CANADIAN WOODS
Their properties and uses

Canadian Woods

THEIR PROPERTIES AND USES

Edited by E.J. Mullins and T.S. McKnight

THIRD EDITION
Published by University of Toronto Press TORONTO BUFFALO LONDON
in cooperation with the Canadian Forestry Service, Environment Canada
and the Canadian Government Publishing Centre, Supply and Services Canada

First edition
King's Printer for Canada, Ottawa 1935

Second edition
King's Printer for Canada, Ottawa 1951

Third edition

Printed in Canada

ISBN 0-8020-2430-0

Government catalogue number FO44-6/1981E

Canadian Cataloguing in Publication Data

Main entry under title:
Canadian woods
Second ed. published: Ottawa: Forestry Branch, 1951.
Bibliography: p.
Includes index.
ISBN 0-8020-2430-0
1. Wood – Canada. 2. Wood-using industries – Canada.
I. Mullins, E.J. II. McKnight, T.S. (Theodore
Samuel), 1927– III. Canada. Environment Canada.
TA419.C38 1981 674'.00971 C81-094571-1

Cette publication est disponible en français sous le titre
Les bois du Canada: leurs propriétés et leurs usages

The forest is a peculiar organism of unlimited kindness and benevolence that makes no demands for its sustenance and extends generously the products of its life activity; it affords protection to all beings, offering shade even to the axeman who destroys it.

Gautama Buddha

Contents

12 *Other Uses and Processes* 285

13 *Pulp and Paper* 303

NFB Photothèque. Photo by Ted Grant

Preface

Canada's forests are one of its most important natural resources. Since the arrival of the earliest settlers, the forests have played a major role in the country's development, supplying the raw material for the necessities of life and for much of our industry. Continuing research has resulted in better use and increased applications of forest products, which now account for a large part of Canada's gross national product.

The purpose of this book is to provide information for engineers, designers, educators, students, and others who require knowledge of the general properties and uses of Canadian woods. The book is divided into sixteen chapters, which discuss wood itself, its processing, and products, and how these products are used today in Canada. Each chapter is written in the simplest practical form, but some use of technical terms is unavoidable. These terms are defined in the glossary at the end of the book. The aim is to produce a book on the science and technology of wood and wood processing that is sufficiently informative for all general purposes. It is impossible, of course, to cover in one book all the technical detail and knowledge about woods that has been accumulated over the years. For the reader who is interested in pursuing any of the topics in more depth, a list of references or a bibliography indicating sources of further reading can be found at the end of each chapter.

Canadian Woods: Their Properties and Uses was first published in 1935, and a second edition followed in 1951. This new edition is based on the earlier publications and incorporates additional knowledge gathered over the last three decades. It reflects the vast accumulation of information that has resulted from nearly 70 years of research on wood and wood products by forest-research organizations in Canada. Most of the material was prepared by members, or former members, of the scientific and technical staff of the Eastern and Western Forest

Products Laboratories, when these establishments were part of the Canadian Forestry Service of Environment Canada. Both laboratories are now units of Forintek Canada Corp., under whose auspices the work of research goes on, seeking solutions to industrial problems and investigating possibilities for better and more complete economic use of the Canadian forest resource.

ACKNOWLEDGMENTS

This book could not have been completed without the cooperation and assistance of many individuals and organizations. The editors are indebted to Mr F.L.C. Reed and Dr R.J. Bourchier of the Canadian Forestry Service for their support and encouragement of the project, and to Mr P.L. Northcott, now retired from the Canadian Forestry Service, for his invaluable help in initiating the preparation of the book. We also wish to thank the authors for giving their time and talents to writing the text and for their cooperation and forbearance with our numerous editorial demands. Our thanks are extended to the reviewers for their helpful and constructive comments.

Special thanks are due to several persons whose contributions helped bring this work to completion: to Mr R. Joly of the Canadian Forestry Service Scientific Publishing Unit and Professor Jean Poliquin of Laval University for their evaluation and editing of the French text and associated materials in preparation for publication; and to Mrs T. Charbonneau, who handled the enormous task of typing the many drafts of both the English and French texts.

Finally, we gratefully acknowledge the contributions made by the many individuals and organizations who provided data, photographs, advice, and other assistance that have greatly aided the successful completion of the book.

E.J. MULLINS
T.S. MCKNIGHT

CANADIAN WOODS
Their properties and uses

1

P.L. NORTHCOTT
Retired, formerly with
Canadian Forestry Service, Ottawa

Canada: A Forest Nation

Figure 1.1 Douglas-fir stand on Vancouver Island, British Columbia. (Photo by British Columbia Forest Service)

Canada is known as a forest nation. The growing stock of Canadian forests, which is estimated at 19 billion m³ (681 billion cu ft), is about 7% of the total of the world's forests. More than half of the land area in Canada's ten provinces is covered with forests and there is additional forest land in the Yukon and Northwest Territories. Canada has the second largest forest reserves of softwood species in the world, after the USSR, and is the world's major exporter of forest products.

THE FORESTS OF CANADA

The forests of Canada are divided into eight regions as shown in Figure 1.2. Each region is a major geographic zone characterized by broad uniformity in physical features and in the composition of the dominant tree species (Rowe 1972).

About 140 native tree species are found in Canadian forests (Hosie 1969). All trees fall into one of two groups – the coniferous (or softwood) trees, which hold their needle-like leaves for two seasons or longer, and the deciduous (or hardwood) trees, whose broad leaves change color in autumn and are shed from the tree, usually before winter. Only 31 of the native species in Canada are conifers, yet they dominate Canada's forests, accounting for about 80% of the total volume of merchantable timber (excluding the forests of the Yukon and Northwest Territories, for which no accurate species breakdown is available). The approximate percentages of merchantable timber provided by the major species are shown in Figure 1.3. A wide variety of coniferous and deciduous species makes up the remainder.

Canada has 342 million hectares (845 million acres) of forest land. By legislation about 10 million hectares (24 million acres) are reserved for primary uses other than timber production, such as national and provincial parks, wildlife refuges, ecological reserves, and wilderness areas. Of the nonreserved forest lands, 79% are provincial Crown

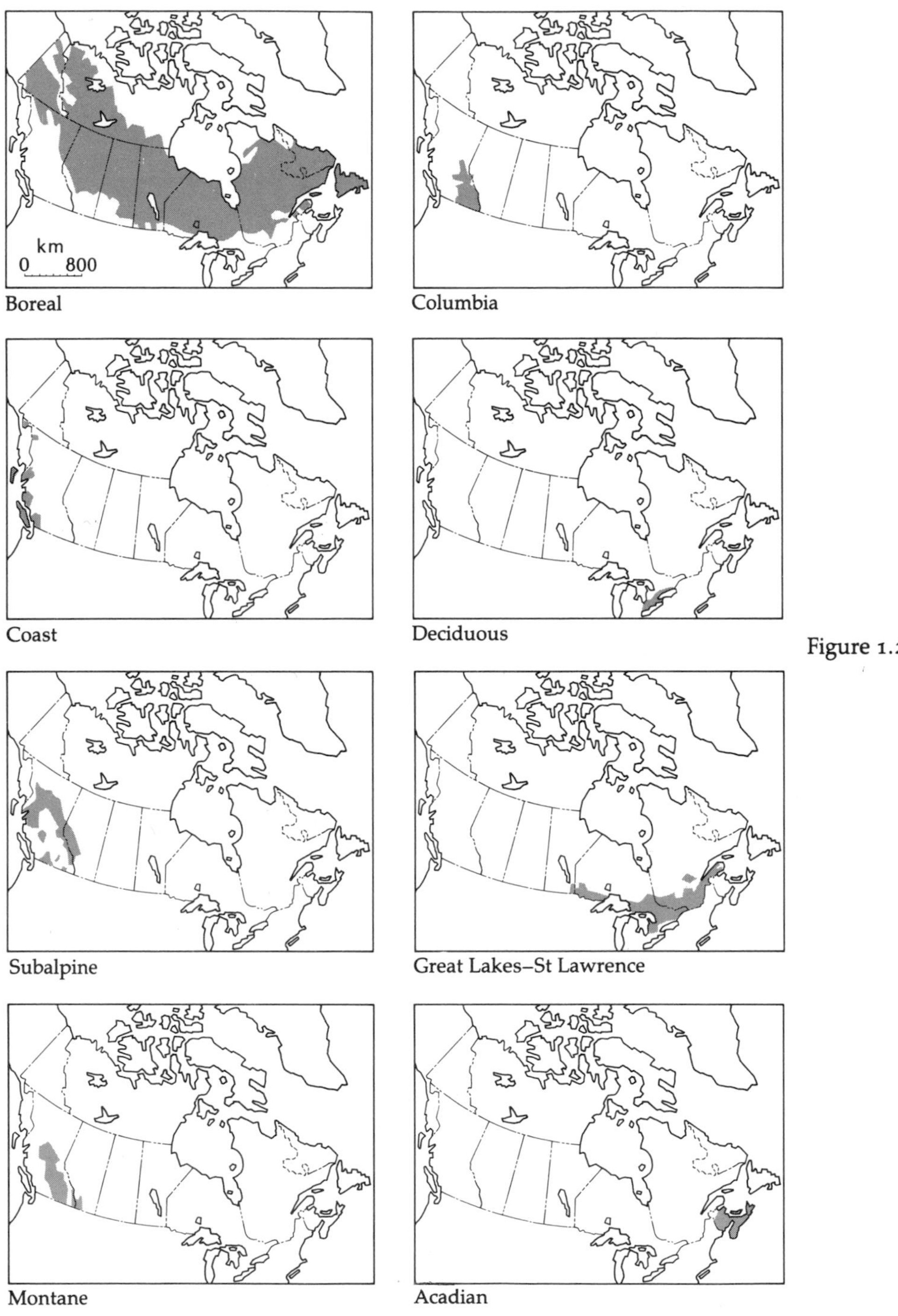

Figure 1.2 Forest regions of Canada.

forests, 16% are under federal jurisdiction, and 5% are privately owned.

Each of the provincial governments is responsible for the management of the public forest land within its boundaries through implementation of its Crown Timber Act, Forest Act, or similar legislation. Although the details of forest legislation vary from one province to another, all provide protection of the forest (e.g., protection against forest fires) and management of the forest for timber, recreation, wildlife, and other purposes in accordance with the public interest. In many cases, management responsibilities are delegated under long-term forest tenures to companies who have the capability of managing their own timber-harvesting areas. The federal government administers the forests of the Yukon and Northwest Territories as well as those of military reserves, national parks, national wildlife areas, and the federal research forests.

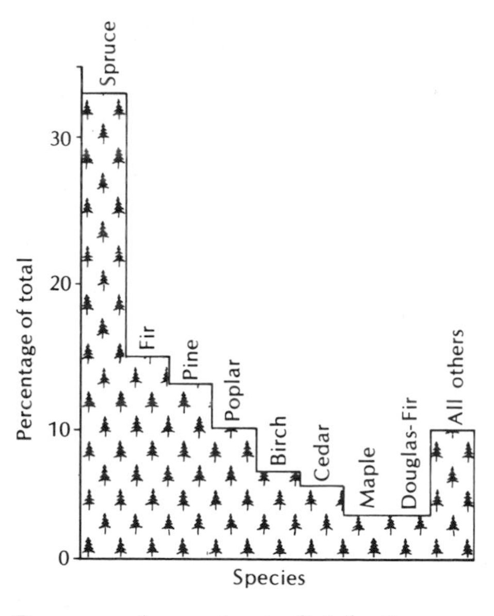

Figure 1.3 Approximate distribution of merchantable timber in Canadian forests.

THE FOREST: A COMPLEX COMMUNITY

Canadians have never been more keenly aware of the ecological importance of the forest than they are today. Usually people think of trees when they think of the forest, but trees are only one part of a complex ecosystem of many organisms and physical elements. Herbs, shrubs, mosses, grasses, and other plant families are as much a part of the forest as trees are. Added to these are the countless varieties of insects, birds, mammals, fish, and other organisms that share the forest environment. Forming a base for this complex system are the physical elements of soil, air, water, geography, and climate. All are inseparable parts of the total ecosystem.

Man too is part of this community and has interacted with the forest environment since prehistoric times. The forest provided early man with many of his basic needs – a hunting ground, a source of fuel, and shelter. Ironically, with all of our sophisticated technology, we still use the forest to help satisfy many of these same needs. At the same time, we now enjoy its other benefits. As our leisure time increases, we are able to use the forest for a wide variety of recreational activities – camping, fishing, hiking, and escape from an increasingly complex, mechanized, and synthetic world. The 19.4 million Canadians and tourists from other countries who visited the national parks in 1978 attest to the popularity of our forested park lands.

Education has made us aware of the pervading influence of the forest on our own environment. We have become aware of two important aspects of the forest's role on this planet: its ability to recycle carbon dioxide back to oxygen and its influence on the recycling of water. The ability of forest lands to store water and to control runoff has benefited man immeasurably; yet it is only in recent times that we have come to appreciate the importance of this ability. The forest also plays a vital part in moderating climate by influencing relative humidity and temperature.

Although the forest influences mankind in many ways, its most direct and tangible benefit is the provision of an adequate supply of the world's major natural, renewable, and versatile raw material – wood.

THE HARVESTING OF FOREST PRODUCTS

Forest products have played a major role in the country's development since the arrival of the earliest settlers. The forest provided construction materials for the log cabins of the pioneers, it enabled an export trade of oak and pine timbers for the British navy to develop and flourish, and it has continued to contribute greatly to Canada's

Figure 1.4 Canoeists setting up camp in Algonquin Park, Ontario. (NFB Photothèque. Photo by Ted Grant)

Table 1.1 *Output and export of selected Canadian forest products, 1979**

Commodity	Output	Export
Lumber – production (1 000 m³)	44 809	31 558
Veneer – shipments (1 000 m³)	na	265
Plywood – shipments (1 000 m³)	2 511†	501
Particleboard – production (1 000 m³)	1 279	345
Rigid insulation board – production (1 000 m² [12.7 mm basis])	52 045	5 322
Hardboard – production (1 000 m²) (3.175 mm basis)	64 796	9 196
Wood pulp – production (1 000 tonnes)	20 408	7 090
Paper and paperboard – production (1 000 tonnes)	13 736	9 619

Sources: Canadian Pulp and Paper Association (1980); Statistics Canada (1978); Statistics Canada (1979a, b, c, d, e, f)
*See Table 1.2 for the same data in units of imperial measure.
†Construction type plywood only.

Table 1.2 *Output and export of selected Canadian forest products, 1979**

Commodity	Output	Export
Lumber – production (million fbm)	18 989	13 374
Veneer – shipments (million cu ft)	na	9.4
Plywood – shipments (million cu ft)	88.7†	17.7
Particleboard – production (million cu ft)	45.2	12.2
Rigid insulation board – production (million sq ft [½ in. basis])	560.2	57.3
Hardboard – production (million sq ft)	697.5	99.0
Wood pulp – production (1 000 tons)	22 496	7 815
Paper and paperboard – production (1 000 tons)	15 141	10 603

Sources: Canadian Pulp and Paper Association (1980); Statistics Canada (1978); Statistics Canada (1979a, b, c, d, e, f)
*See Table 1.1 for above data in SI metric units.
†Construction type plywood only.

prosperity. Today forest products constitute a major segment of Canada's export and domestic economy.

In 1978 the pulp and paper industry and the sawmill and planing mill industries ranked third and sixth, respectively, among Canadian manufacturing industries in the value of shipments of their own manufacture. In the same year the harvesting and processing of timber directly generated work for over 300 000 people, whose salaries and wages amounted to $5.08 billion. The total value added for all forest industries was $9.76 billion on a value of shipments of $18 billion. This value added represents 15% of that for all manufacturing industries.

The forest industries comprise the most important single manufacturing sector in Canada in terms of employment, dollar value of the final products, and foreign exchange earnings. Although total wood-product exports are expected to continue increasing, they make up a diminishing proportion of total Canadian exports (down from 35% in 1950 to 19% in 1979), owing to the increased export of such high-value manufactured goods as automobiles. However, the forest industries still produce the largest foreign exchange earnings. Softwood lumber, newsprint, and wood pulp were Canada's most important exports in 1979.

Production and export data for Canada's forest products for 1979 are shown in Tables 1.1 and 1.2. In addition to the commodities shown in the tables, 891 thousand m^3 (31.5 million cu ft) of logs and bolts, 810 thousand m^3 (28.6 million cu ft) of pulpwood, and 2755 thousand m^3 (97.3 million cu ft) (solid wood basis) of pulp chips were also exported. The total value of forest product exports in 1978 was $9.6 billion, and the figure rose to $11.9 billion in 1979. Approximately 70% of Canada's forest products are exported to the United States; most of the remainder are exported to the United Kingdom, Japan, and western Europe.

THE RENEWABLE RESOURCE

Unlike many of our natural resources, our forests are a renewable resource. With careful management, we can ensure that they will thrive and continue to provide the many benefits to which we have become accustomed.

Foresters can calculate an 'allowable cut' of trees per year for any given forested area that will ensure a sustained yield in perpetuity. Based on various assumptions, including a high level of forest management and a stable forest land base, the maximum allowable cut of both hardwood and softwood species in Canada has been estimated at 276 million m^3 (9747 million cu ft). At present the volume of wood harvested per year is roughly 57% of this amount, since in 1978 the cut was 156 million m^3 (5505 million cu ft). However, the harvest of economically accessible softwood species is very near the upper limit. Shortages are emerging in every region.

The real challenge to forest managers and researchers throughout Canada is not merely to sustain the forest at the level of growth to meet expected wood requirements, but to increase it to levels far beyond current predictions. There are many ways to increase the sustained yield.

Nature depletes the forest at a rapid rate. The annual loss due to fire, insects, and disease is estimated at 85 million m^3 (3000 million cu ft). This figure is not far below the present annual commercial harvest. Preventing this natural loss of wood has tremendous potential for the future.

Commercial thinning, fertilization, and advanced reforestation techniques are just a few of the silvicultural treatments that may help to increase forest yields. Geneticists can also provide us with faster-growing and disease-resistant species with which to stock our future forests.

The trend today is to find uses for the entire tree. Thus, portions of trees left in the woods as forest residues, as well as mill residues at processing plants, will more and more become the raw material of tomorrow's expanded requirements.

The efforts of forest and forest products research at all levels – universities, provincial forest services, forest products companies, and the Canadian Forestry Service – are aimed at increasing the productivity and usefulness of Canada's forests. Wise and careful management of Canadian forests will offer future generations of Canadians the scenic splendor, relaxation, abundant wildlife, forest products, employment, and economic prosperity that we enjoy today.

Figure 1.5 Multipurpose harvester unloading pulpwood logs, Lac Cing Doigts, Quebec. (NFB Photothèque. Photo by George Hunter)

REFERENCES

Canadian Pulp and Paper Association. 1980. *Reference tables 1980*. Montreal

Hosie, R.C. 1969. *Native trees of Canada*. 7th ed. Can. For. Serv., Environment Canada, Ottawa

Rowe, J.S. 1972. *Forest regions of Canada*. Can. For. Serv. Publ. 1300. Environment Canada, Ottawa

Statistics Canada. 1978. Cat. 25-202 Annual. *Canadian forestry statistics 1978*. Ottawa

– 1979a. Cat. 36-001. *Hardboard*. Vol. 27 (12), Dec. Ottawa

– 1979b. Cat. 36-002. *Rigid insulating board*. Vol. 33 (12), Dec. Ottawa

– 1979c. Cat. 36-003. *Particleboard*. Vol. 15 (12), Dec. Ottawa

– 1979d. Cat. 35-001. *Plywood*. Vol. 27 (12), Dec. Ottawa

– 1979e. Cat. 35-002. *Lumber*. Vol. 34 (12), Dec. Ottawa

– 1979f. Cat. 35-003. *Lumber*. Vol. 33 (12), Dec. Ottawa

2

Commercial Woods

E. PEREM

Retired, formerly with Eastern Forest Products Laboratory, Ottawa

C.F. MCBRIDE

Retired, formerly with Western Forest Products Laboratory, Vancouver

C.T. KEITH

Eastern Forest Products Laboratory, Ottawa

About 140 native tree species are described in the Canadian Forestry Service publication *Native Trees of Canada* (Hosie 1969). Of these, 31 are coniferous species (or softwoods) and the others are deciduous species (or hardwoods). Many species are of limited commercial importance because they occur only locally in small quantities, or they are small in size. Nomenclature of the Canadian wood species used in this edition of *Canadian Woods* is generally the common tree and botanical name published in *Native Trees of Canada*. Additional vernacular names that are employed in the wood-using industries, which often vary according to locality, are listed for each species in the following discussion.

This chapter describes the commercially important woods from trees native to Canada. Information given under the heading 'Resistance to decay' applies exclusively to the heartwood* of the species, as do the remarks on ease of penetration with preservatives. Sapwood of all species lacks natural decay resistance, but in general it is easily penetrated by preservative solutions. The term 'degrade' used frequently in describing the seasoning characteristics of the different timbers refers to a reduction in the quality of a wood due to the development of such defects as checks, splits, and distortions in shape.

The characteristics that influence the commercial applications of wood include a wide range of mechanical, biological, and chemical properties. Mechanical properties determine the suitability of wood for structural purposes, veneer production, furniture manufacture, and many other uses. Biological properties affect paper-making characteristics and applications in which resistance to decay, penetration of preservatives, and similar considerations are important. Chemical extractives in wood affect its capacity to take paint and bleaches, its suitability for pulping, its resistance to insects, and other characteristics far out of proportion to the minor quantities that are usually present. More detailed information on the properties of different woods is provided in Chapters 3, 4, and 5.

* For the definition of technical terms see the Glossary at the back of the book. Many of the terms are discussed at appropriate places in the text (see the Index).

SOFTWOODS (coniferous species)

Eastern White Cedar – *Thuja occidentalis* L.
Other names: northern cedar, arbor-vitae, white cedar

The Tree
The tree is usually 13 m (45 ft) high and 30 cm (12 in.) in diameter, but it is occasionally taller. It is found in all the eastern provinces except Newfoundland and extends westerly to eastern Manitoba and approximately as far north as James Bay.

The Wood
General description The heartwood is light brown; the sapwood is nearly white. The texture is fine and even, the grain is usually straight, and the growth-ring figure is not prominent. The wood has a pleasant, aromatic odor and weathers to a silvery sheen with inconspicuous checks. Eastern white cedar is a comparatively small tree and its wood contains many small knots, but otherwise is relatively free of natural defects.
Weight The wood is lighter in weight than that of any other commercial Canadian species at about 335 kg/m^3 (21 lb/cu ft) when air-dry.
Strength Strength, stiffness, hardness, and shock resistance are low.
Processing The wood seasons readily without warping and shrinks very little in drying. It has good working qualities, is easy to glue and stain, and holds paint well. It is difficult to bend without excessive breakage. Its nail- and screw-holding power is low.
Resistance to decay Decay resistance of the heartwood is high, superior to that of other commercial species of eastern Canada.
Uses Eastern white cedar is one of the most durable woods in Canada and is especially good for applications in which it is exposed to the hazard of decay. It is used for shingles, boats and canoes, greenhouses, outdoor furniture, tanks and vats, fence posts, exterior siding, log cabins, floats, poles, and piling.

Figure 2.1 Interior of Vancouver house with western red cedar ceiling and beams. (Photo: Council of Forest Industries of British Columbia)

Western Red Cedar – *Thuja plicata* Donn
Other name: giant arbor-vitae

The Tree
The largest cedar native to North America, western red cedar occasionally attains heights of 60 m (200 ft) and diameters of 450 cm (15 ft). Usually it is 30–45 m (100–150 ft) in height and 100–200 cm (3–6 ft) in diameter. The stem is buttressed at the base. The tree is found all along the west coast and also occurs in the wetter areas of the interior of British Columbia.

The Wood
General description The sapwood is thin with a light yellow tinge, and the heartwood varies in color from light straw to pinkish-red to deep, warm brown. The wood is straight-grained and soft and splits readily. It has a very distinctive aromatic odor. Considerable amounts of clear wood are found in the larger trees.

Weight The wood is light in weight at 385 kg/m³ (24 lb/cu ft) when air-dry.
Strength The wood is not strong but performs satisfactorily when used for round poles.
Processing Western red cedar dries readily with very little shrinkage. It has exceptionally good working qualities and gives a smooth, satiny finish. It takes stains and paints well and has good gluing properties. The heartwood is very difficult to penetrate with preservatives.
Resistance to decay The wood of this tree is classed as highly resistant. The heartwood contains large amounts of extractives, which account for its color, odor, and high durability. The lighter-colored areas of the heartwood, usually nearest the sapwood, contain the largest amount of fungicidal substances and thus are the most resistant to decay.
Uses Western red cedar has found many uses because of its excellent properties, especially its durability in contact with the soil and in situations favorable to decay. Since its early use by the Indians and then by the early settlers for house construction and other purposes, it has proved valuable for shingles and shakes for roofing or weather-boarding, poles, fence posts, and fencing. The clear lumber is preferred for siding and decking. Other uses are for doors, interior finish, greenhouses, boats and canoes, light construction, and kraft pulp.

Yellow Cedar – *Chamaecyparis nootkatensis* (D. Don) Spach
Other names: yellow cypress, Alaska cedar

The Tree
Like the western red cedar, the yellow cedar has a buttressed and usually a fluted base. The foliage, however, is lacier and more pendulous. It grows on the coast of British Columbia above 900 m (3000 ft) in the south and gradually at lower elevations until it is found at sea level in the north. It is a medium-sized tree, 18–24 m (60–80 ft) in height and 60–90 cm (2–3 ft) in diameter, although occasionally trees more than 30 m (100 ft) high and more than 150 cm (5 ft) in diameter occur. It is usually very slow growing, and trees as old as 1200 years have been found.

The Wood
General description The wood is light yellow, straight, and even-grained, and has a characteristic pungent odor. The sapwood is very narrow and white to yellowish-white and is not clearly distinct from the heartwood.
Weight The weight is moderate at 480 kg/m³ (30 lb/cu ft) when air-dry.
Strength The wood is of medium strength among the softwoods.
Processing Yellow cedar seasons well and, because of its low shrinkage and natural extractives, is very stable. Because it is easily worked and does not tend to splinter, it is prized for joinery and carving.
Resistance to decay It has high natural durability.
Uses Because it wears to a smooth finish without splintering and has high durability, yellow cedar is a popular choice for stadium seats and bridge decking. It is used extensively in boat building and sash and door manufacture, and for roof decking and panelling, where its light

color and fine texture are advantageous. Other uses are in light framing, tanks, flumes, patio decks, and carving.

Douglas-fir – *Pseudotsuga menziesii* (Mirb.) Franco

The Tree

Douglas-fir is found throughout the southern half of British Columbia and extends into southwestern Alberta. On the coast the tree reaches its greatest size, mature trees commonly reaching heights of 60 m (200 ft) and diameters of 200 cm (6 ft) or more. On good sites with deep soil where soil and atmospheric moisture are plentiful, trees over 350 cm (12 ft) in diameter and 1200 years old have been found. In the interior the tree grows up to 30 m (100 ft) in height on average sites, and may reach a height of 45 m (150 ft) on the better sites in the wet belt areas.

The Wood

General description The sapwood is light in color and narrow, usually less than 5 cm (2 in.) wide. The heartwood ranges from yellowish to reddish-brown. Earlywood and latewood have a pronounced difference in color, the latewood having darker, more sharply defined bands. This difference in color results in a distinctive grain pattern when a log is flat-sawn or rotary-peeled for veneer. The stems of large trees grown on the coast are clear of branches for most of their length. The wood is therefore available in long lengths and large sizes as clear lumber and as large structural timber.

Weight The weight is moderate at 545 kg/m³ (34 lb/cu ft) when air-dry.

Strength Next to western larch, Douglas-fir is the strongest commercial softwood in Canada and thus is used extensively for structural

Figure 2.2 Preserved wood foundation of Douglas-fir plywood. (Photo: Council of Forest Industries of British Columbia)

Figure 2.3 Glued-laminated beams of Douglas-fir in the Mt Orford Playhouse, Quebec. (Photo: Canadian Wood Council)

purposes. Its hardness makes it suitable for use where wear is a factor.
Processing The wood dries easily and quickly because the heartwood generally has a moisture content below 40%. Wood of Douglas-fir from the British Columbia coast is relatively easy to impregnate with preservatives, but that from the interior region of the province is difficult to treat because natural growth characteristics of the trees from that region make it less permeable to the preservatives. The wood works readily with hand and power tools and is easily glued.
Resistance to decay Douglas-fir is moderately resistant to decay and thus is used untreated in many situations. Under conditions favoring decay, it should be pressure-treated with preservatives for long service life.
Uses Douglas-fir is one of the best-known timber-producing trees in the world. It is used for more purposes than any other wood in Canada, although the volume of this timber cut is being exceeded by that of both spruce and hemlock. The most important use of Douglas-fir is as lumber and plywood in building and construction. Other important uses are in structural timbers, piling, railroad ties, crossarms, sash and doors, flooring, interior finish, wood-stave tanks and pipes, and kraft pulp. It is sometimes used in masts and spars, and in boat and barge construction.

Balsam Fir – *Abies balsamea* (L.) Mill.
Other names: white fir, balsam, Canadian fir

The Tree
Balsam fir is usually 15–21 m (50–70 ft) high and 30–60 cm (1–2 ft) in diameter. It is widely distributed in Canada, from Newfoundland to Alberta.

The Wood
General description The wood is light in color with no contrast between heartwood and sapwood. It has practically no odor or taste when dry. The number of knots in the lumber is usually large, but the knots themselves are small. The wood is straight-grained and medium-textured and has a moderate growth-ring figure. It weathers to a gray color with little sheen.
Weight The wood is light in weight at 385 kg/m^3 (24 lb/cu ft) when air-dry.
Strength The wood is fairly low in bending and compressive strength, in stiffness, and in resistance to impact.
Processing The wood seasons readily, except when it contains areas of 'wet wood' that may result in the development of defects during drying. Changes in moisture content result in moderate shrinkage or swelling. The wood glues readily and holds paints satisfactorily, but its machining properties are below average. It is moderately resistant to impregnation with preservatives.
Resistance to decay Balsam fir is not resistant when exposed to conditions favoring decay.
Uses Balsam fir is used for many of the purposes for which spruce is used. As lumber, it is marketed with the spruces in the Spruce-Pine-Fir grade. The lumber is employed in building construction (light structural framing, sheathing, subflooring, scaffolding, concrete forms, and

interior woodwork) and in the manufacture of particleboard, construction plywood, container veneer, and boxes and crates. Large quantities of balsam fir are used for pulp.

Figure 2.4 Balsam fir is widely used for light framing in residential construction. (Photo: Dept. of Industry, Trade & Commerce)

Amabilis Fir – *Abies amabilis* (Dougl.) Forbes
Other name: Pacific silver fir

The Tree
Amabilis fir is confined to the coast of British Columbia except the Queen Charlotte Islands. It occurs on wetter sites, usually in association with western hemlock, western red cedar, and Sitka spruce. The mature tree commonly reaches heights of 45 m (150 ft) and diameters of 60–120 cm (2–4 ft).

The Wood
General description The wood is white to light buff in color and generally straight- and even-grained. It is odorless when dry.
Weight The wood is light in weight at 445 kg/m^3 (28 lb/cu ft) when air-dry.
Strength The strength properties are similar to those of western hemlock, and for lumber-grading purposes the two species are grouped. Amabilis fir has a higher than normal stiffness-to-density ratio.
Processing Amabilis fir seasons moderately well and has satisfactory machining properties. The heartwood is moderately difficult to penetrate with preservatives.
Resistance to decay Durability is low.
Uses The two important uses of amabilis fir are for pulp and as lumber for general construction. Other uses are for boxes and crating.

Grand Fir – *Abies grandis* (Dougl.) Lindl.
Other name: lowland fir

The Tree
Grand fir is confined to the relatively low elevations on the southern coast of British Columbia and the wet-belt areas of the Kootenay Valley (part of the Columbia Forest Region). It is the largest of Canada's true firs, normally reaching 38–45 m (125–150 ft) in height and 60–120 cm (2–4 ft) in diameter. The small amount of grand fir growing in British Columbia makes it a relatively unimportant commercial species.

The Wood
General description The wood is similar to that of other true firs.
Weight The wood is light in weight at 445 kg/m^3 (28 lb/cu ft) when air-dry.
Strength Grand fir has strength properties similar to those of amabilis fir and is also marketed with western hemlock.
Processing Grand fir is similar to amabilis fir in processing qualities.
Resistance to decay It is not resistant.
Uses Grand fir is used for the same purposes as amabilis fir.

Alpine Fir – *Abies lasiocarpa* (Hook.) Nutt.
Other names: subalpine fir, white fir

The Tree
Alpine fir is found in the interior of British Columbia and in western Alberta. In the forest it normally attains heights of 20–30 m (65–100 ft) and diameters of 30–75 cm (1–2½ ft).

The Wood
General description The wood is light in color with very little distinction between sapwood and heartwood. Some knots have a yellowish tinge.
Weight The wood is very light in weight at 385 kg/m³ (24 lb/cu ft) when air-dry.
Strength Alpine fir has strength properties similar to those of white spruce and is grouped with white and Engelmann spruce and lodgepole pine for marketing in the Spruce-Pine-Fir grade.
Processing Wet pockets in the wood make it difficult to dry uniformly. It machines satisfactorily to a good finish. The heartwood is very difficult to penetrate with preservatives.
Resistance to decay The wood is not resistant to decay.
Uses The principal use of alpine fir is as lumber for residential frame construction. It is also used for pulpwood and for boxes and crates.

Eastern Hemlock – *Tsuga canadensis* (L.) Carr.

The Tree
Eastern hemlock is usually 15–21 m (50–70 ft) high and 30–90 cm (1–3 ft) in diameter. It is found in Canada from Nova Scotia to southwestern Ontario.

The Wood
General description The wood is buff in color with a reddish-brown tinge, although the outermost annual rings are usually of lighter hue. The wood is moderately coarse and uneven in texture, and the growth-ring figure is distinct but not conspicuous. It is odorless when dry and weathers to a light gray with a moderate sheen. The occurrence of various defects – such as shake and cross grain – is fairly common.
Weight The wood is light in weight at 465 kg/m³ (29 lb/cu ft) when air-dry.
Strength The wood is moderate in bending and compression strength, fairly low in stiffness, and moderate in resistance to impact.
Processing Eastern hemlock shrinks moderately in drying but is prone during seasoning to degrade associated with ring shake and cross grain. It glues easily with all common types of adhesives and holds paint moderately well. It has intermediate nail-holding properties and a tendency to split when nailed; machining properties are below average. The wood is moderately difficult to impregnate with preservatives.
Resistance to decay Eastern hemlock is not resistant when exposed to conditions favorable to decay.
Uses It is mainly used in building construction for framing, roofing, siding, sheathing, subflooring, scaffolding, and concrete forms. When

treated with preservatives, it is used for piling, railway ties, and culverts. Other uses include crates and boxes, construction plywood, and pulpwood.

Western Hemlock – *Tsuga heterophylla* (Raf.) Sarg.

The Tree

Western hemlock is a large, graceful tree that frequently grows to heights of 50 m (160 ft) and diameters exceeding 100 cm ($3\frac{1}{2}$ ft). In dense stands the stem is often clear for three-quarters of its length. Western hemlock occurs on the coast of British Columbia and in the interior valleys where rainfall is plentiful.

The Wood

General description The wood is light in color with a pinkish to reddish-brown tinge; there is little difference between the sapwood and the heartwood. The sapwood is 70–130 mm (3–5 in.) in width. The wood has a sour odor when fresh and is straight- and even-grained, with a medium to fine texture. The annual rings are distinct, and the transition from earlywood to latewood is subtle. Dark streaks, caused by bark maggots, occur frequently along the annual rings and are used as an identifying feature. Western hemlock and amabilis fir lumber are difficult to distinguish visually.

Weight The weight is moderate at 480 kg/m^3 (30 lb/cu ft) when air-dry.

Strength Western hemlock is strong, although not as strong as Douglas-fir. Nevertheless, it can be used for many of the same purposes.

Processing Western hemlock dries slowly because of its high moisture content. With care, it can be dried satisfactorily either in kilns or by air drying. It shrinks considerably as it seasons and holds its shape well if dried close to the moisture content it will acquire under the atmospheric conditions in which it will be used. It planes easily and machines well. It is moderately difficult to treat with wood preservatives.

Resistance to decay The wood is only slightly resistant to decay. It should be treated with a preservative when used in moist situations.

Uses Western hemlock is widely used in all types of construction, both residential and industrial. It is the most important pulpwood species in British Columbia. Other uses are in ladder stock, boxes, interior finish, and treated railway ties. The bark contains a high percentage of tannin, which is extracted and used as a mud additive in oil-well drilling; it is also used in agriculture as a trace-metal complexing agent to help the soil retain nutrient elements. Additional information on the use of tannin extracted from western hemlock bark can be found in Chapter 5.

Mountain Hemlock – *Tsuga mertensiana* (Bong.) Carr.

The Tree

Mountain hemlock is an alpine or subalpine species occurring at higher elevations throughout the coastal region and the inland mountains of British Columbia. It is normally a small tree, up to 15 m (50 ft) in height and 25–50 cm (10–20 in.) in diameter. On wind-swept mountains and in exposed wetter areas on the northern coast, however, it is dwarfed and scrubby. On certain sheltered sites it will grow to merchantable size, up to 30 m (100 ft) tall and 90 cm (3 ft) in diameter, in association with western hemlock.

The Wood
The wood of mountain hemlock is similar to that of western hemlock and is used for the same purposes.

Western Larch – *Larix occidentalis* Nutt.
Other name: western tamarack

The Tree
In a forest, the tree develops a long, slightly tapering stem, often free of branches for most of its height. It grows in the southern interior of British Columbia, and occasional trees are found in southwestern Alberta. Heights of 30–55 m (100–180 ft) and diameters of 60–150 cm (2–5 ft) are common. Western larch and eastern larch (tamarack) are the only Canadian conifers having foliage that is deciduous.

The Wood
General description The sapwood is usually narrow and whitish to pale straw brown in color, while the heartwood is russet to reddish-brown. The wood has an oily appearance, feels greasy, and is straight-grained.
Weight The wood is heavy in weight at 640 kg/m^3 (40 lb/cu ft) when air-dry.
Strength The wood is hard, the strongest of the commercial conifers in Canada.
Processing The wood is somewhat difficult to work but takes a smooth, hard finish. It seasons fairly well, but checking causes some difficulty. The heartwood is difficult to penetrate with preservatives.
Resistance to decay The wood is of moderate durability.
Uses Because it is a hard, strong wood, western larch is used for heavy structural purposes interchangeably with Douglas-fir. Other uses are in construction lumber, ties, sash and doors, piling, and tanks.

Tamarack – *Larix laricina* (Du Roi) K. Koch
Other names: eastern larch, hackmatack, Alaska larch

The Tree
The tamarack is a small to medium-sized tree, 9–21 m (30–70 ft) in height and 30–60 cm (1–2 ft) in diameter. In Canada it has a transcontinental range that includes the forests of the northern regions.

The Wood
General description The heartwood is yellowish-brown to russet brown in color; the sapwood is whitish and narrow. The growth-ring figure is prominent, and the transition from earlywood to latewood is abrupt. The wood is fairly coarse in texture. Spiral grain is common.
Weight The weight is moderate at 560 kg/m^3 (35 lb/cu ft) when air-dry.
Strength The wood is fairly high in bending and compressive strength, and low in resistance to impact.

Processing The wood shrinks moderately in seasoning but has a tendency to warp. It requires above average care in machining and is difficult to penetrate with preservatives.
Resistance to decay The heartwood is moderately durable, but treatment with preservatives is necessary under conditions favoring decay.
Uses Tamarack is used for railroad ties (treated), poles, posts, and piling. It is also employed as lumber in building construction, for boat building, tanks, crates and boxes, and pulpwood.

Figure 2.5 White pine used for interior finish of a kitchen. (Photo: Canadian Wood Council)

Eastern White Pine – *Pinus strobus* L.
Other names: white pine, Quebec pine, Weymouth pine

The Tree
Eastern white pine is the tallest conifer in eastern Canada, commonly reaching heights of 27–38 m (90–125 ft) and diameters of 45–75 cm (1½–2½ ft). Its range extends from Nova Scotia to southeastern Manitoba, the most important stands being in the St Lawrence River drainage area.

The Wood
General description The heartwood varies from a straw brown to a light reddish-brown, and the sapwood is almost white. The wood is straight-grained and has a uniform texture, and the growth-ring figure is inconspicuous. The wood weathers to a light gray with moderate sheen.
Weight The wood is light in weight at 415 kg/m^3 (26 lb/cu ft) when air-dry.
Strength This pine is weaker and softer than other eastern pines and has fairly low resistance to impact.
Processing The wood of eastern white pine is very highly regarded. It has low shrinkage and seasons readily and uniformly. It is worked very easily by hand and machine tools, is easy to glue, and has good nailing and screw-holding properties. The paint-retention properties are good. It is treated with preservatives fairly easily.
Resistance to decay The heartwood is moderately durable, but treatment with preservatives is required under conditions favorable to decay.
Uses Because of its low shrinkage and uniform texture, eastern white pine is used extensively for window sashes and frames, doors, shelving, cabinet work, and other items in which dimensional stability is important. It is also used for light and medium construction, exterior and interior finish, boat building, and caskets. It is a popular wood for toys, wood carvings, and other woodenware.

Western White Pine – *Pinus monticola* Dougl.
Other names: Idaho white pine, silver pine

The Tree
The tree grows in the southern coastal area of British Columbia and in the wet-belt areas of the southern interior. It seldom grows in pure stands but usually forms a small part of a mixed stand. It is normally a tall tree with a clear stem that has very little taper. It usually grows to 23–38 m (75–125 ft) in height and up to 90 cm (3 ft) in diameter.

Occasionally trees 55 m (175 ft) high and up to 150 cm (5 ft) in diameter are found.

The Wood

General description The wood is similar to that of eastern white pine. The heartwood is cream-colored to light reddish-brown. The sapwood is narrow and yellowish-white. Knots are considerably darker; this feature puts the wood in demand for panelling.
Weight The wood is light in weight at 415 kg/m^3 (26 lb/cu ft) when air-dry.
Strength Moderately low.
Processing The wood dries readily and is worked easily with hand or machine tools to a very smooth finish. It has good nail-holding properties and glues well.
Resistance to decay The wood has some natural decay resistance, but treatment with preservatives is recommended before use in situations favorable to decay.
Uses Principal uses are in interior finish, millwork, pattern stock, drawing boards, and wood flour. Lower grades are used in light construction.

Ponderosa Pine – *Pinus ponderosa* Laws.

Other names: yellow pine, bull pine

The Tree

Mature trees have straight stems with little taper and are often free of branches for much of their length. The range in Canada is confined to dry areas in the southern interior of British Columbia. On good sites the tree reaches 30–45 m (100–150 ft) in height and 90–150 cm (2–5 ft) in diameter. On very dry open sites the tree is usually very short. At the lower elevations ponderosa pine grows in pure, parklike stands; at elevations up to 1200 m (4000 ft), it is found mixed with Douglas-fir and lodgepole pine.

The Wood

General description Mature trees have a very thick sapwood that is pale yellow. The heartwood is darker, ranging from yellow to reddish-brown. As the growth rate is usually slow, the wood has a very uniform texture. It has a characteristic pleasant pine odor.
Weight The weight is moderate at 515 kg/m^3 (32 lb/cu ft) when air-dry.
Strength Medium.
Processing Ponderosa pine seasons readily, but care must be taken to avoid blue stain in the sapwood, particularly in the spring. It has good nail-holding properties and works well without splintering. It can take preservative treatments easily.
Resistance to decay Durability is low.
Uses Ponderosa pine is used principally in pattern stock, sash and doors and millwork, boxes and crates, interior and exterior finishing, and as lumber for general construction.

Red Pine – *Pinus resinosa* Ait.
Other name: Norway pine

The Tree
Red pine is usually 18–24 m (60–80 ft) in height with a diameter of 30–60 cm (1–2 ft). Under favorable conditions, it may reach a height of 38 m (125 ft) and a diameter of up to 90 cm (3 ft). It is found from Newfoundland to southern Manitoba.

The Wood
General description The heartwood is pale brown to reddish-brown; the sapwood is yellowish-white and wide. The growth-ring figure is fairly conspicuous, the grain is usually straight, and the texture is medium.
Weight The wood is light in weight at 450 kg/m^3 (28 lb/cu ft) when air-dry.
Strength The wood is moderately strong in bending and compression and in resistance to impact.
Processing The wood seasons readily with very little degrade. It has good machining properties, is easy to glue, and has good nail- and screw-holding properties. Its painting characteristics are satisfactory, except when the wood is excessively resinous. It is easily impregnated with preservatives.
Resistance to decay The wood is not resistant when exposed to conditions favorable to decay.
Uses Red pine is used mainly for building construction, siding, and interior and exterior millwork. It is also employed for crates and boxes and as pulpwood. When treated with preservatives, it is used for poles, piling, and railroad ties.

Jack Pine – *Pinus banksiana* Lamb.
Other names: princess pine, gray pine, banksian pine

The Tree
Jack pine is usually 12–21 m (40–70 ft) in height and 20–30 cm (8–12 in.) in diameter. It is found from Nova Scotia to northern Alberta.

The Wood
General description The heartwood is light brown, and the sapwood is nearly white. The wood has a resinous odor. The growth-ring figure is conspicuous, and the wood is rather uneven in texture.
Weight The weight is moderate at 495 kg/m^3 (31 lb/cu ft) when air-dry.
Strength Jack pine is moderately strong in bending, less strong in compression, and fairly resistant to impact.
Processing The wood seasons without difficulty. Dimensional changes caused by changes in the moisture content are relatively small. It has satisfactory machining properties and is easy to glue with all commercial types of adhesive. The wood holds nails well and takes paint satisfactorily. It is moderately resistant to impregnation with preservatives.
Resistance to decay The wood is not resistant under conditions favoring decay.
Uses Jack pine lumber is graded with the spruces in the Spruce-Pine-

Fir group. It is used in building construction as framing, sheathing, scaffolding, and interior woodwork. When treated with preservatives, it is used for railroad ties, poles, and posts. It is also used for boxes and crates, and to a large extent as pulpwood.

Lodgepole Pine – *Pinus contorta* Dougl.
Other name: shore pine

The Tree
The tree grows in the interior of British Columbia and in western Alberta; it is also found in adjacent areas in the southern Yukon and southwestern Northeast Territories. It is noted for its tall, straight form and usually grows in pure stands, attaining heights up to 30 m (100 ft) and diameters up to 60 cm (2 ft). A variety *Pinus contorta* var. *contorta* called shore pine grows on the coast of British Columbia, but it is usually a scrubby tree of little commercial importance.

The Wood
General description The wood is light in color, the sapwood being almost white and the heartwood very light yellow. It is soft and straight-grained with a uniform texture. The flat-sawn wood often has a dimpled pattern.
Weight The wood is light in weight at 465 kg/m^3 (29 lb/cu ft) when air-dry.
Strength Medium.
Processing Lodgepole pine seasons readily, takes a good finish, and holds paints well. It is moderately difficult to treat with preservatives.
Resistance to decay The wood has low durability and should be treated with preservatives if it is to be used in situations favorable to decay.
Uses Lodgepole pine is marketed with white spruce and alpine fir. Its principal use is in construction lumber for the housing market. It is used extensively for railroad ties as well as for boxes, crates, and poles and as pulpwood.

Eastern Spruces – *Picea* spp.
White Spruce – *Picea glauca* (Moench) Voss
Black Spruce – *Picea mariana* (Mill.) B.S.P.
Red Spruce – *Picea rubens* Sarg.
Because the woods of the three eastern spruces cannot be distinguished with certainty on the basis of their structure, all three are usually marketed as eastern spruce. Differences in the appearance and properties of the spruces are associated primarily with differences in the typical growth rates of these species, which reflect the variation in their ecological requirements and tolerances.

The Tree
White spruce and red spruce usually attain a height of approximately 24 m (80 ft) and a diameter of 30–60 cm (1–2 ft). Black spruce, which is normally a slow-growing tree, averages approximately 9–15 m (30–50 ft) in height and 15–25 cm (5–10 in.) in diameter. White and black spruces have a transcontinental range in Canada; the natural distribution of red spruce is limited to Nova Scotia, New Brunswick, Prince

Edward Island, and parts of southern Quebec and eastern Ontario.

The Wood

General description The wood is nearly white, with little or no contrast between sapwood and heartwood. It is usually straight-grained and fairly fine textured. The growth-ring figure is faint; it is somewhat more conspicuous in black and red spruce than in white spruce. The wood weathers to a light gray with a silvery sheen.

Weight The average air-dry weights recorded for the three species are as follows: white spruce, 415 kg/m^3 (26 lb/cu ft); red spruce, 450 kg/m^3 (28 lb/cu ft); black spruce, 480 kg/m^3 (30 lb/cu ft).

Strength The wood of black spruce and red spruce tends to be stronger than that of white spruce. Generally, eastern spruce is classified as medium in strength but above average in stiffness.

Processing The wood seasons fairly easily, and its shrinkage in drying is moderate. Its machining properties are fairly good, and it holds nails well. It is moderately easy to glue and holds paint satisfactorily. The wood is very resistant to impregnation with preservatives.

Resistance to decay Eastern spruce is not resistant in situations favorable to decay.

Uses The wood of eastern spruce is highly valued both as lumber and as pulpwood. As lumber it is used in building construction (framing, sheathing, roofing, scaffolding, subflooring) and for general millwork, ladder rails, and sounding boards for musical instruments. It is also used for the manufacture of construction plywood. The wood is widely used for containers, particularly for food, because it is almost odorless and tasteless.

Engelmann Spruce – *Picea engelmannii* Parry

The Tree

Engelmann spruce grows on the mountains of the southern interior of British Columbia at elevations of 1200 m (4000 ft) and higher. Farther north it is found at lower elevations, where it grows mixed with white spruce. Hybrids are found where the ranges of white and Engelmann spruce overlap. The trees normally grow to 35 m (120 ft) in height and to 90 cm (3 ft) in diameter; on good sites, however, they may reach 55 m (180 ft) in height and up to 180 cm (6 ft) in diameter.

The Wood

General description The wood is similar to that of eastern spruce and cannot be distinguished from it by its structure. It is nearly white in color, with little or no contrast between the sapwood and heartwood.

Weight The wood is light in weight at 415 kg/m^3 (26 lb/cu ft) when air-dry.

Strength The wood is intermediate in strength but noted for its high resilience.

Processing The wood dries readily and is easily worked to a smooth finish. It has good nail-holding properties. It is difficult to treat with preservatives.

Resistance to decay Its durability is low.

Figure 2.6 Newsprint made from spruce pulpwood being wound onto large rolls, which are later cut into smaller rolls for shipment. (NFB Photothèque. Photo by Bill Brooks)

Uses The lumber is used for light construction in housing for studs, floor joists, and rafters. It is used extensively for sheathing grade plywood and as pulpwood. It is also used for boxes because of its lack of odor and for crating. Because it has good resonance properties, it is used for sounding boards for musical instruments.

Sitka Spruce – *Picea sitchensis* (Bong.) Carr.
Other names: tideland spruce, silver spruce

The Tree
Sitka spruce grows on moist sites on the coast of British Columbia, attaining its best growth on the Queen Charlotte Islands. It is normally 90–180 cm (3–6 ft) in diameter and 45–60 m (150–200 ft) high, but many larger trees, up to 350 cm (12 ft) in diameter and up to 90 m (300 ft) in height, are found. Sitka spruce is best known for its long, clear stem, which produces considerable amounts of clear wood.

The Wood
General description The wood is not as light in color as the other spruces, but there is little difference between the sapwood and heartwood. The color ranges from a creamy white to near white with a light pinkish tinge. The wood is usually straight-grained, and large

amounts of clear lumber are produced because of the size of the trees.
Weight The wood is light in weight at 430 kg/m^3 (27 lb/cu ft) when air-dry.
Strength The wood has moderate strength properties but a high strength-to-weight ratio.
Processing Sitka spruce dries readily. It is easily worked to a smooth finish, takes paint well, and holds nails well. The heartwood is difficult to penetrate with preservatives.
Resistance to decay Sitka spruce has low durability.
Uses The principal uses are for general construction, shipbuilding, plywood, boxes and crates, and pulp. It is also used for millwork, musical instruments, and ladder rails. Sitka spruce was once an important wood in airplane construction when wood was used for this purpose, because of its high strength-to-weight ratio and the availability of clear lumber in large sizes.

HARDWOODS (deciduous species)

Red Alder – *Alnus rubra* Bong.
Other names: western alder, Oregon alder

The Tree
Red alder is the most plentiful hardwood species on the British Columbia coast. It grows in pure stands on moist bottom land in coastal valleys. The tree grows to 24 m (80 ft) in height and to 60 cm (2 ft) in diameter. Occasionally trees are found up to 30 m (100 ft) in height and 120 cm (4 ft) in diameter. The tree is relatively short-lived, and sawlog-size trees are produced in 35–50 years.

The Wood
General description The wood is generally white when freshly cut, but after exposure it quickly changes color, varying from a flesh shade to a light reddish-brown. There is little distinction between the heartwood and the sapwood. The grain is usually straight, and occasional pieces have prominent aggregate rays that show a pleasing pattern when the wood is quarter-cut.
Weight The wood is light in weight at 455 kg/cm^3 (29 lb/cu ft) when air-dry.
Strength Medium.
Processing Red alder dries easily with little degrade. It works easily and takes stain readily. It has good gluing qualities and good nail- and screw-holding properties.
Resistance to decay Red alder has low durability in moist conditions.
Uses The principal use is for furniture and kitchen cabinets. It is also used for woodenware and novelties, firewood, and as pulpwood. It was formerly used for charcoal manufacture.

White Ash – *Fraxinus americana* L.
Wood of commercial white ash may include minor amounts of several closely related species of *Fraxinus* other than *F. americana*.

The Tree
Usually 18–21 m (60–70 ft) in height and up to 60 cm (2 ft) in diameter, white ash is found in eastern Canada from Nova Scotia to southwestern Ontario. In addition, red ash (*Fraxinus pennsylvanica* Marsh.), which is a smaller tree, is found in southern Manitoba and Saskatchewan.

The Wood
General description The heartwood is light brown, and the sapwood is nearly white. The growth-ring figure is conspicuous, and the grain is usually straight, with coarse texture.
Weight The wood is heavy at 690 kg/m^3 (43 lb/cu ft) when air-dry. The wood of red ash is slightly lighter.
Strength White ash is a strong and stiff wood, and its impact strength is high. Because of its lower wood density, red ash is somewhat weaker.
Processing The wood seasons readily with little degrade. Shrinkage in drying is moderate. The wood is above average in most machining characteristics, but it is moderately difficult to glue. Its nail-holding power is high, but it shows some tendency to split in nailing. The wood of white ash bends easily.
Resistance to decay The wood is not resistant under conditions favoring decay.
Uses The wood is used for furniture, cabinets, and handles for spades, shovels, rakes, axes, hammers, and similar tools as well as for sporting goods (baseball bats, hockey-stick handles, oars, tennis-racket frames, snowshoes, skis). It is also used for woodenware and novelties.

Black Ash – *Fraxinus nigra* Marsh.
Other names: swamp ash, brown ash

The Tree
Usually 12–18 m (40–60 ft) in height and 30–60 cm (1–2 ft) in diameter, black ash grows in Newfoundland, Nova Scotia, Prince Edward Island, New Brunswick, Quebec, Ontario, and southeastern Manitoba.

The Wood
General description Black ash is similar to white ash in appearance but the heartwood is usually a darker brown. The band of pores in the early part of each annual ring is wider than in white ash, and the annual rings themselves are usually narrower.
Weight The weight is moderate at 560 kg/m^3 (35 lb/cu ft) when air-dry.
Strength The wood is moderately strong and stiff, and its resistance to impact is fairly high.
Processing The seasoning and working properties are similar to those of white ash.
Resistance to decay The wood is not resistant under conditions favoring decay.
Uses Black ash is used for furniture, cabinets, interior finish, boxes, crates, and pallets. Wood of black ash of higher than average density is used for the same purposes as white ash.

Basswood – *Tilia americana* L.
Other names: linden, whitewood, lime

The Tree
Usually 18–21 m (60–70 ft) in height and 45–75 cm ($1\frac{1}{2}$–$2\frac{1}{2}$ ft) in diameter, the tree is found in the southern areas of New Brunswick, Quebec, Ontario, and Manitoba.

The Wood
General description The wood is light in color, shading from a creamy white to a light brown, with little contrast between the sapwood and the heartwood. The grain is usually straight, the texture fine and even, and the growth-ring figure indistinct. The wood has practically no taste or odor.
Weight The wood is light in weight at 465 kg/m^3 (29 lb/cu ft) when air-dry.
Strength Basswood is soft and not strong, and its resistance to impact is low.
Processing The wood seasons readily without significant degrade. The shrinkage is high, but once dry the wood does not change dimensions unduly. It is worked easily by hand tools, although care is required in some machining operations, such as mortising, turning, and shaping. The wood is easily glued with most types of adhesives and has satisfactory nail- and screw-holding power. It holds paint well and is readily treated with preservatives.
Resistance to decay Basswood is not resistant when exposed to conditions favoring decay.
Uses Basswood lumber is used for the manufacture of dimension stock, furniture (mainly concealed parts), millwork, caskets, mobile homes, advertising displays, models, picture frames, drawing boards, wood carvings, toys, and novelties. The veneer is used for baskets and similar containers as well as for the manufacture of plywood. Smaller amounts of basswood are used for excelsior and, in mixture with other hardwoods, for pulp.

Beech – *Fagus grandifolia* Ehrh.
Other name: American beech

The Tree
Usually about 15 m (50 ft) in height and 45–60 cm ($1\frac{1}{2}$–2 ft) in diameter, occasionally taller, beech is found in Nova Scotia, New Brunswick, Prince Edward Island, southern Quebec, and southern Ontario.

The Wood
General description The sapwood of beech is nearly white, and the heartwood is usually reddish-brown. The growth-ring figure is faint, but the rays are distinctly visible, appearing as flecks or flakes on quarter-sawn surfaces. The texture is fine and even. The wood has no distinctive taste or odor when dry.
Weight The wood is very heavy at 750 kg/m^3 (47 lb/cu ft) when air-dry.
Strength The wood is very strong, stiff, and hard, and its resistance to impact is high.

Processing Beech is difficult to season because the wood shows a tendency to check and distort. Shrinkage is high. Resistance to cutting is high, although machining properties are generally good, particularly in mortising, turning, and planing. Beech has good nail- and screw-holding characteristics but tends to split in nailing unless it is predrilled. It is moderately difficult to glue. Its bending properties are good. The heartwood is moderately difficult to treat with preservatives.
Resistance to decay The wood is not resistant when exposed to conditions favoring decay.
Uses Beech lumber is used for flooring, furniture parts, tool handles, brush backs, sporting goods, pallets, and woodenware. The veneer is used as container veneer and for the manufacture of plywood. Considerable quantities of beech are treated with preservatives and used for railway ties.

Yellow Birch – *Betula alleghaniensis* Britton.
Other names: hard birch, curly birch

The Tree
Yellow birch is the largest of the birches native to Canada, averaging about 18–23 m (60–75 ft) in height and about 60 cm (2 ft) in diameter. It is found from Newfoundland westward to Lake Superior in Ontario and beyond along the international boundary.

The Wood
General description The sapwood is whitish, pale yellow, or light reddish-brown; the heartwood is light to dark brown or reddish-brown. Most of the timber is reasonably straight-grained, although the occurrence of curly grain and wavy grain is not uncommon. The texture is fine and even.
Weight The wood is heavy at 670 kg/m^3 (42 lb/cu ft) when air-dry.
Strength The wood is strong and hard, and its resistance to impact is high.
Processing Yellow birch seasons at a slow rate with little degrade, although its shrinkage is fairly high. It takes a smooth finish and has good machining properties. Its nail- and screw-holding power is high, and it resists splitting by nails and screws. It is a fairly good bending wood but requires care in gluing. The wood can readily be treated with preservatives.
Resistance to decay The wood is not resistant when used in contact with the ground or in other situations where the hazard of decay is high.
Uses Yellow birch is used extensively for furniture, flooring, doors, interior finish, and cabinetwork. It is particularly in demand for veneers and plywood. When treated with preservatives, it is used for railway ties. It is also used for containers, both as lumber and as veneer.

White Birch – *Betula papyrifera* Marsh.
Other names: paper birch, canoe birch, silver birch

The Tree
White birch is usually about 16 m (52 ft) in height and 25–35 cm (10–15 in.) in diameter. The tree is distributed throughout Canada. It

has many botanical varieties, including western paper birch (*B. papyrifera* var. *commutata* [Reg.] Fern.). The wood of the western varieties of white birch and that of Alaska birch (*B. neoalaskana* Sarg.) are known under the general designation 'western birch.'

The Wood

General description The wood is usually creamy white, but a brownish central core is often present. The growth-ring figure is faint; the texture is fine and uniform.
Weight The weight is moderate at 640 kg/m^3 (40 lb/cu ft) when air-dry.
Strength White birch is somewhat weaker than yellow birch and much lower in resistance to suddenly applied loads.
Processing The wood seasons satisfactorily, with relatively high shrinkage. The working properties are very similar to those of yellow birch. The wood is moderately easy to glue.
Resistance to decay White birch is not resistant when exposed to conditions favoring decay.
Uses The wood is used for veneer and plywood, interior finish, furniture, woodenware, toys, brushes, dowels, clothespins, pallets, and crates. It is also used as pulpwood.

Black Cherry – *Prunus serotina* Ehrh.

The Tree

Black cherry is usually about 20 m (65 ft) in height and 45–60 cm (1½–2 ft) in diameter. Scattered trees are found in Nova Scotia, New Brunswick, in the valley of the St Lawrence in Quebec and Ontario, and in the southernmost corner of Ontario.

The Wood

General description The heartwood has a rich reddish-brown color; the sapwood varies from nearly white to light brown. The wood has a fine, even texture, and the growth-ring figure is fairly distinct. Dark red gum streaks are occasionally present in the wood.
Weight The weight is moderate at 610 kg/m^3 (38 lb/cu ft) when air-dry.
Strength The wood is fairly strong, stiff, and hard.
Processing Black cherry seasons readily and has comparatively low shrinkage. It holds its dimensions well after seasoning. It is readily worked by tools and has good machining properties. It glues satisfactorily and takes paints and stains well.
Resistance to decay The wood is not resistant under conditions favoring decay.
Uses Black cherry is used mainly for fine furniture and cabinetwork. It is also used for patterns, instruments, toys, and woodenware. The engraving industry formerly used this wood for mounting engravings, electrotypes, and etchings.

Rock Elm – *Ulmus thomasii* Sarg.
Other names: cork elm, hard elm

The Tree

Rock elm is usually 15–21 m (50–70 ft) in height and 60 cm (2 ft) in diameter. In Canada, it is confined to the southern parts of Ontario and to a few localities in Quebec.

The Wood

General description The heartwood is light brown, and the sapwood ranges from nearly white to pale brown. The growth rings are fairly distinct, and the grain is sometimes interlocked. The wood has no characteristic odor or taste.

Weight The wood is very heavy at 735 kg/m^3 (46 lb/cu ft) when air-dry.

Strength The wood is stronger, stiffer, and harder than that of other Canadian elms. It has very high shock resistance and high resistance to wear.

Processing Rock elm is difficult to season because of its tendency to check and warp. It has high shrinkage in drying. The wood is fairly difficult to work, mainly because of its high density. It has good bending, nail-holding, and screw-holding properties. It is easy to glue with all types of adhesives.

Resistance to decay The heartwood shows some resistance to decay.

Uses Rock elm is used for blades of hockey sticks and for other sports equipment, ladder rungs, components of boats, tool handles, bentwood, and wharf and dock fenders.

White Elm – *Ulmus americana* L.

Other names: American elm, soft elm

The Tree

White elm is one of the largest hardwood trees in Canada, sometimes reaching 38 m (125 ft) in height although it is usually 18–24 m (60–80 ft) in height and 60–90 cm (2–3 ft) in diameter. It is found in all provinces except British Columbia, Alberta, and Newfoundland.

The Wood

General description The heartwood varies from pale brown to medium brown; the sapwood is nearly white. The annual rings are distinct; the texture is coarse and the grain is sometimes interlocked.

Weight The weight is moderate at 625 kg/m^3 (39 lb/cu ft) when air-dry.

Strength The wood is of medium strength, considerably less strong, stiff, and hard than rock elm. Resistance to impact is high.

Processing The wood seasons readily with a tendency to warp. It has medium shrinkage in drying. It is fairly difficult to plane, turn, and work with hand tools. It takes nails and screws without splitting, and its nail- and screw-holding properties are very good. It is easy to glue with all types of adhesives. White elm is a good bending wood.

Resistance to decay The wood is not resistant under conditions favoring decay.

Uses White elm is used for furniture, boxes and crates, pallets, caskets, hockey sticks, handles, lobster traps, and ladder rungs. Veneer is made into containers, baskets, and plywood that is used in interior finishing and panelling and as construction plywood. The wood is also used for pulp.

Slippery Elm – *Ulmus rubra* Mühl.
Other names: red elm, soft elm

The Tree
Slippery elm is usually about 16 m (52 ft) in height and 30–60 cm (1–2 ft) in diameter. It is found in the valley of the St Lawrence as well as in the southernmost parts of Ontario.

The Wood
General description The heartwood is brown, and the sapwood is nearly white. The wood is generally similar to white elm in appearance.
Weight The wood is heavy at 670 kg/m^3 (42 lb/cu ft) when air-dry.
Strength Since it has a slightly higher density than white elm, the wood is somewhat stronger, stiffer, and harder.
Processing Slippery elm's processing qualities are similar to those of white elm.
Resistance to decay The wood is not resistant under conditions favoring decay.
Uses Slippery elm has the same uses that white elm does.

Hickory – *Carya* spp.
Shagbark Hickory – *Carya ovata* (Mill.) K. Koch
Big Shellbark Hickory – *Carya laciniosa* (Michx. f.) Loud.
Mockernut Hickory – *Carya tomentosa* Nutt.
Pignut Hickory – *Carya glabra* (Mill.) Sweet
Bitternut Hickory – *Carya cordiformis* (Wang.) K. Koch

The Tree
The tallest of the Canadian hickories is mockernut hickory, with heights up to 27 m (90 ft), closely followed by the big shellbark and shagbark hickories. The approximate height of pignut and bitternut hickories is 12–18 m (40–60 ft). The average diameter of hickories at breast height ranges from 30 to 90 cm (1 to 3 ft). Hickory is not widespread in Canada, and three species are confined mainly to the southernmost parts of Ontario. Shagbark hickory and bitternut hickory have wider natural distributions in southern Ontario and in the valley of the St Lawrence.

The Wood
General description The wood of the different species of hickories is very similar. The heartwood is pale brown to brown, frequently with a reddish tinge; the sapwood is white and wide. The annual rings are distinct, and the grain is usually straight.
Weight The wood of all hickories is very heavy. The average weight for shagbark hickory is 800 kg/m^3 (51 lb/cu ft) when air-dry; for bitternut hickory, 750 kg/m^3 (47 lb/cu ft) when air-dry.
Strength There are slight differences in the strength of the different species of hickory, but all hickories are very strong, stiff, and hard. The resistance of this wood to impact is exceptionally high.
Processing The wood dries slowly and shrinks considerably in drying. It is difficult to work, but its turning and shaping characteristics are good. The nail- and screw-holding power is high, but predrilling is

necessary. Care is required to achieve satisfactory gluing. The wood holds its shape well when bent by steaming it.

Resistance to decay Hickory is not resistant under conditions favoring decay.
Uses The main use of hickory is for tool handles and other handles requiring high shock resistance. Hickory is also used for sporting goods, ladder rungs, machine parts, and turnery.

Ironwood – *Ostrya virginiana* (Mill.) K. Koch
Other name: hop-hornbeam

The Tree
Usually small, 7–12 m (25–40 ft) high and 15–25 cm (6–10 in.) in diameter, the tree is found from Nova Scotia to Ontario and in southeastern Manitoba.

The Wood
General description The heartwood is light brown, and the sapwood is nearly white. The wood has a fine texture, and the growth-ring figure is inconspicuous.
Weight The wood is very heavy at 800 kg/m^3 (51 lb/cu ft) when air-dry.
Strength The wood is very strong, stiff, and hard. Its resistance to impact is very high.
Processing Ironwood shrinks considerably in drying. It is difficult to cut.
Resistance to decay Ironwood is not resistant under conditions favorable to decay.
Uses Because of its small size and limited availability, ironwood is of little commercial importance. However, it is used to some extent for tool handles, ladder rungs, mallets, and the like because of its great strength and toughness.

Hard Maple – *Acer* spp.
Sugar Maple – *Acer saccharum* Marsh.
Black Maple – *Acer nigrum* Michx. f.
The wood of the two species has similar structure and properties and is not segregated commercially.

The Tree
Both species commonly reach heights of 24–27 m (80–90 ft) and diameters up to 75 cm (2½ ft). Sugar maple is the principal species of hard maple. Its range includes the Great Lakes–St Lawrence area, in southern Ontario and Quebec and the Maritime provinces (excluding Newfoundland); the range of black maple is limited mainly to the southernmost corner of Ontario.

The Wood
General description The wood varies from creamy white to light brown, often with a reddish tinge. Some trees may have a darker central core. Dark streaks and patches ('mineral stain') caused by injuries to the growing trees, are common in lumber from some regions. The grain is

Figure 2.7 Hard maple floor of a gymnasium. (Photo: Canadian Wood Council)

usually straight but may be wavy or curly. The texture is fine, and the growth-ring figure relatively distinct. The occurrence of a bird's-eye figure is more common in hard maple than in other species.

Weight The wood is very heavy at 735 kg/m^3 (46 lb/cu ft) when air-dry.

Strength The strength, stiffness, and shock resistance of hard maple are great, as are the surface hardness and resistance to wear.

Processing The wood dries slowly, without difficulty. Shrinkage is high, but the tendency to warp is only moderate. Hard maple is relatively easy to shape and carve without splintering and chipping, although its resistance to cutting is high. It takes paints and stains satisfactorily but requires care for satisfactory gluing. High in nail- and screw-holding power, it tends to split if not predrilled. It is a moderately good bending wood.

Resistance to decay Hard maple has low resistance under conditions favoring decay. It should be pressure-treated with preservatives for such uses, although the wood is somewhat resistant to impregnation.

Uses Hard maple is one of the most important commercial species in Canada. It is used for furniture, interior finish, plywood, flooring, bowling alleys, sporting goods, chopping blocks, tool handles, turnery objects, container veneer, musical instruments, pulpwood, and preservative-treated crossties.

Soft Maple – Acer spp.
Silver Maple – *Acer saccharinum* L.
Red Maple – *Acer rubrum* L.
The wood of the two species has similar structure and properties and is not differentiated commercially.

The Tree

Silver maple usually attains a height of 24–27 m (80–90 ft) and a diameter of up to 90 cm (3 ft). Red maple is not as large, usually reaching 15–23 m (50–75 ft) in height and 60 cm (2 ft) in diameter. Silver maple is

Figure 2.8 Children's furniture made of hard maple. (Photo: Simmons Limited)

found in New Brunswick and throughout southern Quebec and southern Ontario. The natural distribution of red maple is wider, comprising Newfoundland, Prince Edward Island, Nova Scotia, New Brunswick, southern Quebec, and southern Ontario.

The Wood

General description The wood is usually creamy white to light brown. The grain is usually straight, and the growth-ring figure faint.

Weight The wood of red maple is heavier than that of silver maple. The average weight of red maple when air-dry is 610 kg/m³ (38 lb/cu ft), and of silver maple 530 kg/m³ (33 lb/cu ft).

Strength The wood is considerably weaker than that of hard maple. It is usually rated as fairly low in strength and stiffness. Red maple is stronger than silver maple, particularly in bending. The resistance of soft maple to impact is relatively low.

Processing The wood seasons satisfactorily and shrinks moderately in drying. The machining properties are not as good as those of hard maple. Nail-holding is good, but the wood shows some tendency to split if not predrilled. Its bending properties are slightly better than those of hard maple.

Resistance to decay Soft maple is not resistant under conditions favorable to decay.

Uses The wood is used for furniture, interior finish, and some other purposes for which hard maple is used. It is also used for boxes, crates, and pallets. The veneer is used for containers and for plywood. In addition, soft maples are used for pulp.

Bigleaf Maple – *Acer macrophyllum* Pursh
Other name: broadleaf maple

The Tree

The tree is the only commercial maple in British Columbia. It grows on Vancouver Island and the adjacent mainland and islands. It normally reaches 15–21 m (50–70 ft) in height and 60–90 cm (2–3 ft) in diameter. On good sites it may grow to a height of 30 m (100 ft).

The Wood

General description The sapwood is pinkish, and the heartwood is pinkish-brown. The wood has a uniform texture. Occasional trees have a wavy grain.

Weight The weight is moderate at 560 kg/m³ (35 lb/cu ft) when air-dry.

Strength This maple is similar in strength to eastern soft maple.

Processing The wood seasons well, with little or no degrade. It finishes easily, takes stains well, and has good gluing and nailing properties.

Resistance to decay Bigleaf maple is not durable under conditions favorable to decay.

Uses The main use of bigleaf maple is for furniture. Other uses are for turning, musical instruments, kitchen cabinets, and interior finish.

White Oak – *Quercus alba* L.

Five other species of Canadian oak produce wood of similar structure and properties that is marketed under the name of white oak. These species, all of less commercial importance, are: bur oak – *Quercus macrocarpa* Michx.; Garry oak – *Quercus garryana* Dougl.; swamp white oak – *Quercus bicolor* Willd.; Chinquapin oak – *Quercus muehlenbergii* Engelm.; chestnut oak – *Quercus prinus* L.

The Tree

White oak is commonly 15–30 m (50–100 ft) in height and 60–120 cm (2–4 ft) in diameter. It is found in southern Ontario and southern Quebec. Other oaks of the white oak group are usually smaller and have more limited distribution. An exception is bur oak, a small tree that grows not only in southern Ontario and Quebec but also in New Brunswick, Manitoba, and the southeastern corner of Saskatchewan. Garry oak is confined to southwestern British Columbia.

The Wood

General description The heartwood is light brown or grayish-brown; the sapwood is nearly white. The wood has clearly defined annual growth rings separated by bands of large vessels. It is characterized by broad rays, which are particularly conspicuous on edge-grain surfaces. The wood weathers to a dark gray with conspicuous weather checks.
Weight The wood is very heavy. The average air-dry weight of white oak is 750 kg/m^3 (47 lb/cu ft), and that of bur oak is 735 kg/m^3 (46 lb/cu ft).
Strength The wood is strong, stiff, and hard. The strength of bur oak is somewhat lower than that of white oak. Resistance to impact is very high.
Processing The wood dries relatively slowly and tends to check in seasoning. Its shrinkage is fairly high. It is easily worked, particularly in planing, shaping, and mortising operations. The nail- and screw-holding power is high, but predrilling is usually recommended to prevent splitting. The wood is moderately difficult to glue. It is an excellent bending timber. The sapwood is moderately resistant to impregnation; the heartwood is extremely resistant.
Resistance to decay The heartwood is resistant to decay.
Uses The wood is valued for furniture, cabinetwork, and flooring. It is also used for tight cooperage, as in whiskey casks, and for boats, interior finish, doors, and decorative plywood.

Red Oak – *Quercus rubra* L.

Two other Canadian oaks of less commercial importance than white oak produce wood of similar structure and properties, which is marketed under the name of red oak. They are: black oak – *Quercus velutina* Lam.; pin oak – *Quercus palustris* Muenchh.

The Tree

Red oak is usually 18–24 m (60–80 ft) in height and 30–90 cm (1–3 ft) in diameter. On a superior site it may exceed 40 m (100 ft) in height and 120 cm (4 ft) in diameter. Red oak is found in southern Ontario, southern Quebec, and the Maritimes. Black oak and pin oak are similar in size but are found only in the southernmost corner of Ontario.

The Wood

General description The heartwood is darker than that of white oak,

Figure 2.9 High-quality furniture of hard maple solids and selected sugar maple veneers. (Photo: Kaufman of Collingwood)

often with a reddish-brown tinge, and the sapwood is nearly white. The wood is not so fine in texture as white oak and the broad rays are not as prominent on flat-grain surfaces. It weathers to a dark gray with conspicuous checks.

Weight The wood is heavy at 690 kg/m^3 (43 lb/cu ft) when air-dry.

Strength The wood is strong and hard although somewhat weaker than white oak. Its resistance to impact is high.

Processing The wood seasons relatively slowly with moderate shrinkage. It works fairly easily and has good machining properties. It holds nails and screws well but tends to split unless predrilled. It is moderately easy to glue and has very good bending properties. The heartwood is moderately resistant to preservative treatment.

Resistance to decay Red oak is not resistant under conditions favorable to decay.

Uses The uses of the wood are similar to those of white oak: furniture, cabinetwork, flooring, interior finish, doors, and decorative plywood. It is not suitable for tight cooperage. For uses requiring wood of high durability, pressure treatment with preservatives is required.

Aspen – *Populus* spp.
Trembling Aspen – *Populus tremuloides* Michx.
Largetooth Aspen – *Populus grandidentata* Michx.
The woods of the two species cannot be distinguished from each other and are marketed together as aspen.

The Tree
Aspen is usually 12–18 m (40–60 ft) in height and up to 60 cm (2 ft) in diameter. Trembling aspen is the most widely distributed hardwood species in Canada, ranging from Newfoundland and Labrador to British Columbia and the Yukon. Largetooth aspen is found in Nova Scotia, New Brunswick, Prince Edward Island, southern Quebec, and Ontario.

The Wood
General description The wood is light in color, ranging from nearly white to grayish-white. It is usually straight-grained with fine and even texture and a faint growth-ring figure. It weathers to a light gray with a silvery sheen.
Weight The wood is light in weight at 450 kg/m^3 (28 lb/cu ft) when air-dry.
Strength It is comparable to white spruce in strength. Resistance to wear is strikingly high for a low-density wood.
Processing The wood seasons satisfactorily, with moderate shrinkage. Care is required in machining to produce quality surfaces, particularly if tension wood is present. The wood can be readily cut into veneer. It holds nails satisfactorily, with little danger of splitting. It is moderately easy to glue.
Resistance to decay Aspen is not resistant under conditions favorable to decay.
Uses Aspen is widely used for the manufacture of plywood and composite board products and as pulpwood. It is also used for pallets, boxes and crates, furniture parts, lumber core plywood used in the manufacture of furniture, interior trim, and dimension framing lumber.

Balsam Poplar – *Populus balsamifera* L.
Other names: tacamahac, black poplar

The Tree
Balsam poplar is usually 18–24 m (60–80 ft) in height and 30–60 cm (1–2 ft) in diameter. It is found in all provinces of Canada.

The Wood
General description The heartwood zone in balsam poplar is not clearly defined; it is usually grayish-brown, often with a reddish-brown tinge. The sapwood is nearly white. The wood has a faint growth-ring figure. The texture is somewhat coarser than that of aspen, and there is a higher incidence of wet pockets.
Weight The wood is light in weight at 465 kg/m^3 (29 lb/cu ft) when air-dry.
Strength The strength properties are similar to those of aspen. Resistance to impact is low.

Processing The processing characteristics are similar to those of aspen.
Resistance to decay Balsam poplar is not resistant under conditions favorable to decay.
Uses Balsam poplar is used for plywood, particleboard, and pulp in addition to other products for which aspen is also employed.

Black Cottonwood – *Populus trichocarpa* Torr. & Gray
Other name: western balsam poplar

The Tree
Black cottonwood grows throughout British Columbia west of the Rocky Mountains in low-lying damp areas, principally along river bottoms. It also extends through the mountain passes into Alberta. It is the largest of the native poplars and the largest broadleaf tree in British Columbia. The tree is very similar to balsam poplar in form and appearance. It commonly reaches 24–38 m (80–125 ft) in height and 90–150 cm (3–5 ft) in diameter.

The Wood
General description The wood is white to light grayish-brown and odorless when dry, but it has a sour odor when moist. The annual rings are distinct but inconspicuous. The sapwood and the heartwood are not clearly defined. A tree can produce considerable amounts of clear lumber and veneer because of its large size.
Weight The wood is very light in weight at 350 kg/m^3 (22 lb/cu ft) when air-dry.
Strength Low.
Processing The wood seasons easily but slowly because of the very high moisture content of the green or undried wood. It has fairly good machining properties and takes stains and paints readily. It does not split with nailing.
Resistance to decay The durability of the wood is low.
Uses Black cottonwood is used for veneer, plywood, and furniture parts. Other uses are for pulp, berry boxes, excelsior, other boxes, and crates.

Eastern Cottonwood – *Populus deltoides* Bartr.

The Tree
Eastern cottonwood is usually 23–30 m (75–100 ft) high and 60–100 cm (2–3½ ft) in diameter. It is found in limited quantities in southern Quebec and southeastern Ontario. A western variety of the tree, plains cottonwood (*Populus deltoides* var. *occidentalis* Rydb.), grows in southern Manitoba, Saskatchewan, and Alberta.

The Wood
General description The heartwood is grayish to pale brown, and the sapwood is nearly white. The growth-ring figure is faint, and the texture rather coarse.
Weight The wood is light in weight at 430 kg/m^3 (27 lb/cu ft) when air-dry.
Strength The wood is weak, soft, and low in shock resistance.

Processing The processing characteristics are similar to those of aspen although eastern cottonwood is generally more difficult to machine. Its capacity to hold nails is also lower than that of aspen.
Resistance to decay Eastern cottonwood is not resistant under conditions favorable to decay.
Uses The wood is used for pulp, plywood, particleboard, boxes and crates, concealed furniture parts, and excelsior.

Black Walnut – *Juglans nigra* L.
Other names: walnut, American walnut

The Tree
Usually 15–27 m (50–90 ft) in height and 60 cm (2 ft) or more in diameter, black walnut grows in the warmest parts of Ontario, bordering Lakes Ontario, Erie, and St Clair.

The Wood
General description The color of the heartwood varies from light brown to rich dark brown. The sapwood is nearly white. The wood is usually straight-grained, with occasional irregularities of grain.
Weight The wood is heavy at 655 kg/m^3 (41 lb/cu ft) when air-dry.
Strength The wood is strong, stiff, and hard. Its resistance to impact is high.
Processing The wood dries slowly but holds its shape well after seasoning. It has excellent machining properties. It takes and holds paints and stains well and glues satisfactorily.
Resistance to decay The heartwood is highly decay resistant.
Uses Black walnut is used for high-quality furniture and for television, phonograph, and radio cabinets. It is used either in the form of solid wood or as face veneer over plywood or other core stock. The wood is employed for interior finish in homes, restaurants, and public buildings. Other special uses include gunstocks, piano cases, and novelties such as ornaments.

Butternut – *Juglans cinerea* L.
Other name: white walnut

The Tree
Usually 12–15 m (40–50 ft) high and 30–75 cm (1–2½ ft) in diameter, butternut is found in New Brunswick, the valley of the St Lawrence, and southern Ontario.

The Wood
General description The heartwood is a light chestnut brown, and the sapwood varies from almost white to light brown. The wood somewhat resembles black walnut in general appearance and texture. It is without characteristic odor or taste.
Weight The wood is light in weight at 430 kg/m^3 (27 lb/cu ft) when air-dry.
Strength The wood is weak and soft, and its resistance to impact is low.
Processing The wood has low shrinkage and seasons readily. It works well and can be finished to resemble black walnut.

Resistance to decay Butternut is not resistant under conditions favoring decay.
Uses Butternut is used for interior finish, furniture, cabinetwork, boxes, toys, and household woodenware.

Table 2.1 *Common and scientific names of Canadian species*

Common names of Canadian species described in this chapter are given here in English and French. Additional local names by which a species may be identified within a specific locality are excluded but may be found in the description of that species in the text under the heading 'Other names.'

English	French	Scientific
SOFTWOODS		
Eastern white cedar	Thuya occidental	*Thuja occidentalis*
Western red cedar	Thuya géant	*Thuja plicata*
Yellow cedar	Cèdre jaune	*Chamaecyparis nootkatensis*
Douglas-fir	Douglas taxifolié	*Pseudotsuga menziensii*
Balsam fir	Sapin baumier	*Abies balsamea*
Amabilis fir	Sapin gracieux	*Abies amabilis*
Grand fir	Sapin grandissime	*Abies grandis*
Alpine fir	Sapin subalpin	*Abies lasiocarpa*
Eastern hemlock	Pruche du Canada	*Tsuga canadensis*
Western hemlock	Pruche occidentale	*Tsuga heterophylla*
Mountain hemlock	Pruche subalpine	*Tsuga mertensiana*
Western larch	Mélèze occidental	*Larix occidentalis*
Tamarack	Mélèze laricin	*Larix laricina*
Eastern white pine	Pin blanc	*Pinus strobus*
Western white pine	Pin argenté	*Pinus monticola*
Ponderosa pine	Pin ponderosa	*Pinus ponderosa*
Red pine	Pin rouge	*Pinus resinosa*
Jack pine	Pin gris	*Pinus banksiana*
Lodgepole pine	Pin tordu	*Pinus contorta*
White spruce	Épinette blanche	*Picea glauca*
Black spruce	Épinette noire	*Picea mariana*
Red spruce	Épinette rouge	*Picea rubens*
Engelmann spruce	Épinette d'Engelmann	*Picea engelmannii*
Sitka spruce	Épinette de Sitka	*Picea sitchensis*
HARDWOODS		
Red alder	Aulne rouge	*Alnus rubra*
White ash	Frêne blanc	*Fraxinus americana*
Red ash	Frêne rouge	*Fraxinus pennsylvanica*
Black ash	Frêne noir	*Fraxinus nigra*
Basswood	Tilleul d'Amérique	*Tilia americana*
Beech	Hêtre à grandes feuilles	*Fagus grandifolia*
Yellow birch	Bouleau jaune	*Betula alleghaniensis*
White birch	Bouleau à papier	*Betula papyrifera*
Alaska birch	Bouleau d'Alaska	*Betula neoalaskana*
Black cherry	Cerisier tardif	*Prunus serotina*
Rock elm	Orme liège	*Ulmus thomasii*
White elm	Orme d'Amérique	*Ulmus americana*
Slippery elm	Orme rouge	*Ulmus rubra*
Shagbark hickory	Caryer ovale	*Carya ovata*
Big shellbark hickory	Caryer lacinié	*Carya laciniosa*
Mockernut hickory	Caryer tomenteux	*Carya tomentosa*
Pignut hickory	Caryer glabre	*Carya glabra*
Bitternut hickory	Caryer cordiforme	*Carya cordiformis*
Ironwood	Ostryer de Virginie	*Ostrya virginiana*
Sugar maple	Érable à sucre	*Acer saccharum*

Table 2.1 *(concluded)*

English	French	Scientific
Black maple	Érable noir	*Acer nigrum*
Silver maple	Érable argenté	*Acer saccharinum*
Red maple	Érable rouge	*Acer rubrum*
Bigleaf maple	Érable grandifolié	*Acer macrophyllum*
White oak	Chêne blanc	*Quercus alba*
Bur oak	Chêne à gros fruits	*Quercus macrocarpa*
Garry oak	Chêne de Garry	*Quercus garryana*
Swamp white oak	Chêne bicolore	*Quercus bicolor*
Chinquapin oak	Chêne jaune	*Quercus muehlenbergii*
Chestnut oak	Chêne châtaignier	*Quercus prinus*
Red oak	Chêne rouge	*Quercus rubra*
Black oak	Chêne noir	*Quercus velutina*
Pin oak	Chêne palustre	*Quercus palustris*
Trembling aspen	Peuplier faux-tremble	*Populus tremuloides*
Largetooth aspen	Peuplier à grandes dents	*Populus grandidentata*
Balsam poplar	Peuplier baumier	*Populus balsamifera*
Black cottonwood	Peuplier occidental	*Populus trichocarpa*
Eastern cottonwood	Peuplier deltoïde	*Populus deltoides*
Black walnut	Noyer noir	*Juglans nigra*
Butternut	Noyer cendré	*Juglans cinerea*

REFERENCE

Hosie, R.C. 1969. *Native trees of Canada*. 7th ed. Can. For. Serv., Environment Canada, Ottawa

BIBLIOGRAPHY

Bergin, E.G. 1964. The gluability of various eastern Canadian wood species. *Can. Wood Prod. Ind.* 64 (9): 17

Cantin, M. 1965. *The machining properties of 16 eastern Canadian woods*. Dep. For. Publ. 1111. Ottawa

Coleman, D.G. 1966. *Woodworking factbook*. New York: R. Speller & Sons

Farmer, R.H., ed. 1972. *Handbook of hardwoods*. 2nd ed. Princes Riseborough Lab., Build. Res. Establ., London

Kennedy, E.I. 1965. *Strength and related properties of woods grown in Canada*. Dep. For. Publ. 1104. Ottawa

McElhanney, T.A. 1951. Commercial timbers of Canada. In *Canadian woods: Their properties and uses*. 2nd ed. For. Prod. Lab. For. Branch, Ottawa

Panshin, A.J., and de Zeeuw, A.J. 1970. *Textbook of wood technology*, Vol. 1. 3rd ed. Toronto: McGraw-Hill

U.S. Forest Products Laboratory. 1974. *Wood handbook: Wood as an engineering material*. U.S. Dep. Agric. Handb. No. 72 (rev.). Washington, DC

3

C.T. KEITH

Eastern Forest Products Laboratory, Ottawa

R.M. KELLOGG

Western Forest Products Laboratory, Vancouver

The Structure of Wood

Wood is a material of beauty. Most people are attracted to its warmth, color, and shape. Beneath these surface attributes lies a structure of great diversity and complexity, which is seldom seen or appreciated. One cannot help but marvel at the beauty and intricacy revealed under magnification (Figure 3.1). But our interest goes far beyond the aesthetic, for the properties, and therefore the usefulness, of a wood depend on these details of structure and cell-wall composition. In order to use wood wisely, one must understand its biological origin, its structure and composition, its abnormalities, and its natural variability.

Figure 3.1 Scanning electron micrograph (200×) of aspen showing cross-section of surface.

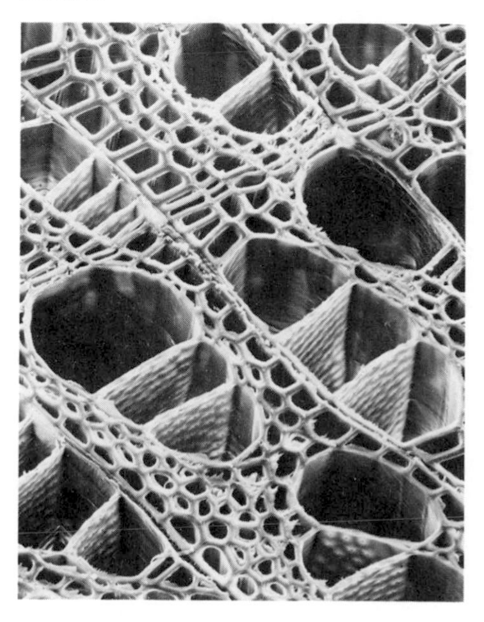

Woods are grouped into two general classes, commonly called softwoods and hardwoods. These terms are not entirely accurate, because some softwoods are actually harder than some hardwoods. The two classes are distinguished by differences in physical properties that are the result of differences in the structure of the wood and make them suitable for different uses.

Botanically, the softwoods are conifers, classified in a group called gymnosperms, which are plants having exposed seeds, usually borne in cones. Pines and spruces are well-known examples of this group. Canadian softwoods have needlelike or scalelike leaves, which, except in the larches, persist throughout the winter.

The other botanical group, which consists of the various orders of hardwoods, is known as the angiosperms. These have true flowers and seeds enclosed in a fruit. The broad leaves of principal Canadian hardwoods are deciduous, although in certain cases the dead leaves may remain on the trees for considerable periods before falling.

TREE GROWTH

When a tree seed germinates under favorable conditions, it sends up a young shoot, which forms a layer of wood around a soft, central core of pith. From this stage onward, as long as the tree remains alive, it

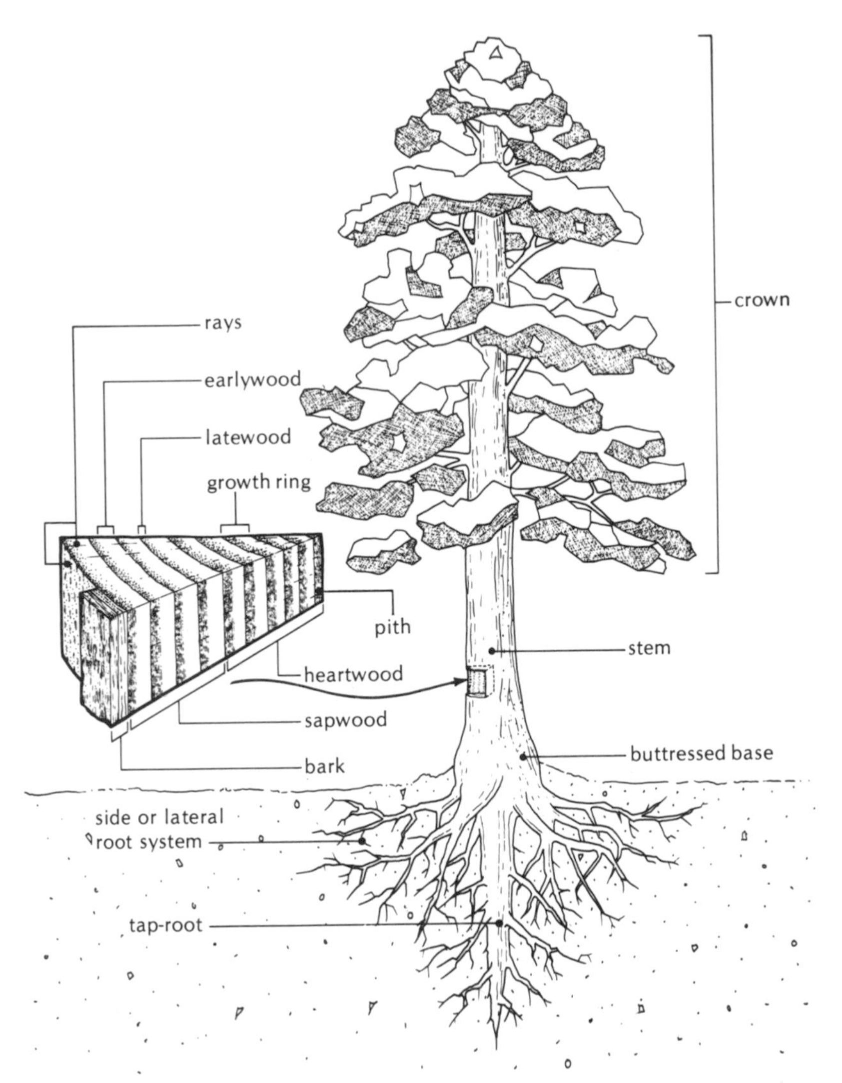

Figure 3.2 Terminology of the parts of a tree and structure of wood.

continues to extend its crown and its root system and to increase the diameter of its stem, limbs, and roots by the addition of new layers of wood and bark (Figure 3.2). Sap, which is a dilute solution of minerals derived through the roots, is conducted through the tubelike cells of the outer layers of wood, or sapwood, to the crown of the tree. Here in the leaves, through the process known as photosynthesis, the sap combines with gases from the air, in the presence of sunlight and chlorophyll, to produce the food required for growth. This food is transported through the conductive cells of the inner bark to the growing tissues.

Growth in Height

Trees grow in length by cell division in apical meristems, or growing points, located at the tips of all twigs and roots. Through this cell division and elongation of newly formed cells during the growing season, increments of new tissue are added to extend the height of the tree and the length of its branches and roots.

Because growth in trees is accomplished by periodic additions to the top (and to the branch tips), it is apparent that an object or marker attached to the stem of a young tree will not move upward as the tree grows older but will remain essentially at the same position.

Figure 3.3 Cross-section of pine showing cells of cambium region, wood, and bark.

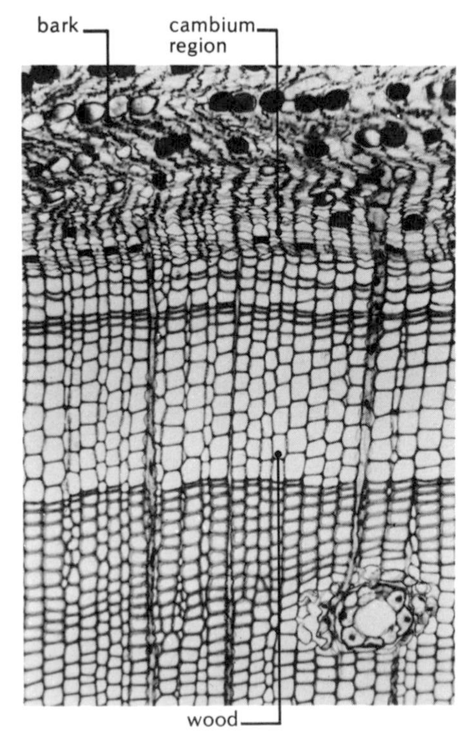

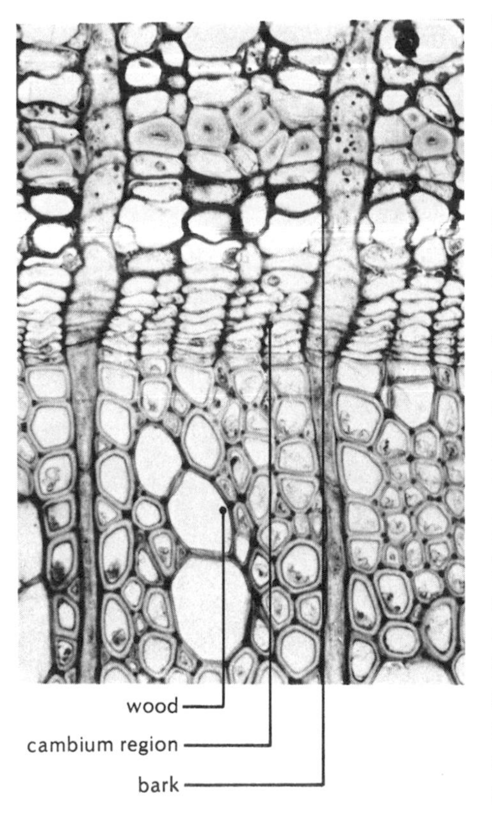

Figure 3.4 Cross-section of poplar showing cells of cambium region, wood, and bark.

Growth in Diameter

Tree stems (as well as branches and roots) increase in diameter through cell division of the cambium, a thin layer of growth-producing tissue located between the wood and bark (Figure 3.3). The cambium envelops the wood of the entire body of the tree and is progressively extended into the new increments of height growth as these are produced. Through the process of cell division, the cambium produces new wood from its inner surface and new bark from its outer surface, thus increasing the diameter of the stem, branches, and roots.

Because of their common origin from a single cambial cell, wood cells tend to be aligned in radial files. These can be seen more clearly in softwoods (Figure 3.3) than in the hardwoods (Figure 3.4), because softwood cells are relatively uniform. Through the activity of the cambium, a new layer of wood develops (each year, in temperate latitudes) outside the previous wood of the stem and branches (Figure 3.5). This concentric arrangement of growth layers is visible on the ends of logs of most of the common kinds of timber. These growth layers or growth rings are called annual layers or, more commonly, annual rings, because they generally consist of the wood added during one year. The annual rings are most readily seen in species where there is a well-defined boundary between earlywood and latewood. In softwoods, the visibility of the rings is due to the contrast between the thick-walled latewood cells, produced at the end of the last growing season, and the thin-walled earlywood cells, produced when growth resumed the following spring.

CHARACTERISTICS OF WOOD

Sapwood and Heartwood

When first formed, layers of wood are light-colored and function as sapwood, conducting sap and storing reserve food. The complete band of light-colored sapwood is normally composed of several years' growth, its thickness depending in general on the age of the tree, the species, and the rate of growth.

After a number of years, each annual layer undergoes certain physical and chemical changes, ceases to conduct sap, and becomes

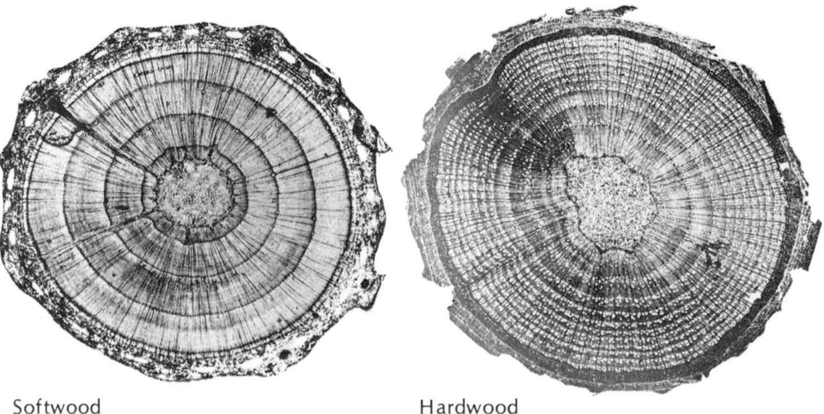

Figure 3.5 Cross-sections of young trees showing the outer bark and several annual layers of wood surrounding the central pith.

heartwood (Figure 3.6). Infiltration of the tissues by extraneous organic substances, commonly called heartwood extractives, normally occurs during this transformation and results in a conspicuous darkening in the color of the wood in some species, such as black walnut and western red cedar. The spruces and balsam fir of eastern Canada are examples of species that have a low content of extractives and therefore show virtually no difference in color between heartwood and sapwood. The high decay resistance of the heartwood of certain species, such as white oak and western red cedar, is due to the presence of substances in the heartwood extractives which are toxic to decay fungi. However, it is not uncommon to find trees in which the heartwood has decayed and the stem consists of a shell of sapwood. In these trees the flow of sap and the protective mechanisms of the living tissues have prevented fungi from successfully attacking the sapwood.

Figure 3.6 Cross-section of white elm showing heartwood, sapwood, and bark.

As heartwood forms, the cell walls and pit membranes may become encrusted and vessel cavities plugged with ingrowths of small cells known as tyloses, causing a significant reduction in permeability. The heartwood of white oak, for example, is impermeable because its vessels are closed with tyloses, and thus it is a choice material for staves for tight cooperage. In some species, such as spruce, the low permeability of the heartwood may be an advantage, by maintaining the buoyancy of pulpwood bolts during water driving, or a disadvantage by resisting penetration by preservative treating solutions for protection against decay.

Because of the deposition of resins, gums, tannins, and other secretions, the heartwood of some species may have considerably greater density than the sapwood. Extractives are sometimes present in sufficient quantities to cause slight increases in compressive strength and hardness. The sapwood of freshly felled trees, especially conifers, is likely to be much heavier than the heartwood because of its higher moisture content. In some hardwoods, however, the heartwood may have the higher moisture content.

Some woods are characterized by a thin layer of sapwood; others have sapwood that may be several centimetres thick. The width of sapwood varies considerably even within one species. Apparently, no specific length of time is required for sapwood to change into heartwood. The transition to heartwood seems to be governed partly by the time that has elapsed since the wood was formed and partly by the distance that separates it from the periphery of the tree. Thus, the sapwood of trees with wide annual rings is likely to change into heartwood somewhat sooner than the sapwood of trees with narrow annual rings. Fast-growing trees therefore tend to have fewer annual rings in the sapwood than slow-growing trees do, but the sapwood is generally wider than that of slow-growing trees. Slow growth tends to produce a narrow zone of sapwood containing a relatively large number of annual layers.

Grain, Texture, and Figure

These three terms are frequently used in describing timber. Unfortunately, in popular use, they have so many different meanings as to be confusing, if not contradictory, and therefore require careful definition in any discussion of lumber.

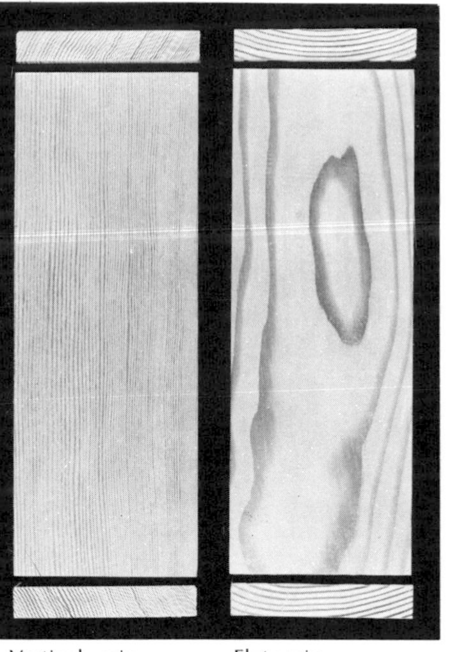

Figure 3.7 Edge- or vertical-grain and flat-grain boards of Douglas-fir, showing how figure is determined by plane of cutting. Edge-grain boards are cut radially to the annual growth rings or parallel to the rays; flat-grain boards are cut tangentially to the annual growth rings or essentially at right angles to the rays.

Grain

According to widely accepted terminology, grain refers to the direction of the cellular elements of wood that determines a plane of cleavage. A straight-grained piece of timber is one that splits parallel to the axis of the piece.

Grain is sometimes used to signify the size of the constituent cells of wood; thus, some of the hardwoods that contain cells of sufficiently large diameter to be conspicuous are called coarse-grained or open-grained, particularly by painters and finishers, whereas woods composed of small-diameter cells are described as fine-grained or close-grained. Some woodworkers use the term grain to refer to a characteristic of the annual rings; wood with wide, conspicuous annual layers is called coarse-grained, and wood with narrow layers is called fine-grained.

The terms edge grain and flat grain (Figure 3.7) have achieved such wide use that they are necessary to any wood user's vocabulary. An edge-grain or vertical-grain board is one in which the annual rings are oriented at approximately right angles with the wide face. The faces of these boards are approximately in the same plane as the rays, which in some woods, such as oak and beech, show a characteristic figure, sometimes called silver grain or flake.

Flat-grain (or flat-sawn) boards have faces that are approximately parallel to the annual rings. When they are cut from the periphery of very large trees, the annual layers show very little curvature. Such use of the term grain arises from the conception of grain as synonymous with annual layer.

Texture

In common practice the terms texture and grain are sometimes used interchangeably, a practice that may lead to confusion and misunderstanding. In its generally accepted meaning, texture describes the prevailing size of the cellular components of wood. To a lesser extent, texture may also be used to describe the relative uniformity in size of the structural components of wood. Timbers such as white pine or basswood, for example, might be described as fine-textured or even-textured because their cells are uniformly small in diameter. This property makes them suitable for pattern making and carving. Bands of large vessels in such hardwoods as ash and hickory result in wood of coarse texture that requires the application of a filler during the finishing operation of furniture.

Figure

This term is used to describe any characteristic pattern visible on the surface of wood. Figure may originate from a variety of causes, including growth increments, grain deviations, and variations in color.

The flat sawing of lumber and, similarly, the rotary peeling of veneer produce the irregular, concentric, and nested parabolic patterns characteristic of the flat-grain figure. This figure is particularly prominent in those softwoods, such as Douglas-fir, with contrasting colors of earlywood and latewood. The distinctive flat-grain figure in such hardwoods as ash and elm is associated with a contrast in texture between the large earlywood pores and the relatively small fibers of the latewood.

Quarter sawing of lumber and quarter slicing of veneer produce a typical edge-grain or vertical-grain figure, in which the annual layers are visible as a series of parallel lines varying in prominence according to the amount of contrast between the earlywood and latewood. Timbers of sugar maple, beech, and sycamore, with large rays, sometimes called flecks, are likely to display a more or less conspicuous ray fleck figure on edge-grain surfaces. In oaks and similar woods, the rays, which are both broad and deep, produce several special types of figure, called silver grain, flake, or rift, on edge-grain surfaces.

Figure in wood may also originate from deviations in the grain direction. Curly or wavy grain, for example, which is fairly common in yellow birch, results from wavelike undulations in the orientation of the cell elements (Figure 3.8). Light is reflected at various angles from the surface of such wood, depending on the inclination of the fibers, so as to present a pleasing effect of alternating light and dark bands that change according to the direction from which the wood is observed. Undulating grain is often present in sugar maple in waves of shorter length and much closer spacing than in curly or wavy grain. This pattern in maple is sometimes known as fiddleback figure.

Figure 3.8 Curly grain in sugar maple.

It is not uncommon for wood fibers to develop in a spiral pattern around the stem. In some timbers the direction of the spiral grain tends to reverse periodically, resulting in a condition known as interlocked grain. Such wood typically displays a pattern of vertical stripes on its edge-grain surface because light is reflected differently from the different zones. Such a ribbon-stripe figure is common in some members of the mahogany family.

Several special types of ornamental figure in wood are associated with grain deviations that occur at the sites of certain growth structures. Stump, crotch, and burl figures are examples of this type of figure. Logs containing these features are usually sliced into veneers to obtain the greatest yield of highly figured wood. In addition, special techniques of veneer cutting (back cutting, cone cutting, etc.) can be used to augment the natural figure to achieve special effects.

A highly figured Canadian wood is the so-called bird's-eye maple. The bird's-eye structure is a result of numerous radial series of conical depressions in the annual layers that form rounded areas of the grain remotely resembling small eyes. Although no one knows why, subsequent growth layers conform to the initial indentations and create long series of conical depressions extending radially throughout the stem. One widely held theory on the formation of bird's-eye figure is that the depressions are initiated by the activity of parasitic fungi in the cambial tissue. The resultant damage to the cambium prevents development of the wood in local areas, and although the cambium appears to recover shortly afterward, the depressions established at such sites continue to exist for many years.

Similar series of growth-ring depressions are found in many other hardwoods but seldom in sufficient numbers to produce an ornamental effect. Conical growth depressions also occur in certain softwoods, notably lodgepole pine; the figure produced is known as dimpled grain. The dimples are attributed to local pressure on the cambium exerted by the expansion of resin cavities in the inner bark. Radial series of depressions also commonly occur in Sitka spruce and in white pine and probably occasionally exist in most softwoods. These depressions, however, are usually not conical but may extend several centimetres along the grain.

Figure in wood also arises from variations in color, especially in species with dark-colored heartwood, such as western red cedar and black walnut. The distribution of these variations tends to bear little relation to growth increments or other structural features of the wood. In some instances, figure may be the result of contrasting color between heartwood and sapwood, as in eastern red cedar.

Color and Luster

Color

Color is a variable feature in wood and must be used with care in species identification. As mentioned earlier, the sapwood of most species is typically pale; the familiar dark colors of woods, such as those associated with cedar, walnut, and mahogany, are the colors of the heartwood only. Heartwood of the tropical timber ebony is probably one of the darkest-colored woods found in nature. Perhaps the most unusual natural pattern of color variation in wood is found in the prominently striped zebrawood.

The color of a freshly machined wood surface tends to change with the passage of time and exposure to light. Most timbers darken with age. The color of certain timbers, however, tends to fade or bleach out after prolonged exposure.

Abnormal colorations in wood are often related to the presence of fungi that cause stains and decay in timber. Abnormal colorations may also result from chemical reactions that take place between extractive substances in wood and certain materials. The blue-gray stain, which develops very quickly when moist oak is in contact with a ferrous metal, is an example of this type of coloration. Wood of the hard maples and basswood may develop olive-green to black streaks, known as mineral stain. These stains are discussed in detail in Chapter 5.

Luster

This term describes the natural light-reflecting qualities of a wood surface. A wood lacking in luster appears dull; a lustrous wood has a natural gloss or sheen. Luster is a somewhat subjective quality and varies in relation to the character of the surface being examined. To the experienced eye, this feature can sometimes be useful in wood identification.

Odor and Taste

The basic cell-wall substance that makes up wood has no perceptible odor or taste. When present in wood, odor and taste are normally associated with extractive substances, frequently those formed at the time of heartwood formation. A characteristic odor may provide an immediate clue to the identification of certain woods. A typical natural odor is usually most pronounced in freshly cut material and tends to diminish with the passage of time. It is recommended, therefore, that a few cuts be made to expose some fresh surface when checking a specimen for characteristic odor.

Sometimes a strong odor may be a disadvantage in wood. Woods such as poplars and true firs are chosen for food boxes and containers because they are free of chemical substances that could impart an odor or taste to the food. Conversely, because the strong odor of eastern red cedar is believed to have a repellent effect on moths, this wood is widely used to line clothes closets and chests. Some highly scented oriental woods are burned as incense.

Unpleasant odors sometimes develop in green logs, especially during storage in warm weather. Such odors are not 'natural' but are due to the action of microorganisms (molds, bacteria, etc.) that are breaking down carbohydrate, protein, and fatty materials in the wood and bark.

STRUCTURE OF WOOD

Softwoods

The annual rings of a softwood such as white pine or western larch may be examined closely if a small area on the end of a sample is cut cleanly with a sharp knife. Each annual layer may be seen to have a

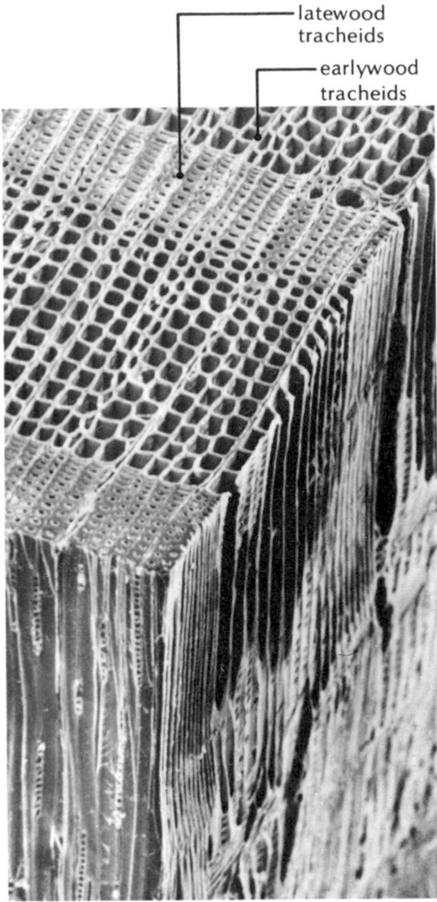

Figure 3.10 Scanning electron micrograph (75×) of Douglas-fir showing abrupt transition between earlywood and latewood tracheids.

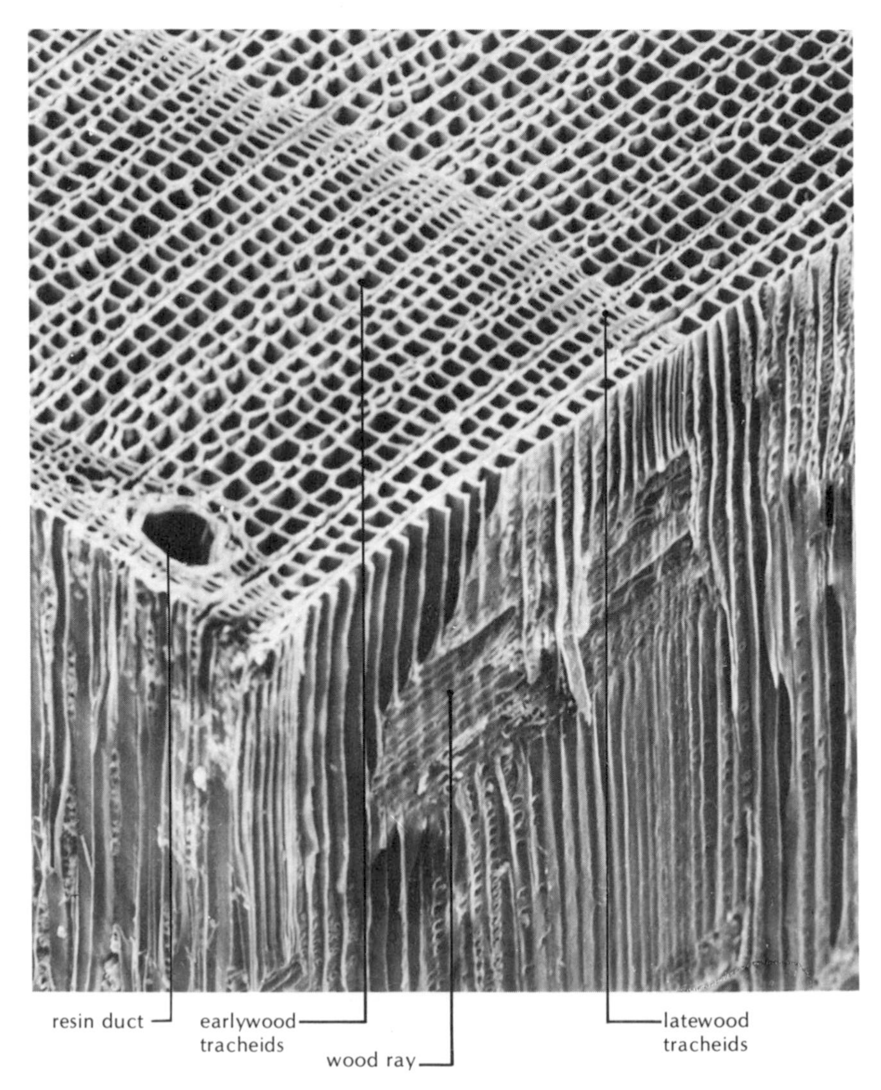

Figure 3.9 Scanning electron micrograph (75×) of lodgepole pine showing resin duct, wood ray, and gradual transition between earlywood and latewood tracheids.

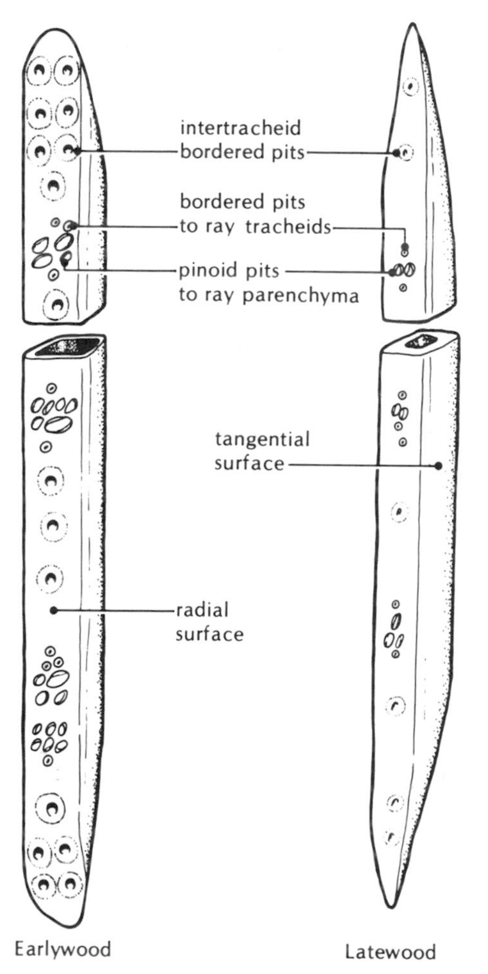

Figure 3.11 Earlywood and latewood tracheids illustrating various types of pitting: intertracheid bordered pits; bordered pits to ray tracheids; pinoid, or small, pits to ray parenchyma. (Adapted from *Utilization of the Southern Pines* by P. Koch [USDA Agr. Handbk. 420, Vol. 1], p. 96)

comparatively light and soft portion on the inside, toward the pith, and a darker and harder region at the outside, toward the bark. Examined with a hand lens (10–14×), each annual layer will show a regular, honeycomblike structure (Figures 3.9 and 3.10). Softwoods are composed largely of cells shaped like tiny tubes, frequently 100 times longer than they are wide, with closed, pointed ends (Figure 3.11). Packed closely together, these long cells are arranged parallel to the stem. A section cut across these tubular structures reveals their regular arrangement in radial rows – that is, in a series of generally straight lines from the pith towards the bark. These tubular cells are popularly known as fibers, although this term is properly applied only to a special type of cell with a very small cavity, or lumen, found in hardwoods. Botanically, these softwood cells are referred to as tracheids. The average length of tracheids in eastern Canadian softwoods is 3–4 mm, while some western Canadian softwoods have tracheids half as long again. Tracheids have overlapping ends, which communicate with each other through pits in the walls. Although these pits are distributed along the entire length of the tracheids, they are more frequent at the ends.

The difference in hardness between the inner earlywood and outer latewood portions of the annual layer is due to clearly discernible differences in the structure of the cells of the two regions. The inner region is composed of cells with relatively large cavities, or lumens, and thin walls, whereas the outer cells have smaller lumens and thicker walls. The growing period in the spring produces the light inner region, known as earlywood or springwood, and subsequent growth later in the growing season produces the denser outer region, which is called latewood or summerwood. The growing period ends with the latewood; the next year's growth adds a new layer of earlywood to the latewood. The contrast between the light earlywood and the darker and denser latewood sufficiently distinguishes the separate annual increments to give timber its characteristic layered appearance.

Examination of cleanly cut end surfaces of softwoods with a hand lens discloses numerous fine, light lines called rays. Rays are minute, ribbonlike structures formed of short cells which are grouped into strands that cross the annual layers at right angles in a radial direction. Although rays are present in both softwoods and hardwoods, they are of diagnostic importance chiefly in identifying hardwoods.

In some softwoods, resin ducts, or canals, appear on the end surface, either as minute openings or as small, light-colored spots, and contain and transmit resinous materials. These vertical ducts, which extend in the direction of the tracheids, are often visible as fine streaks, like scratches, on planed longitudinal surfaces, especially in pine. In dry wood resin ducts may be open and empty or closed, either because they are full of resin or because in heartwood the cavity generally becomes filled with a growth of small wood cells known as tylosoids. Some resin canals extend radially, but they are difficult to detect unless they have become darkened.

Hardwoods

Hardwoods differ from softwoods in possessing vessels that are contained in a matrix of small, thick-walled fibers. On examination of the clean-cut end of a hardwood board, the vessels are seen in transverse section and appear as great numbers of small round holes or pores in the wood. The vessels are composed of relatively short tubular cells with more or less open ends, which fit together longitudinally like minute lengths of pipe and form long, continuous channels especially adapted for conducting sap. These cells are of much larger diameter than the fibrous cells, which compose the greater portion of the wood. The fibrous cells are shorter than the tracheids of softwoods and are often so small in diameter that the cell lumens are indistinguishable when examined with a hand lens.

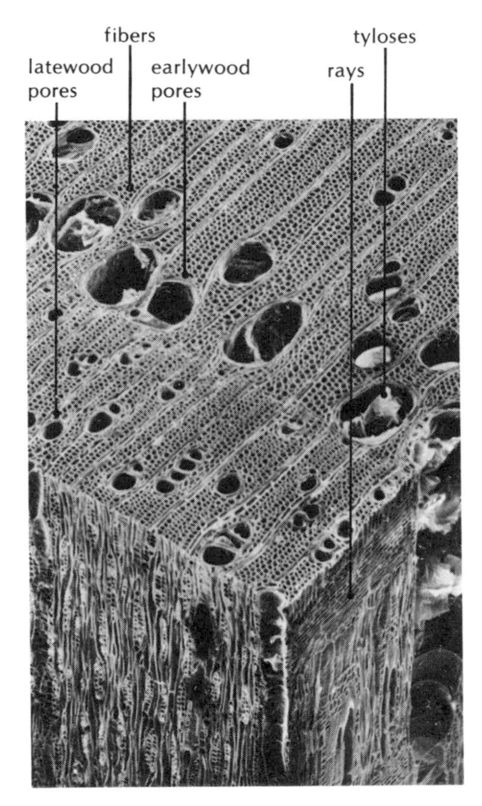

Figure 3.12 Scanning electron micrograph (40×) of a ring-porous hardwood (ash). Earlywood pores are much larger than latewood pores. Fibers are relatively uniform across the annual ring. Rays of ash are typically 1–3 cells wide. Tyloses infiltrate the earlywood pores in the heartwood.

For purposes of identification, the hardwoods may be divided into two general classes on the basis of pore arrangement. Hardwoods in the *ring-porous* class contain pores, that not only are very much larger in the earlywood than in the latewood (Figure 3.12) but are arranged close together in a more or less continuous layer in the earlywood region of the annual ring. The *diffuse-porous* hardwoods contain pores of more nearly uniform size distributed fairly uniformly throughout the annual layers (Figure 3.13). Ash and oak are examples of ring-porous woods; birch, maple, and poplar are typical diffuse-porous species. The annual layers of some diffuse-porous hardwoods may be difficult to distinguish because these woods do not show the distinct contrast in texture between earlywood and latewood found in some of the softwoods. The structure of such hardwoods is often so uniform that a fine line is the only perceptible sign of the boundary of the annual rings. Certain tropical woods, both hardwood and softwood, grow without clearly defined seasonal layers.

In the sapwood of both classes of hardwoods, the vessels are usually open. As the growing tree ages, however, the cavities of the heartwood vessels of many species become filled with ingrowths of small cells known as tyloses. In some woods the cavities become filled with deposits of gummy substances. Where vessels are mostly unobstructed, as in the outermost sapwood of white oaks or even in the heartwood of red oaks, it is possible to blow air through the wood. By contrast, where vessels are closed with tyloses, as in the heartwood of white oaks, the wood is relatively impermeable except at very high pressures. Tyloses are visible through a hand lens in the large vessels of white oaks and certain other hardwoods as shiny structures frequently resembling tiny soap bubbles.

The rays in hardwoods are much more strongly developed and diverse than in softwoods. Although in some hardwoods, as in softwoods, the rays are so narrow as to be invisible without the aid of a lens, in oak, for example, they may reach a millimetre in diameter and several centimetres in height (measured in the direction of the fibers). The rays function both as reservoirs for the storage of food material and as pathways for transverse conduction of sap. They are composed of cells called parenchyma, which are much shorter than the fibers. The rays of certain conifer species also contain, at their upper and lower margins, cellular elements similar in size to ray parenchyma called ray tracheids.

When wood with wide rays is quarter-sawn, the rays lie on the wide faces of the boards, exposing their bandlike character as attractive streaks across the grain. The rays in such boards often have a wavy or

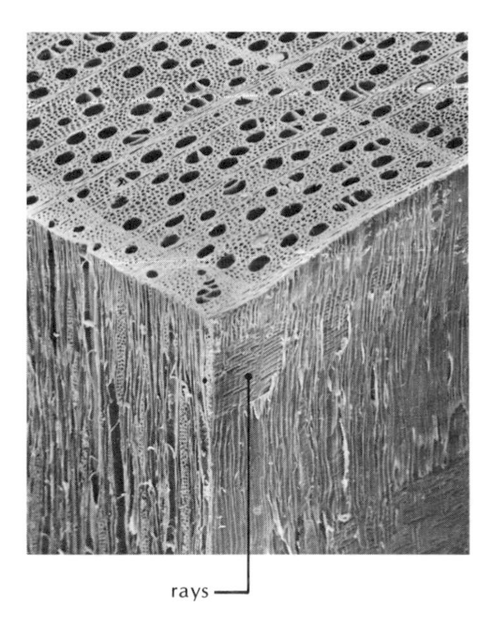

Figure 3.13 Scanning electron micrograph (60×) of a diffuse-porous hardwood (cherry). Pores are relatively uniform across the annual ring. Rays of cherry are mostly 3–4 cells wide.

curved outline because in commercial practice the boards are unlikely to have a truly radial face, and the large rays are rarely perfectly straight. On tangential surfaces of the boards, the rays are usually less conspicuous because of their small cross-sectional area.

Another characteristic of some hardwood species is the visibility of wood parenchyma. When present, wood parenchyma often appears in whitish patches, lines, or even fine dots running parallel to the grain. Such tissue may occur at the outer edge of annual rings, scattered diffusely, grouped about the pores, in well-defined bands, or in various combinations and gradations of development. When light-colored, wood parenchyma may be invisible or difficult to distinguish in the light-colored hardwoods; in other cases, however, it may be of diagnostic importance.

Careful observation of the various structural characteristics thus far enumerated may be sufficient to identify the common species of wood. Although such properties as weight, color, odor, taste, and even size may be decidedly important in species identification, these properties may become considerably altered during aging or as a result of various treatments. It is plain that observation of the structural features of wood presents the surest diagnostic method. Although many experienced woodworkers have learned by long practice to identify the woods they use by general appearance, their methods are sometimes difficult to describe satisfactorily. Novices will probably do well to begin by observing the fundamental structure of common woods, even though their object is to devise methods of rapid identification by general appearances. With a preliminary grounding in structural characteristics, close observation of wood will soon demonstrate the simple relationship between the structure and the general appearance of different kinds of wood.

A key for microscopic identification of woods commonly used in Canada is included at the end of this chapter to illustrate the procedure used in identifying woods by their structural differences. This key should be used with care, for mistakes are easily made by the inexperienced. The forest products laboratories of Forintek Canada Corp., the Canadian Forestry Service, university faculties of forestry, and some institutes of technology have facilities for identifying woods on request.

CELL-WALL STRUCTURE

One must delve deeply into the structure of wood to understand its basic nature. Examination with a light microscope will reveal the details of pit and wall structure essential to making identifications at the species level. Even greater detail of the wall structure is revealed by the higher resolving power of the electron microscope. These observations are not only useful for identification but also essential to our understanding of the relationships between wood structure and its properties and behavior.

Examination at the higher levels of magnification, as shown in Figure 3.14, reveals the cell wall to be composed of a number of layers. The neighboring cells are joined together by a lignin-rich layer known as the middle lamella. This layer is dissolved in a chemical pulping

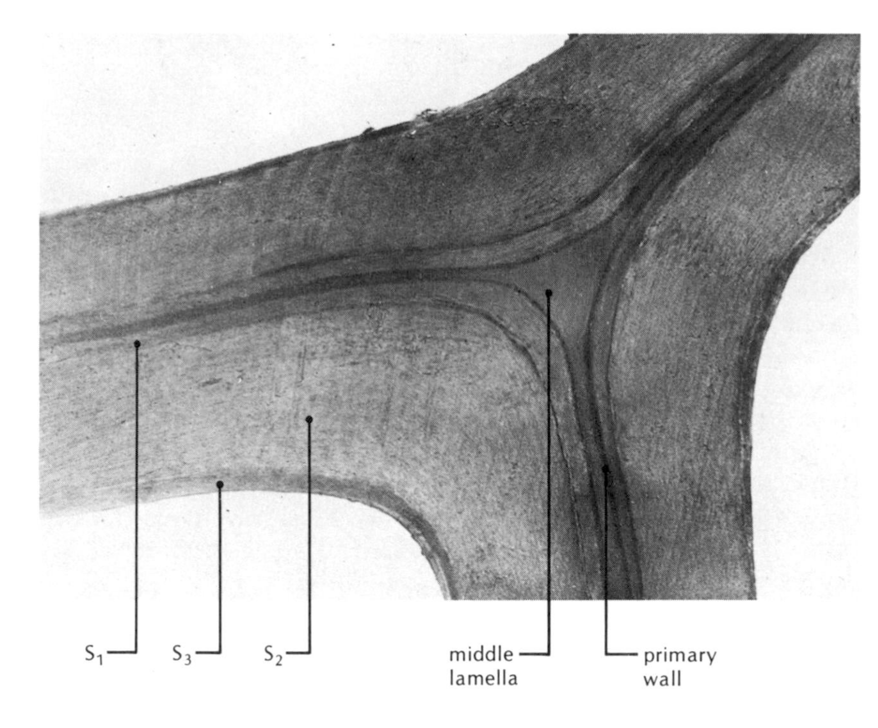

Figure 3.14 Transmission electron micrograph (11 000×) of a coniferous cell wall cross-section showing layered structure, including middle lamella, primary wall, and the three layers of the secondary cell wall (s_1, s_2, and s_3).

process to separate the wood into pulp fibers. The main structural components of the cell wall are threadlike elements called microfibrils, which are organized and oriented differently in each layer of the cell wall.

Primary Wall
The thin, outermost layer of the cell wall, which is formed first during the development of the cell, is known as the primary wall (P). The microfibrils in this wall are arranged irregularly in flat helixes. As the cell reaches full size, the wall is thickened by the addition of a three-layered secondary wall laid down from inside the cell. The layers of the secondary wall are numbered consecutively; s_1 is closest to the outside primary wall, s_2 makes up the bulk of the cell wall, and s_3 is nearest the inside lumen of the elongated tracheid cell (Figure 3.15).

S_1 Layer
The s_1 layer is relatively thin and is composed of several laminae, or layers. The microfibrils in successive laminae exhibit an alternating right- and left-handed, almost flat helical arrangement.

S_2 Layer
This thick layer contains many laminae and makes up the bulk of the cell wall. The microfibrils are arranged in a steep right-handed helical arrangement so that they are nearly parallel to the cell axis.

S_3 Layer
The innermost layer of the secondary wall, the s_3 layer, is similar to the s_1 layer, being thin and having microfibrils oriented in flat helixes with

Figure 3.15 Layered cell-wall structure illustrating the orientation of the microfibrils within each layer: middle lamella, primary wall, three layers of the secondary wall (s_1, s_2, and s_3).

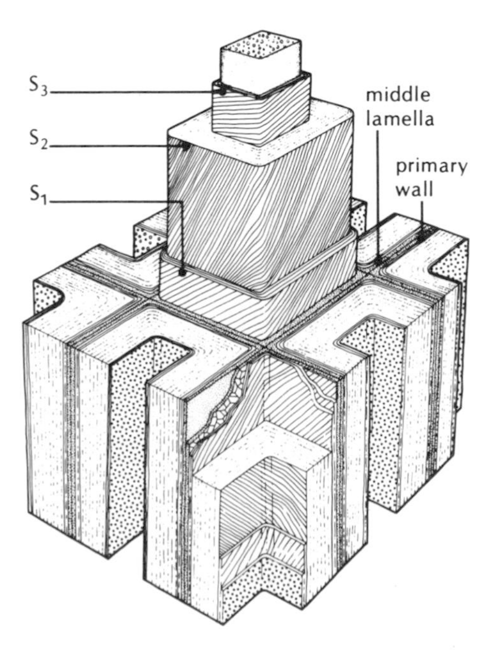

both right- and left-handed orientations. An even thinner layer known as the warty layer may line the lumens of the tracheids and fibers of many species.

Microfibrils
The importance of the orientation of the microfibrils to many properties of wood becomes clearer when the structure of the microfibril is considered.

Each microfibril consists of linear cellulose molecules arranged in strands which pass through phases of parallel and nonparallel order, known as crystalline and amorphous regions. In the crystalline regions, the molecules are called crystallites, after the crystal-like form they assume, and are oriented essentially parallel to the microfibril axis. In the amorphous regions, the cellulose molecules exhibit a lack of parallel order and there may be a higher proportion of shorter chain molecules. Thus, the crystallites are separated from each other by amorphous regions. The forces bonding the crystallites together along the microfibril axis are much stronger than those that bond them together laterally. As a result, strength and stiffness of wood are influenced by the orientation of the microfibril elements.

Water and Shrinkage
In the growing tree, the cell walls are saturated with water, which occupies all the interstices in and around the cellulose microfibrils and the interlamellar spaces. Varying amounts of water are usually present in the cell cavities as well. When wood is seasoned by exposure to dry air, it gradually gives up its moisture, beginning with the water from the cell cavities, known as 'free water', because this water is held by much lower attractive forces than those holding the 'bound water' located within the cell walls. The moisture content (the weight of water in proportion to the weight of dry wood substance) of the wood is slowly diminished until it reaches the critical point where all the free water is lost from the cell cavities. This point, which occurs at a moisture content of about 30% for most native species, is known as the fiber saturation point (fsp).

If wood is dried to any level above its fiber saturation point, loss of water from the cell cavities does not result in any shrinkage in the wood or any change in its physical or mechanical properties. When wood is dried below this point, however, water is removed from the cell walls, allowing attractive forces to draw the cellulose microfibrils laterally together. The result is shrinkage, which occurs mainly at right angles to the direction of the microfibrils' orientation. Since the predominant orientation of the microfibrils is essentially parallel to the axis of the cells, the cells tend to shrink in diameter but not in length; that is, the wood tends to shrink across the grain but not along it.

Wood's affinity for moisture is so great that dry wood will absorb water from moist air and swell. If the air becomes sufficiently dry, moisture will be withdrawn from the wood and it will shrink. As a result, those cupboard doors and dresser drawers that appear to be so loose during the dry winter season can eventually start to pinch and bind during the humid months of summer.

Pits

The pits are important structural details of wood visible only under microscopic examination. Pits are unthickened areas, or discontinuities, in the secondary wall of the cells. In the living tree, the pits function as pathways for liquid movement between the cells. Normally they are found in adjacent pairs in neighboring cells so that together they form what is known as a pit pair, although the term pit is used loosely to designate the complete structure. The two halves of the pit pair are usually separated by a membrane consisting of the middle lamella and the primary walls of the adjacent cells.

Pits are designated as either simple or bordered. In bordered pits the secondary wall overarches the pit membrane (Figure 3.16) and constricts the pit chamber toward the lumen. In the bordered pits of softwood tracheids the pit membrane may be thickened in the center of the chamber to form a disklike structure called a torus. The surrounding membrane is composed predominantly of radially oriented microfibrils and is called the margo. As the developing cell wall matures, the margo usually develops a porous nature so that open pathways between cells are established. In simple pits, the membrane is a simple, flat structure (Figure 3.17). These pits are found in some cells.

The bordered pit serves as a valve in the control of both liquid movement within the tree and the passage of liquids in either drying or preservative treating processes. When the torus is suspended in the center of the pit chamber, liquids may pass from one cell to the next through the openings in the margo. However, the torus may be displaced within the pit chamber, particularly when the wood is dried, so that it irreversibly closes the pit aperture. This condition is known as pit aspiration and can make certain species very difficult to impregnate with preservative liquids. Even without aspiration, the pits, particularly in the heartwood, may become occluded with extraneous materials, and the pores in the margo may become greatly reduced in size and effectiveness as pathways for flow.

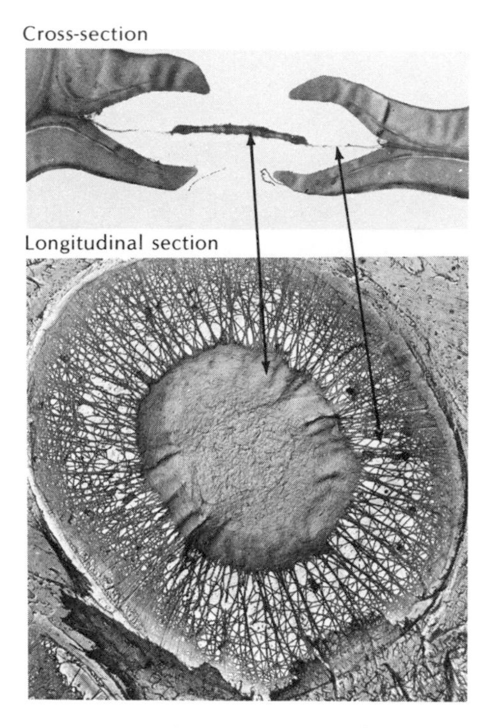

Figure 3.16 Electron micrographs (upper, 2100×; lower, 2800×) of softwood bordered-pit structure.

Figure 3.17 Electron micrograph (4800×) of a simple pit membrane.

NATURAL DEFECTS AND ABNORMALITIES

Knots

A knot is simply a portion of a tree branch that has become embedded in the wood of the stem. A tight or intergrown knot (Figure 3.18) is one in which the annual layers of wood are continuous with those of the stem. Such knots are formed when branches are living parts of the tree and the cambium constitutes a continuous sheet over the stem and branches. In dense forest stands the lower branches of a tree do not get enough sunlight and eventually start to die. When a branch dies, its cambium ceases to function and its growth is terminated. Growth of the main stem continues and gradually encases more and more of the basal portion of the dead branch. There is now no continuity of wood between the stem and branch, and the knots in lumber sawn from wood produced at this stage are known as loose, encased, or sometimes black knots (Figure 3.19). These knots frequently become loose and fall out. Should branches be removed through natural or artificial pruning, their stubs may eventually be completely overgrown by the enlarging stem and any external evidence of the branches will disappear.

Because the annual layers of wood are continuous between the stem and living branches, considerable distortion in structure occurs at these locations within the tree. Lumber with knots is characterized by considerable grain deviation, (or cross grain), in the vicinity of the knots. The faces of vertical-grain boards are more or less parallel to the long axis of embedded branches. An intergrown knot exposed on the surface of such a board is elliptic, with the small end connected to the pith, and is commonly known as a spike knot (Figure 3.20). The faces of flat-grain boards, in contrast, are more or less perpendicular to the long axis of embedded branches so that knots appear generally roundish to oval in such lumber.

The number, type, size, and distribution of knots influence both the appearance and the properties of timber and are taken into account in classifying lumber into grades. The characteristics of the knots, in turn, are related to the branching habits and natural pruning qualities of the trees. These characteristics appear to be under fairly strong genetic control, but they may also be influenced by environmental factors such as the availability of light, which is determined by how densely the trees are stocked.

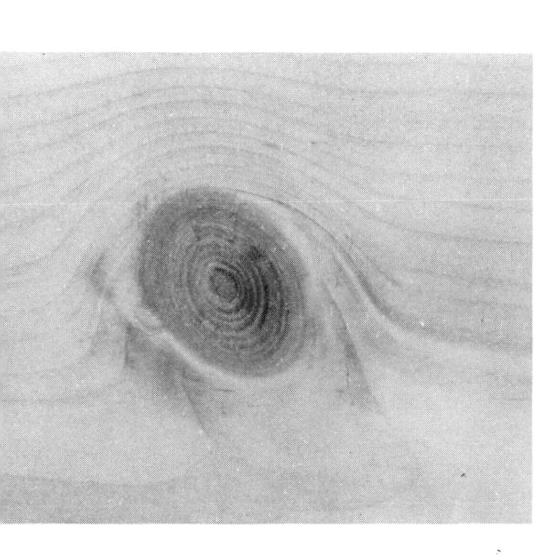

Figure 3.18 Cross-section of a tight or intergrown knot.

Pitch Streaks and Pockets

Pitch streaks or patches are areas of excessively high resin content that sometimes occur in those softwood timbers that contain normal resin ducts – spruces, pines, larches, and Douglas-fir. Although the normal cellular structure appears to be unchanged in these areas, resin fills the

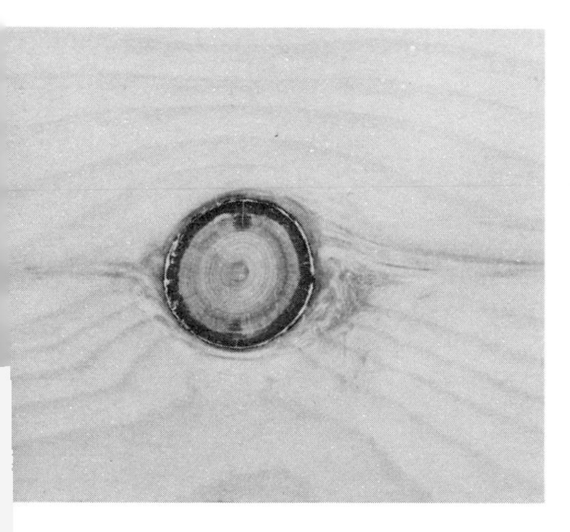

Figure 3.19 Cross-section of an encased knot.

Figure 3.20 Intergrown spike knot in eastern white pine.

cell cavities and apparently saturates the cell walls. It is believed that some type of injury is usually responsible for initiating the process of resin infiltration.

Pitch pockets (Figure 3.21) are actual cavities in the wood, usually filled with liquid or sometimes solid resin. These cavities are variously described as lenticular or planoconvex in shape (based on their appearance on radial or transverse surfaces) and tend to occur in the same softwoods as do pitch streaks. Small pitch pockets may extend only a few centimetres (one inch) or less in length and in tangential breadth, and less than a few millimetres in the radial direction. Pitch pockets may reach very large dimensions, occasionally containing several litres of resin in mature Douglas-fir.

The cambium layers of certain trees are susceptible to damage by insects (often the larvae of certain flies or small species of beetles that bore tunnels in the cambium). Several series of irregularly shaped cells with brownish contents are produced before the cambium is able to reestablish normal growth. These may extend for a few centimetres or more along the grain and, when present on the surface of planed lumber, are sometimes called pith flecks because of their superficial resemblance to the brownish pith. In the natural process of repair to a damaged area of cambium, portions of bark may become embedded in the wood. The term bark pockets is applied to such blemishes when they appear on the surface of lumber.

Cross Grain

Cross grain is a general term describing any condition in which the fibers are not aligned parallel to the axis of a piece of wood. Those special categories of cross grain known as curly, wavy, and interlocked grain have already been discussed in the section on grain, texture, and figure because of their importance in the production of highly figured material for ornamental uses.

Spiral grain is a cross-grain condition found in both hardwoods and softwoods in which the fibers are oriented in a helical pattern around the stem (Figure 3.23). The helix may be oriented toward the left or the right, and its orientation may reverse periodically throughout the stem. The spiral pattern of growth is so common that it is considered the normal pattern in some species. It can often be seen in poles, fence posts, or other situations where bark-free logs have weathered and developed surface seasoning checks. Some specialists believe the tendency to follow a spiral pattern of growth is a genetic characteristic.

Diagonal grain describes a cross-grain condition arising when logs are sawn at an angle to the fiber direction. This condition may occur when severely tapered logs are sawn parallel to the pith or when crooked logs are sawn.

Cross grain can be detected on machined wood surfaces if one looks closely at the orientation of vessel lines or vertical resin canals. It is also useful to observe the orientation of the annual layers on the edge-grain surface and to look for small seasoning cracks or checks on flat-grain surfaces (Figure 3.22). The extent of cross grain in the areas surrounding intergrown knots is approximately proportional to the size of the knots. Tearing along the grain or roughness of machined surfaces may also indicate the presence of cross grain.

Cross-sectioned surface

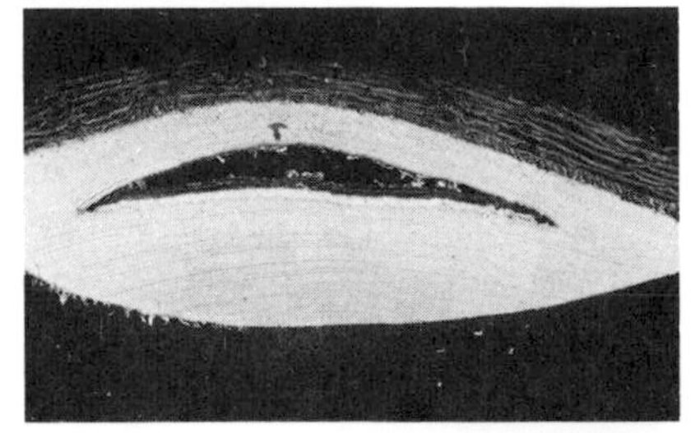

Radial surface

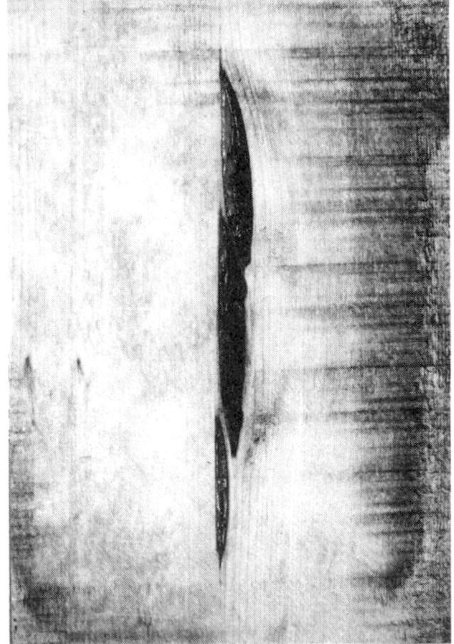

Figure 3.21 Pitch pocket in Douglas-fir as seen on cross-sectional and radial surfaces.

Figure 3.22 Orientation of surface checks indicates cross grain in maple wood board.

Figure 3.23 Spiral-grained and straight-grained forest trees. (Photo: Forest Products Laboratory, Forest Service, USDA)

Cross grain can reduce the stiffness and strength of timber (see Chapter 4) and cause serious problems of checking and distortion of material when it is seasoned. The point at which cross grain becomes a serious problem is determined by the slope of the grain in relation to the edge of the piece of material. A slope of grain greater than 1 in 25, for example, will begin to have a deleterious effect on the tensile strength of wood. Compressive strength is less sensitive to this defect and is not affected until the slope of grain reaches about 1 in 15. Excessively cross-grained material should not be used where strength is a critical factor, as in structural lumber, ladder rails, tool handles, and sporting goods. (See Chapter 4 for a more detailed discussion of slope of grain.)

Shakes and Checks

The following comments apply to defects that occur in the standing tree. Checks and separations that may develop in timber during its processing and seasoning are discussed in Chapters 4 and 7. Some authorities employ the term shake to describe all separations that occur in the standing tree, regardless of the orientation of the plane of failure. Others use this term to refer to ring shakes only (Figure 3.25), or separations oriented in the tangential plane and usually located at the boundary between growth increments. Although ring shakes are sometimes attributed to the release of growth stresses in trees, very often the cause appears to be associated with injury or damage to the stem.

Figure 3.24 Freshly cut spruce log containing small heart check.

The terms heart checks (Figure 3.24), heart shakes, and rift cracks are used to describe separations oriented across the growth rings, extending radially across the tree stem. Star shake refers to a series of such separations radiating from the pith. These defects are more commonly found near the base of relatively large trees and are attributed to such causes as wind sway, growth stresses, desiccation of the heartwood, and frost injury. For further discussion of the effect of shakes and checks on strength and seasoning, see Chapter 4, under 'Shakes and Checks,' and Chapter 7, under 'Drying Defects and Their Causes.'

Figure 3.25 Douglas-fir board containing ring shake.

Brash Wood

Brash wood may be defined as weak wood with low resistance to shock. It cannot withstand suddenly applied loads (impacts), and its bending and tensile strength properties are below normal. Typically, it produces a very brittle fracture, yielding abrupt, relatively smooth fracture surfaces. Brash wood may be the result of several factors.

Low-density wood, or wood that is exceptionally light in weight, is likely to be brash. Low density in wood means that a small proportion of the total volume is composed of cell-wall substance. This condition could be due to a high proportion of thin-walled cells (earlywood tracheids, vessels, parenchyma, etc.) or to a general reduction in the wall thickness of all cells. The narrow-ringed wood produced near the outside of large, overmature trees is often exceptionally light in weight and likely to be brash. Wood with exceptionally wide annual rings, particularly that in certain conifers, frequently has a low proportion of thick-walled latewood tracheids, a low density, and a high incidence of brashness. These characteristics are often found in the annual layers of wood located close to the central pith, which are formed during the juvenile stage of development, giving these layers the name juvenile wood.

Compression wood, an abnormal type of wood that occurs in coniferous trees, is another significant cause of brashness in timber. Although it is of unusually high density, compression wood is low in tensile strength and shock resistance (especially after seasoning) and produces an abrupt fracture. The irregular behavior of compression wood is usually attributed to its below-normal cellulose content and to the excessive inclination of the microfibrils in the cell walls.

Brashness and compression failures are frequently associated. Compression failures appear as minute wrinkles or slip planes in the cell walls (Figure 3.27), which can be readily observed with a polarizing microscope. The failures in individual cells tend to become aggregated into fine lines traversing the wood across the grain direction and may become large enough to be visible to the naked eye. Possible causes of compression failures include snow loads, stresses induced by severe

Figure 3.26 Compression wood in a cross-section of spruce.

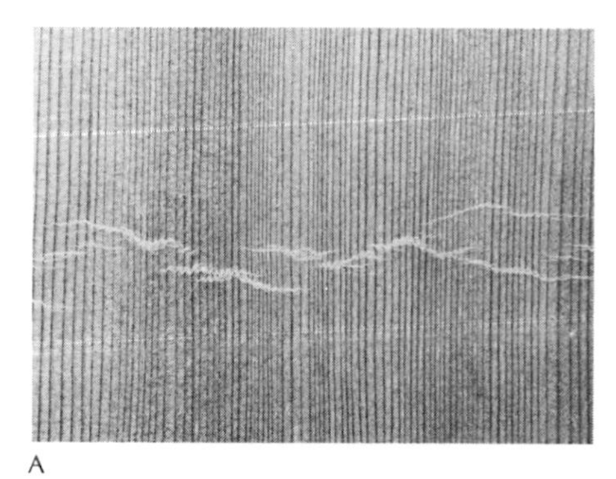

A

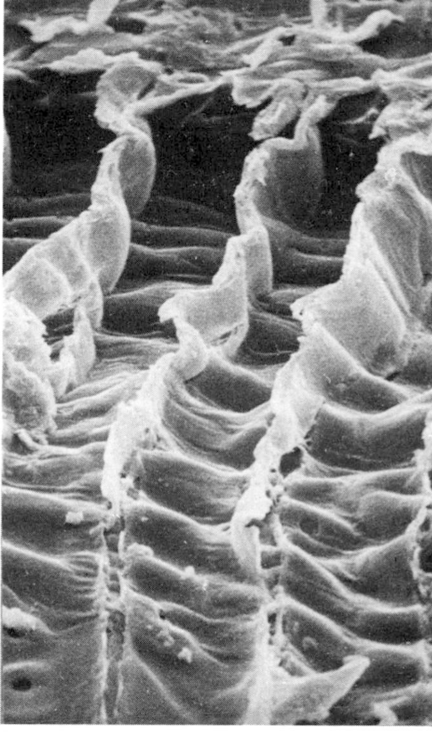

B

Figure 3.27 Compression failure shown (A) as fine lines across grain direction and (B) by scanning electron micrograph as buckling of cell walls in longitudinal direction. (Photo [*upper*]: Forest Products Laboratory, Forest Service, USDA)

wind sway in standing trees, and damage to timber through felling or rough handling during logging or processing operations.

Exposing wood to excessive heat reduces its impact strength; thus, heat might contribute to brashness under special circumstances. Deterioration of wood by decay is another cause of brashness.

Reaction Wood

Reaction wood is an abnormal type of wood that occurs in the stems of crooked or leaning trees and in most branches. Because the growth layers in reaction wood are usually somewhat wider than normal, an eccentric growth pattern may eventually be established in the stem, with the reaction wood located along the long radius. The reaction wood of coniferous trees has been given the name compression wood because it forms on the lower, or compression-stress, side of a leaning stem. In hardwood trees, the reaction wood forms on the upper, or tension-stress, side and is called tension wood. The formation of reaction wood is associated with a straightening response of a tree and probably results from a hormone imbalance rather than from simple stress.

Compression Wood

In freshly felled logs, compression wood can usually be detected by its dark reddish color. The annual layers of compression wood, which are generally wider than normal, give the appearance of an unusually large proportion of latewood (Figure 3.26). The wide layers are due partly to the thickness of the earlywood cell walls and partly to the gradual transition from earlywood to latewood. The high proportion of cell-wall substance in compression wood results in its darker color and also in its abnormally high density. In the dry condition, compression wood tends to have a dull appearance (lacking normal luster), with no clear distinction between earlywood and latewood. Contrast between earlywood and latewood can usually be improved temporarily by moistening the wood surface.

The thick-walled cells of compression wood are rounded in cross-section, unlike the more nearly rectangular or polygonal cells of normal wood. Because the rounded cells do not fit together in close contact as normal wood cells do, intercellular spaces are a characteristic feature of compression wood. What appear to be abundant radial checks in the walls are actually helically arranged cavities in the principal wall layer (S_2). The microfibrils in this wall layer are arranged in a relatively flat helix and thus form a fairly wide angle with the cell axis. This structural organization in the S_2 layer of compression wood contrasts sharply with the steep, helical arrangement of the microfibrils, nearly parallel to the cell axis, usually found in normal wood.

Although many of the strength properties of compression wood are greater than those of normal wood, compression wood is generally weaker than might be expected from its high density. Particular weaknesses include tensile strength parallel to the grain; stiffness in bending, or modulus of elasticity; and impact resistance (especially for wood in the dry condition). In the green condition, the high strength of compression wood may be partially due to the lower percentage of moisture by weight in the saturated cell walls. In other words, compression wood has a lower fiber saturation point than normal wood.

Perhaps the most serious drawback of compression wood is its anomalous behavior in seasoning. In spite of its high density, its shrinkage across the grain is lower than that of adjacent normal wood; however, its longitudinal shrinkage is exceptionally high – up to ten or more times greater than normal. The characteristic shrinkage behavior of compression wood can be explained by the structural organization of its cell walls, particularly the wide microfibril angle in the thick S_2 layer. In practice, bands of compression wood are likely to occur in the same board as bands of normal wood. Because of the different shrinkage properties of the two types of tissue, stresses develop in the board during drying that can result in severe checking and distortion (cupping, twisting, bowing, etc.) of the seasoned lumber.

Compression wood also differs significantly from normal wood in chemical composition, especially in the proportions of lignin and cellulose. It has higher lignin content than normal wood and thus has limited usefulness as pulp furnish. Normal wood of a typical conifer might contain a little over 40% cellulose and perhaps a little under 30% lignin. These figures might be expected to change in compression wood to a little over 30% cellulose and a little under 40% lignin.

Tension Wood

Tension wood, usually found on the upper side of branches and leaning stems in hardwoods, is generally much less conspicuous and therefore less readily detected than is compression wood in conifers. Visual characteristics indicating the presence of tension wood vary somewhat among the different tree species. On freshly cut log ends, tension wood may sometimes be revealed by a slight difference in color or luster. Often a fuzziness or woolliness on the surface texture of green lumber or veneer indicates the presence of tension wood. In seasoned material, tension wood may be the cause of warping and other distortion, including collapse.

Anatomically, the occurrence of tension wood is reflected by changes in the cell walls of the hardwood fibers. Although the structure and distribution of the vessels and other cell types appear to be unaffected, the fiber wall commonly contains a lower proportion of lignin and a higher proportion of cellulose than normal wood. While there are a few exceptions, tension wood in most hardwood species is characterized by the presence of a special wall layer, the gelatinous, or G, layer, located adjacent to the cell lumen inside the other layers of the secondary wall. This gelatinous layer appears to consist mainly of highly crystalline cellulose microfibrils lying almost parallel to the cell axis. It often has a swollen or folded appearance, and because it frequently becomes detached during the cutting of thin sections for microscopic study, it probably is not very firmly attached to the cell wall.

The abnormal properties of tension wood are generally less pronounced than those of compression wood. Nevertheless, both density and longitudinal shrinkage tend to be high in tension wood. Although longitudinal shrinkage seldom exceeds 1%, it is sufficient to cause distortion during seasoning. The strength of tension wood appears to be

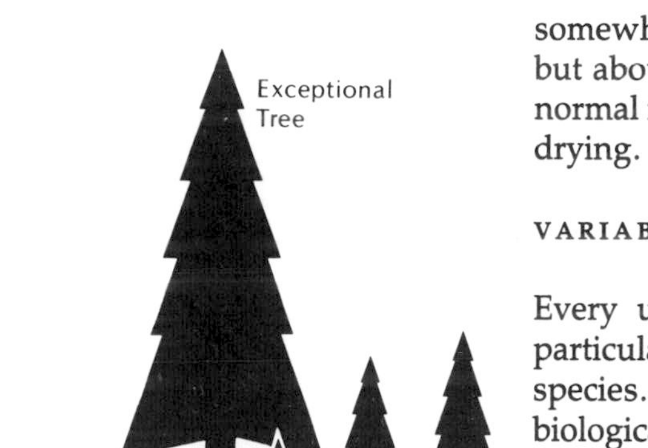

Figure 3.28 Outstanding growth of 27-year-old tree compared with average growth of white spruce trees suggests it may have superior genetic characteristics desirable for propagation.

somewhat below normal in bending and in longitudinal compression but above normal in impact resistance. The tensile strength is below normal in green wood but increases to above-normal values following drying.

VARIABILITY IN WOOD OF THE SAME SPECIES

Every user of wood who selects one species over another for a particular use is taking advantage of the natural differences between species. The user must understand, however, that because of its biological nature, the wood of any one species is not uniform. Pieces of wood originating from different trees, or even from different parts of the same tree, may vary considerably in anatomical structure and in physical, chemical, and mechanical properties.

Variation between Trees

Variations in properties of wood from different stems can generally be attributed either to differences in conditions of growth (site and stand characteristics) or to differences in the genetic makeup of individual trees. In a stand growing in the open, where sunlight is not restricted, trees are likely to have wider annual rings and more stem taper and are more likely to produce knotty timber (because of persistent branches) than trees grown in relatively dense forest stands. Variations in climatic conditions and the availability of moisture tend to influence seasonal growth in height and the proportions and densities of the earlywood and latewood that make up the annual layers of wood. Other differences associated with such factors as latitude and altitude can result in the formation of geographic races within a species. Douglas-fir from coastal forests of British Columbia, for example, tends to produce wood of higher density and with a greater average tracheid length than Douglas-fir from the interior of the province.

Part of the variation in many basic wood characteristics is due to differences in the genetic constitution of individual trees. In studies involving sampling of species over extensive areas, individual trees are generally located with exceptional characteristics for their species. Such trees presumably have a superior genetic composition and may be used in breeding experiments for forest-tree-improvement programs in the hope of producing superior progeny.

Figure 3.28 illustrates the degree to which individual variations may exist in a species. The average dry weight of wood produced at various ages by plantation-grown white spruce is compared with that of a single tree that had produced more than three times as much wood as the average tree of the same age. Not all the superiority is necessarily genetic, but the assumption of genetic superiority is the best basis for initial selection in tree-improvement programs.

Variation within Trees

Radial Variation

At a given height in a tree, a number of significant characteristics of wood show specific patterns of variation across the radius of the stem from the first-formed layers of wood, adjacent to the central pith, to the most recently formed layer of wood just inside the bark. In softwoods, the tracheids, or fibers, are invariably shortest near the pith but become longer in successive growth increments until they eventually reach an essentially constant length, characteristic of the mature wood (Figure 3.29). A similar trend appears to exist in tracheid diameter and, to a lesser extent, in cell-wall thickness, although much less information is available on these trends.

The length of time required to reach maximum cell dimensions varies greatly from species to species. Shorter-lived trees, such as poplars, may reach mature cell dimensions in 10–20 years, whereas long-lived species, such as the giant redwoods in the United States, may take 200 years. In a few species tracheid length apparently never reaches a stable maximum value but continues to increase at a greatly reduced

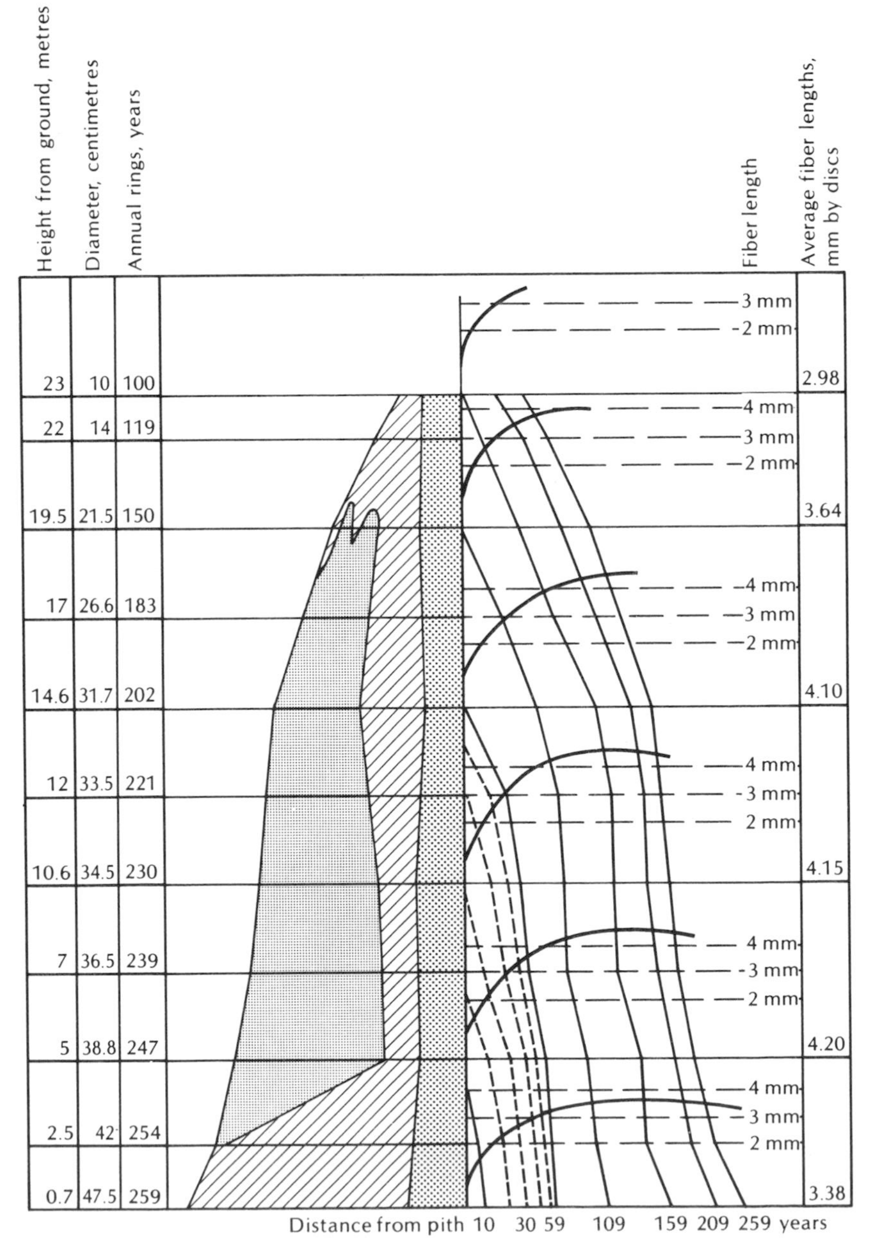

Figure 3.29 Diagram illustrating patterns of variation in length of fibers within a single stem of red pine, *Pinus resinosa*.

Height above ground, diameter, and number of annual layers at each level where fibers are measured is shown at left.

On the tree diagram hatched areas on the left side indicate the characteristic length of the fibers in different regions of the stem. In widest hatched area fibers are 1–3 mm long; in medium hatching, 3–4 mm; in finest hatching, over 4 mm.

On the right side the lines extending upward from the 0.7 m level indicate 50-year increments. Dashed lines between the 10th and 59th annual layers indicate 10-year growth increments. The height of the curves at each level traces the increasing length of fibers from innermost to outer layers.

rate throughout the mature period. Variation in fiber length has been studied extensively for many years because of its importance in pulp and paper manufacture, where it can be an important factor in determining the strength and quality of the final product. The generally lower strength of paper made from hardwoods compared with that produced from softwood fibers can be attributed, to a large degree, to the marked difference in fiber length. Once fibers reach a length of 3½–4 mm, however, further increases in length produce only marginal improvements in paper strength.

During the juvenile period of growth, some hardwoods show characteristic differences in the distribution of such cell types as vessels, parenchyma, and fibers. Furthermore, the vessels in juvenile wood may have a different shape or type of perforation from that of mature wood. The typical ring-porous structure that characterizes the mature wood of some hardwood species may be lacking or incompletely developed in some of the first-formed layers. Ignorance of these facts could cause mistakes in species identification.

In juvenile wood, the microfibrils in the principal layer of the secondary wall (s_2) lie at a fairly wide angle to the cell axis. This microfibril angle,however, decreases sharply (in close agreement with increasing fiber length) and reaches a relatively small constant value for mature wood. This pattern of variation in the microfibril angle can affect the occurrence of distortions in the shape of wood products exposed to significant changes in moisture content.

The density of wood, which indicates the proportion of the total volume occupied by cell-wall substance, may show one of several radial patterns of variation across the stem. Perhaps the most common trend, especially for conifers, is an increasing density from pith to bark. A few species, however, show exactly the opposite trend. The hardwoods appear to be fairly equally divided between the increasing and the decreasing radial patterns. It should be noted that the patterns of variation in density discussed here are those that are likely to be encountered in trees of relatively uniform growth rate (ring width). Because density appears to be correlated with ring width in some instances, normal radial patterns of density variation can be significantly modified by variation in width of the rings. There is considerable evidence that the proportion of latewood and the density (and associated strength properties) of wood tend to decrease noticeably in coniferous trees when they each a condition of extreme old age or overmaturity.

The chemical composition of wood also appears to vary with radial position in the stem. Both cellulose and alpha cellulose content show increases of 5–10% over the juvenile period in many conifers. Lignin content, in contrast, follows a pattern of very gradual reduction from pith to bark, the total decrease amounting to approximately 2–3%. The quantity of extractives present in wood also varies to some extent in relation to radial position in the stem. Many conifers have a significant resin content, which tends to diminish in a radial pattern from pith to bark. This pattern is usually modified to some extent by a large decrease in resin content across the heartwood-sapwood boundary. Decay resistance associated with heartwood extractives usually tends to increase outward from the pith, reaching a maximum in the outer heartwood near the sapwood boundary.

Variation across the Annual Layer

A considerable variation in fiber length exists in different parts of a single growth increment The length of coniferous tracheids tends to decrease sharply at the beginning of seasonal growth and reaches a minimum value just before the production of latewood begins. Tracheid length then increases by 12–25% to reach a maximum value near the end of the increment. The variation in fiber length of hardwood species across the annual layer appears to be relatively large (up to 80%) in short-fibered species and relatively small (up to 15%) in long-fibered species. The vessel elements of hardwoods show similar patterns of variation. In general, the character of the pattern of variation in fiber length across the growth increment appears to be related to the degree of contrast between earlywood and latewood. Plotted values of the radial variation in fiber length show relatively sharp peaks for highly contrasting latewood but relatively gradual peaks when latewood contrast is not so pronounced. These differences in fiber length within a single growth increment are one of the factors that contribute to the differences in paper-making qualities between earlywood and latewood pulp furnish.

Coniferous species growing in temperate climates show a characteristic pattern of decreasing radial diameter of tracheids and increasing wall thickness across the annual layer. In hardwoods, the diameter and volume of vessels may decrease across the growth increment. Variation in density across the annual layer is related to the transition from earlywood to latewood. In conifers, the change to latewood is accompanied by a pronounced increase in wall thickness and reduction in lumen volume. In native softwood species, the density of latewood averages two to three times that of earlywood.

The collection of meaningful information on the proportions of earlywood and latewood and on the pattern of variation in density across the growth increment has been facilitated in recent years by the use of such techniques as x-ray densitometry. In this technique, x-ray negatives of wood cross-sections can be scanned on a densitometer, and intraring density and ring-width data produced by a computerized data-acquisition system. The system can be programmed in a number of ways to produce intraring density plots (like that shown in Figure 3.30) or to integrate the data to derive values for such factors as earlywood width; latewood width; total ring width; minimum, maximum, or total ring density; and mean earlywood or latewood density.

Observations of a number of coniferous species indicate that minor differences exist in the chemical composition of wood within the individual growth increment. In mature stems, the earlywood portions of annual layers show a somewhat lower cellulose content and a somewhat higher lignin content than the latewood portions. In addition, the latewood cellulose is reported to have a higher degree of crystallinity than that of earlywood. Differences in proportions of various chemical components across the annual layer usually do not exceed 2–3%.

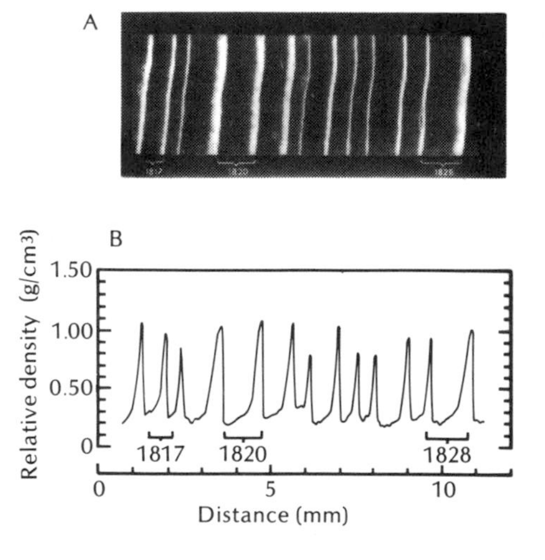

Figure 3.30 Scanning of (A) x-ray negatives of wood cross-sections with a densitometer will produce (B) intraring relative density plots useful in the evaluation of wood quality.

Variation Associated with Height in the Stem

Variations in wood characteristics associated with vertical position in the stem appear to be much less prominent than those associated with radial position. Fiber length, which has been examined extensively in both softwoods and hardwoods, does show a fairly consistent variation along the vertical axis. Within a given growth increment, fiber length tends to increase from the base of the tree to a maximum part way up the stem. From this point to the top of the increment there is a gradual decrease in fiber length. The complete pattern of variation in fiber length within a single red pine is shown in Figure 3.29. This pattern of variation has been observed in a variety of hardwoods as well.

Exceptions have been reported. In some softwood species, the maximum tracheid length, once reached, does not decrease but remains relatively unchanged to the upper tip of the growth increment. In a few species (including both hardwoods and softwoods), fiber length consistently diminishes from the base to the top of the stem in each increment.

Several different patterns of variation in density in relation to height position in the stem have been reported. The most common pattern for softwoods is that of decreasing density from butt to top of the stem. In hardwoods, the opposite pattern (increasing density from butt to top) is usually found. Some hardwood and some softwood species show a trend of decreasing density part way up the stem and a trend of increasing density from that point to the top. As mentioned previously, patterns of variation in density may be influenced significantly by variations in other related factors, such as the rate of growth.

Variation Associated with Growth Rate (Ring Width)

Results of some investigations of coniferous species indicate an inverse relationship between growth rate and fiber length, increasing fiber length being associated with decreasing growth rate (ring width). Other investigations have shown that maximum fiber lengths tend to occur in wood of moderate ring width. When annual rings are significantly wider or narrower than normal, fiber length tends to decrease. Because earlywood fibers tend to be shorter than latewood fibers, a positive correlation appears to exist between fiber length and proportion of latewood. The proportion of latewood in many conifers tends to increase with age and decrease with height in the stem.

Both the frequency of occurrence and the size of wood rays are related to the rate of growth. Positive correlations between ring width and the volume and height of rays have been found in several coniferous woods. A study of several species of ring-porous hardwoods showed that wide-ringed wood has significantly more rays per unit of transverse surface area than narrow-ringed wood. In wide rings, ray width increases as a result of increases in both number and size of ray cells. In some species, rays frequently change from a width of essentially one cell to a width of two cells when growth becomes exceptionally rapid.

Wood technologists have devoted considerable effort over the years to examining relationships between the rate of growth and the density of wood. The rate of growth, as indicated by the width of annual layers, can be readily observed in many kinds of timber. If ring width could be correlated with a technologically important characteristic such as basic wood density, ring width would provide a visual means of appraising that quality. In the heavier softwoods, which normally contain a relatively large proportion of latewood, a moderate rate of growth has usually been found to produce the densest wood. Extremely wide rings or extremely narrow ones tend to produce a lower proportion of latewood and exhibit a lower overall density than rings of moderate width. Woods in this category (Douglas-fir, hard pines, etc.) show the greatest contrast betweeen earlywood and latewood in the annual layer and the greatest normal variation in wood density. Woods such as spruce, which have small or moderate amounts of latewood and a gradual transition from earlywood to latewood, generally show increases in density with diminishing width of annual rings.

In ring-porous hardwoods, the annual layer contains a zone of large earlywood pores that merges fairly abruptly into a latewood zone containing a large proportion of dense, fibrous tissue. The zone of large earlywood pores appears to be relatively unaffected by variation in the rate of growth, which essentially determines the width of the latewood portion of the annual layer (Figure 3.31). The significant positive correlation between ring width and proportion of latewood results in a similar correlation between ring width and basic density of wood. Practical use is made of this correlation in specifications for such products as tool and implement handles; the requirement of a minimum ring width ensures that wood of high density and strength is used. In wood with excessively wide annual layers, the latewood fibers have sometimes been found to be relatively thin-walled and the density of such wood to be somewhat below the expected value.

Diffuse-porous hardwoods lack the row of extremely large pores that characterizes the earlywood of ring-porous species and, in this sense, are more homogeneous in structure. The heavier native woods of this type, such as birch and maple, tend to be relatively independent of density variations associated with the rate of growth. There is some

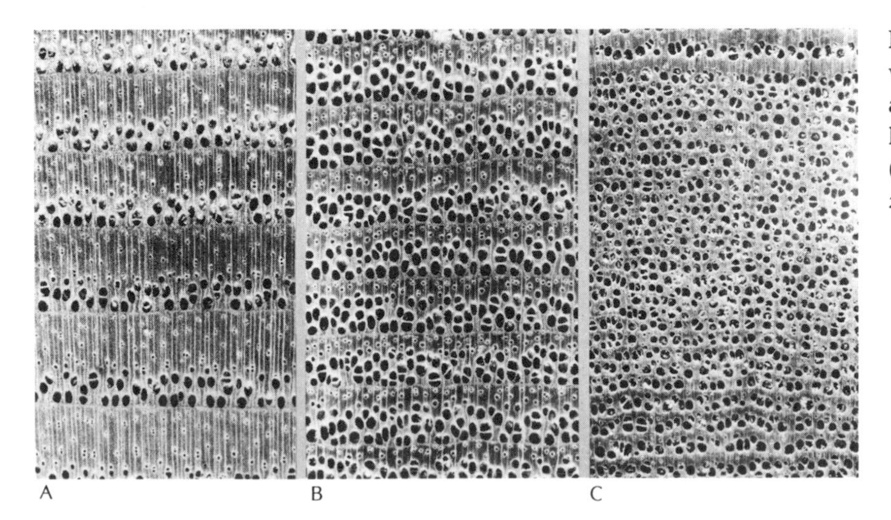

Figure 3.31 Variations in wood density with growth rate in black ash: (A) average weight, 560 kg/m³ – 5 rings/cm; (B) light weight, 430 kg/m³ – 8 rings/cm; (C) extremely light weight, 350 kg/m³ – 20 rings/cm.

evidence, however, that certain of the lighter diffuse-porous hardwoods tend to be denser at moderate growth rates than at extremely fast or extremely slow growth rates.

It should be emphasized that an assessment of the relationship between growth rate and density in both softwoods and hardwoods, must take into account the effect of other variable factors, such as maturity in age or size.

The average consumer may make little conscious use of information on the variable nature of wood. However, this variability leads to differences in grain and appearance, allowing the consumer to select pieces of wood furniture or other products that appeal to his or her personal sense of aesthetics. The more serious user of wood or the producer of wood products must recognize the variability in wood within the same species in order to use it intelligently. The manufacturer of wood pulp must recognize, for example, that decisions to use previously unused tops and branches or to lower the age at which trees are harvested will affect the proportion of juvenile wood in the pulp. This change is apt to significantly affect the average morphological characteristics of the pulp fibers and, in turn, the physical properties of paper produced from the pulp.

IDENTIFICATION OF WOOD SAMPLES

The identification of wood samples is usually more difficult than the identification of trees. Differences between genera and species are based on the structure of flowers, leaves, buds, bark, and so on, and not on wood structure. During evolution wood structure did not differentiate as much as the external features of trees did; hence, in many cases, identification of wood only to a species group is all that is possible, particularly when the examination is carried out solely with the aid of a hand lens. For accurate observation of wood structure, a fresh cross-sectional surface should be cut with a very sharp knife or razor blade. Moistening the cut surface is frequently helpful, and the aid of a low-power (10–14×) hand lens is almost indispensable.

One of the basic tools for wood identification is a dichotomous key of the type provided below. By choosing between contrasting statements about diagnostic features provided in the key, one is led to features that are characteristics of a single genus or species. Descriptions are for normal samples. Wood from branches, roots, and the first few annual rings closest to the pith may differ sufficiently to cause confusion. Reaction wood, either tension wood in hardwoods or compression wood in softwoods, can make proper identification even more difficult.

Proficiency in wood identification, especially of softwoods, requires considerable practice on known samples leading to the development of what has been referred to as a 'seeing eye.' The following key has been developed for the identification of woods commonly used in Canada. Keys for North American, European, and tropical woods important to the North American market are listed in the bibliography.

Key for Microscopic Identification of Woods Commonly Used in Canada

1 Pores absent; general cell structure consisting of distinct radial rows of tracheids visible only with magnification. **2**

Pores present; may be visible with or without aid of magnification. **12**

2 Resin ducts present, mostly confined to latewood. **3**

Resin ducts normally absent. **7**

3 Resin ducts large, relatively distinct and numerous, visible to the eye as light or dark spots on end grain, solitary or in small tangential groups of 2–3. On tangential surface of boards, resin ducts visible as fine longitudinal lines or scratches. **4**

Resin ducts small, relatively inconspicuous, usually not numerous, sometimes absent in some rings, barely visible to the naked eye, solitary or in small tangential groups of 2–5. **5**

4 Transition from earlywood to latewood gradual. Wood practically uniform, soft, light. **Eastern white pine, Western white pine**

Transition from earlywood more or less abrupt. Latewood conspicuous, darker, and denser than earlywood.
Lodgepole pine, Jack pine, Ponderosa pine, Red pine

5 Latewood not very conspicuous; transition from earlywood gradual. Latewood usually occupying less than 1/3 ring width. Wood comparatively soft and light. Heartwood not colored, or little different from sapwood.
White spruce, Red spruce, Black spruce, Engelmann spruce, Sitka spruce

Latewood conspicuous; transition from earlywood abrupt. Latewood usually occupying 1/3 or more of ring width. Wood moderately hard and heavy. Heartwood colored. **6**

6 Heartwood light reddish brown. Sapwood whitish to yellowish or reddish-white. Resin canals numerous, mostly in small tangential groups.
Douglas-fir

Heartwood tending to brown. Sapwood whitish. Wood surfaces feel oily. Resin canals relatively few, mostly solitary or in small tangential groups.
Tamarack, Western larch

7 Wood without aromatic odor. **8**

Wood with aromatic odor. **9**

8 Transition from earlywood gradual; earlywood whitish to light brown in color, latewood brownish.
Alpine fir, Amabilis fir, Balsam fir, Grand fir

Transition from earlywood may be more or less abrupt. Wood with a pale reddish-brown tinge. **Eastern hemlock, Western hemlock**

9 Heartwood dark, sapwood whitish. **10**

Heartwood relatively light colored. **11**

10 Wood moderately hard and heavy. Heartwood reddish-brown to dull red. Odor and taste mild, not spicy. **Eastern red cedar**

Heartwood reddish brown; odor fragrant; taste faintly bitter. Wood moderately soft and light. **Western red cedar**

11 Heartwood yellowish or yellowish-brown; taste somewhat bitter. **Yellow cedar**

Heartwood light brown; odor mild. **Eastern white cedar**

12 Pores with sharp transition in size between earlywood and latewood. Wood ring-porous. **13**

Pores uniform or with gradual transition in size between earlywood and latewood. Wood diffuse-porous or semi-ring-porous. **19**

13 Pores in latewood create radial, flamelike designs. **14**

Latewood without radial, flamelike designs. **15**

14 Small pores of latewood not distinct with hand lens. Large pores of earlywood filled with tyloses in heartwood. Broad rays commonly exceeding 2.5 cm (1 in.) in height. **White oak**

Small pores of latewood distinct with hand lens. Tyloses in large pores of earlywood usually absent or sparse. Broad rays seldom exceeding 2.5 cm (1 in.) in height. **Red oak**

15 Latewood figured with long or short wavy tangential bands of pores. **16**

Latewood not figured by wavy tangential bands of pores. Pores in latewood few, solitary or in small multiples. **18**

16 Pores in earlywood usually in a single row. **17**

Pores in earlywood in more than one row. **Slippery elm**

17 Pores in earlywood regularly spaced with few tyloses. Transition to latewood abrupt. **White elm**

Pores in earlywood variable in size, irregularly spaced, and mostly plugged with tyloses. Transition to latewood more or less gradual. **Rock elm**

18 Earlywood pores closely spaced, usually in several rows. Tyloses present but not abundant. Transition from earlywood to latewood abrupt. **White ash, Black ash**

Earlywood pores not closely but irregularly spaced, usually in a single row. Light-colored tissue (parenchyma) in numerous fine tangential lines plainly visible with lens in latewood. Transition from earlywood to latewood gradual. **Hickory**

19 Pores easily visible on all surfaces without lens and on longitudinal surfaces appearing as minute grooves or scratches. Pores not uniformly distributed throughout growth ring. **20**

Pores invisible or barely distinguishable without aid of lens. Pores more or less uniformly distributed throughout growth ring. **21**

20 Heartwood light chocolate brown or purplish-brown. Wood hard and heavy, with characteristic odor on freshly cut surface. **Black walnut**

Heartwood light chestnut brown. Wood soft and light without characteristic odor. **Butternut**

21 Rays of variable width; some twice (or more) the width of largest pores. Caution: broad rays may be widely spaced. **22**

Broad rays absent; all rays narrow. **24**

22 Rays predominantly broad, appearing as closely packed broken lines on tangential surface. **Sycamore**

Broad rays few relative to fine rays. **23**

23 Broad rays distinct and lustrous; appear on radial surface as brownish flecks, usually not over a few millimeters in depth but rarely 6 mm ($\frac{1}{4}$ in.) or over. Growth ring ends with distinctly darker zone of latewood. **Beech**

Broad rays dull and relatively indistinct. Frequently over 6 mm ($\frac{1}{4}$ in.) in depth. Growth ring does not end with distinct zone of latewood. **Red alder**

24 Rays indistinct to naked eye. **25**

Rays distinct to naked eye. **26**

25 Rays indistinct with hand lens. **Balsam poplar, Black cottonwood, Eastern cottonwood, Largetooth aspen, Trembling aspen**

Rays distinct with hand lens. **27**

26 Pores not crowded; distributed fairly uniformly throughout annual ring. **Sugar maple, Black maple, Red maple, Bigleaf maple**

Pores numerous, usually crowded, particularly in earlywood that is initiated by a relatively inconspicuous single more or less continuous band of pores. **Black cherry**

27 Pores relatively few, not crowded. Wood hard to moderately hard. **28**

Pores numerous, crowded. Wood soft, easily dented with thumbnail. Growth ring delineated by narrow, light-colored line. **Basswood**

28 Pores clearly wider than widest rays. **White birch, Yellow birch**

Width of pores equal to or smaller than width of widest rays. **Sugar maple, Black maple, Red maple, Bigleaf maple**

BIBLIOGRAPHY

Harlow, W.M. 1970. *Inside wood, masterpiece of nature*. Washington: Am. For. Assoc.
Jane, F.W. 1956. *The structure of wood*. London: Black
Kollmann, F., and Côté, W.A., Jr. 1968. *Principles of wood science and technology*, Vol. 1: *Solid Wood*. Berlin: Springer-Verlag
Panshin, A.J., and de Zeeuw, C. 1970. *Textbook of wood technology*, Vol. 1. 3rd. ed. New York: McGraw-Hill
Tsoumis, G. 1968. *Wood as raw material*. Oxford: Pergamon Press
Wilson, B.F. 1970. *The growing tree*. Amherst: Univ. of Mass. Press
Zimmerman, M.H., and Brown, C.L. 1971. *Trees – Structure, function*. New York: Springer-Verlag

4

A.W. PORTER

Porter Engineering Ltd.,

Richmond, British Columbia

Strength and Physical Properties of Wood

Today a wide selection of building materials is available. One of the preferred materials is wood, whose unique characteristics and versatility have long made it the natural choice for building homes and for manufacturing many other products.

Data on the physical and mechanical properties of wood that make it so useful are presented in this chapter for the reference of architects, engineers, and others interested in the use of wood for structural and other applications.

FACTORS AFFECTING THE STRENGTH OF WOOD

Density and Relative Density

Wood is a porous, cellular material, and therefore the amount of solid wood substance in a given volume of wood is a good indicator of its strength properties and, to a lesser degree, its machining, drying, and thermal characteristics. Two terms often used to characterize the porosity of wood are density and relative density. Relative density is the term used in metric practice under the International System of Units (SI), and replaces the traditional term specific gravity.

The density of a body is its weight, or mass, per unit volume. Because both the weight and the volume of a wood sample can vary significantly according to the moisture content, it is important to state the conditions under which the density or relative density is obtained. Usually one of the following three conditions is selected:

Weight	Volume	Name given to density or relative density
Oven-dry	Oven-dry	Oven-dry
Oven-dry	Air-dry, or 12% moisture content	Nominal
Oven-dry	Green, at or above the fiber saturation point	Basic

As an example, Douglas-fir has an average oven-dry density of 510 kg/m^3 (31.8 lb/cu ft), and an average basic density of 450 kg/m^3 (28 lb/cu ft).

There is a 30-fold variation in the density of wood species throughout the world, from 40 kg/m^3 (2.5 lb/cu ft) for balsa wood to 1400 kg/m^3 (87.5 lb/cu ft) for lignum vitae. Studies have shown that the density of the material composing the cell walls of wood fibers is approximately the same for all wood species, namely, 1540 kg/m^3 (96.2 lb/cu ft). This value indicates that the cell-wall substance itself is approximately 50% denser than water, which weighs 1000 kg/m^3. The ability of wood to float is therefore determined by its porosity and moisture content.

The relative density of a solid or liquid substance is the weight, or mass, of any given volume divided by the weight, or mass of an equal volume of water. As we have previously stated, Douglas-fir has an oven-dry density of 510 kg/m^3 (31.8 lb/cu ft). The oven-dry relative density for this species is the ratio of 510 kg to 1000 kg, or 0.51. Expressed in imperial units, the oven-dry relative density, or specific gravity, of Douglas-fir is the ratio of 31.8 lb to 62.4 lb (weight of a cubic foot of water), or 0.51. The range in oven-dry relative density for Canadian species is from 0.31 for eastern white cedar to 0.79 for hickory and ironwood.

The relative density of a species is an excellent indicator of its strength properties. The data in Table 4.1 provide valuable information on the relation between relative density and various strength properties of clear wood. Some properties, such as modulus of elasticity, increase in a linear manner with relative density, while others, such as modulus of rupture, increase as a power function.

Given the relevant relative density, acceptable approximations of

Figure 4.1 Hoisting large glued-laminated timbers into position during construction of the Centennial Pier, Vancouver, British Columbia. (Photo: Council of Forest Industries of British Columbia)

Table 4.1 *Functions relating mechanical properties* to relative density† of clear, straight-grained wood*

Property	Relative density–strength relation‡	
	Green wood	Air-dry wood (12% moisture content)
Static bending		
Fiber stress at proportional limit, MPa	$70.3G^{1.25}$	$115G^{1.25}$
Modulus of elasticity, million MPa	$16\,300G$	$19\,300G$
Modulus of rupture, MPa	$121G^{1.25}$	$177G^{1.25}$
Work to maximum load, kJ/m^3	$245G^{1.75}$	$223G^{1.25}$
Total work, kJ/m^3	$710G^2$	$501G^2$
Impact bending, height of drop causing complete failure, mm	$2\,900G^{1.75}$	$2\,400G^{1.75}$
Compression parallel to grain		
Fiber stress at proportional limit, MPa	$36.20G$	$60.33G$
Modulus of elasticity, million MPa	$20\,100G$	$23\,300G$
Maximum crushing strength, MPa	$46.4G$	$84.1G$
Compression perpendicular to grain, fiber stress at proportional limit, MPa	$20.7G^{2.25}$	$31.9G^{2.25}$
Hardness		
End, N	$16\,600G^{2.25}$	$21\,300G^{2.25}$
Side, N	$15\,200G^{2.25}$	$16\,800G^{2.25}$

*SI metric units are converted from imperial units shown in Table 4.1A (U.S. Forest Prod. Lab. 1974) included in the Appendix at the end of this chapter.

†Formerly specific gravity.

‡The properties and values should be read as equations; for example, modulus of rupture for green wood = $121G^{1.25}$, where G represents the relative density of oven-dry wood, based on the volume at the moisture condition indicated.

the strength properties of species whose strength properties have not been measured can be extrapolated from Table 4.1. Strength properties are known for only one-third of Canada's 140 odd species and only a small fraction of the thousands of tropical woods.

The relative density of wood within any one species varies greatly. For example, the basic relative density of white spruce ranges between 0.246 and 0.462. The cause of this variation within a species was the subject of many early research studies, which reached the following conclusions about influences on the magnitude of the variations:

1. Local site conditions in which a species grows are more important than location within its geographic range.
2. The age of a tree has little effect on the density of wood produced except for wood near the pith.
3. Generally, density decreases with increasing height of a tree.
4. The percentage of latewood is a better indicator of density than rate of growth.

For a more detailed discussion of factors influencing relative density, see 'Variability in Wood of the Same Species' in Chapter 3.

Moisture Content

The moisture content of a piece of wood is the weight (mass) of water it contains, expressed as a percentage of the weight (mass) of oven-dry wood. Moisture in wood exists as free water, or water vapor in the cell lumens, and as bound water, which is adsorbed in the cell walls. When green wood dries, the first water to evaporate is the free water within the cell lumens. When no free water is left in the lumens but the cell walls are still saturated, the fiber saturation point has been attained. This condition usually occurs when the wood reaches a moisture content of 30%, although it can vary from 24% to over 30%, depending on the species.

Until the fiber saturation point is reached, no appreciable change occurs in the strength properties of wood. When wood dries below this point, however, shrinkage occurs and strength and stiffness increase. Various procedures have been developed that enable strength adjustments to be made to compensate for differences in moisture content. For relatively small sizes of clear wood, the following equation can be used (Brown, Panshin, and Forsaith 1952):

$$\log S_3 = \log S_1 + [(M_1 - M_3)/(M_1 - M_2)] \log (S_2/S_1) \qquad (1)$$

where S_1 and M_1 are one pair of corresponding strength and moisture content values, as found from tests; S_2 and M_2 are another pair; and S_3 is the strength value to be adjusted for moisture content, M_3. Should one of the known values be for the green condition – that is, above the fiber saturation point – a value of 25% should be used for M_1, and the strength at M_1 taken as S_1.

If no test results are available, the information provided in Tables 4.2 and 4.3 for two reference conditions – green and 12% moisture content – may be used for M_1, S_1, and M_2, S_2, respectively. For example,[1] to determine an approximate value of the shear strength of a clear sample of white spruce at 19% moisture content, one uses the values from Tables 4.2 and 4.3:

$$S_1 = 4.62 \text{ MPa}, \qquad M_1 = 25\%;$$

$$S_2 = 6.76 \text{ MPa}, \qquad M_2 = 12\%.$$

Then the shear strength S_3 at $M_3 = 19\%$ is found by substituting in equation (1):

$$\log S_3 = \log 4.62 + [(25 - 19)/(25 - 12)] \log (6.76/4.62)$$

and

$$S_3 = 5.51 \text{ MPa}.$$

The increases in strength due to drying shown in Table 4.3 for small, clear specimens may not be achieved for wood in larger, structural sizes, owing to shrinkage and seasoning defects. The actual increases depend on the strength property and moisture content in question and are discussed in more detail in the American Society for Testing and Materials (ASTM) standards D245-70, D2555-73, and D2915-74.

1 In imperial units, the calculation proceeds as follows. From Tables 4.2A and 4.3A (see Appendix), $S_1 = 670$ psi, $M_1 = 25\%$, and $S_2 = 980$ psi, $M_2 = 12\%$. Then the shear strength S_3 at $M_3 = 19\%$ is found by substituting in equation (1): $\log S_3 = \log 670 + [(25 - 19)/(25 - 12)] \log (980/670)$, and $S_3 = 798$ psi.

Table 4.2 *Average clear-wood strength values* for commercial species in green condition*

		Property									
Species	Relative density,† basic	Shrinkage, green to oven-dry based on dimensions when green (%) Radial	Tangential	Volumetric	Modulus of rupture (MPa)	Modulus of elasticity (MPa)	Compression parallel to grain, crushing strength max. (MPa)	Shear strength (MPa)	Compression perpendicular to grain, fiber stress at proportional limit (MPa)	Tension perpendicular to grain (MPa)
SOFTWOODS										
Cedar										
Eastern white	0.30	1.7	3.6	6.4	26.6	3 550	13.0	4.55	1.35	2.26
Western red	0.31	2.1	4.5	7.8	36.5	7 240	19.2	4.80	1.92	1.64
Yellow	0.42	3.7	6.0	9.4	45.8	9 240	22.3	6.07	2.41	2.69
Douglas-fir	0.45	4.8	7.4	11.9	52.0	11 100	24.9	6.36	3.17	2.81
Fir										
Amabilis (Pacific silver)	0.36	4.2	8.9	12.5	37.8	9 310	19.1	4.92	1.61	1.89
Balsam	0.34	2.7	7.5	10.7	36.5	7 790	16.8	4.68	1.68	2.02
Hemlock										
Eastern	0.40	3.5	6.7	11.2	46.7	8 760	23.6	6.30	2.79	2.36
Western	0.41	5.4	8.5	13.0	48.0	10 200	24.7	5.18	2.57	2.69
Tamarack	0.48	2.8	6.2	11.2	47.0	8 550	21.6	6.34	2.85	2.76
Larch, western	0.55	5.1	8.9	14.0	59.8	11 400	30.5	6.34	3.58	2.87
Pine										
Eastern white	0.36	2.5	6.3	8.2	35.4	8 140	17.9	4.28	1.64	2.21
Jack	0.42	4.0	5.9	9.6	43.5	8 070	20.3	5.67	2.31	2.44
Lodgepole	0.40	4.7	6.8	11.4	39.0	8 760	19.7	4.99	1.90	2.29
Ponderosa	0.44	4.6	5.9	10.5	39.3	7 790	19.6	4.96	2.41	2.69
Red	0.39	3.7	6.3	9.6	34.5	7 380	16.3	4.90	1.94	2.41
Western white	0.36	3.7	6.8	10.7	33.3	8 200	17.4	4.50	1.62	1.57
Spruce										
Black	0.41	3.8	7.5	11.1	40.5	9 100	19.0	5.49	2.07	2.34
Engelmann	0.38	4.2	8.2	11.6	39.0	8 620	19.4	4.84	1.85	2.18
Red	0.38	4.0	7.9	11.7	40.5	9 100	19.4	5.56	1.88	2.41
Sitka	0.35	4.6	7.8	11.7	37.4	9 450	17.7	4.37	2.01	2.11
White	0.35	3.2	6.9	11.3	35.2	7 930	17.0	4.62	1.69	2.12
HARDWOODS										
Aspen, trembling	0.37	3.6	6.6	11.8	37.6	9 030	16.2	4.95	1.37	3.04
Birch, yellow	0.56	5.8	7.1	15.1	56.8	10 600	23.4	8.20	3.36	5.21
Maple, sugar	0.60	4.6	8.8	15.7	70.5	11 700	31.4	11.14	5.89	7.18
Oak, red	0.58	3.6	6.7	12.0	64.5	10 800	27.2	9.38	5.44	6.54

*SI metric units are converted from imperial units shown in Table 4.2A included in the Appendix at the end of this chapter.
†Formerly specific gravity.

Table 4.3 *Average clear-wood strength values* for commercial species in air-dry condition*

	Property										
	Relative density†		Shrinkage, green to air-dry based on dimensions when green (%)			Modulus of rupture (MPa)	Modulus of elasticity (MPa)	Compression parallel to grain, crushing strength max. (MPa)	Shear strength (MPa)	Compression perpendicular to grain, fiber stress at proportional limit (MPa)	Tension perpendicular to grain (MPa)
Species	Nominal	Oven-dry	Radial	Tangential	Volumetric						
SOFTWOODS											
Cedar											
Eastern white	0.30	0.31	–	–	3.8	42.3	4 380	24.8	6.93	2.68	2.63
Western red	0.34	0.34	–	–	4.8	53.8	8 270	33.9	5.58	3.43	1.46
Yellow	0.43	0.46	–	–	5.0	79.7	11 000	45.9	9.21	4.74	3.49
Douglas-fir	0.49	0.51	–	–	7.0	88.6	13 500	50.1	9.53	6.01	3.06
Fir											
Amabilis (Pacific silver)	0.39	0.41	–	–	7.5	68.9	11 400	40.8	7.54	3.61	3.06
Balsam	0.35	0.37	1.2	4.3	5.7	58.3	9 650	34.3	6.25	3.14	2.08
Hemlock											
Eastern	0.43	0.45	2.4	4.7	6.2	67.1	9 720	41.0	8.75	4.28	2.06
Western	0.43	0.47	–	–	8.1	81.1	12 300	46.7	6.48	4.53	2.93
Tamarack	0.51	0.54	–	–	7.1	76.0	9 380	44.8	9.00	6.15	3.47
Larch, western	0.58	0.64	–	–	8.0	107.0	14 300	60.9	9.25	7.31	3.62
Pine											
Eastern white	0.37	0.38	–	–	4.5	65.0	9 380	36.2	6.10	3.39	2.63
Jack	0.44	0.45	2.1	3.8	5.7	77.9	10 200	40.5	8.23	5.70	3.65
Lodgepole	0.41	0.46	–	–	6.6	76.0	10 900	43.2	8.54	3.65	3.78
Ponderosa	0.46	0.49	–	–	6.1	73.3	9 510	42.3	7.03	5.22	3.47
Red	0.40	0.42	1.9	4.1	6.5	69.7	9 450	37.9	7.50	4.96	3.54
Western white	0.37	0.40	–	–	6.0	64.1	10 100	36.1	6.34	3.23	2.64
Spruce											
Black	0.43	0.44	1.7	4.0	6.5	78.3	10 400	41.5	8.65	4.25	3.43
Engelmann	0.40	0.42	–	–	6.8	69.5	10 700	42.4	7.55	3.70	2.72
Red	0.40	0.42	–	–	6.2	71.5	11 000	38.5	9.20	3.77	3.70
Sitka	0.39	0.39	–	–	6.0	69.8	11 200	37.8	6.78	4.10	2.48
White	0.37	0.39	1.4	4.0	6.8	62.7	9 930	36.9	6.79	3.45	3.28
HARDWOODS											
Aspen, trembling	0.41	–	2.7	5.7	8.3	67.6	11 200	36.3	6.76	3.52	4.19
Birch, yellow	0.61	–	–	–	9.9	106.0	14 100	52.1	14.67	7.24	7.52
Maple, sugar	0.66	–	2.9	6.4	9.3	115.0	14 100	56.4	16.71	9.72	9.21
Oak, red	0.61	–	–	–	6.9	98.7	11 900	49.8	14.38	8.89	6.52

*SI metric units are converted from imperial units shown in Table 4.3A included in the Appendix at the end of this chapter.
†Formerly specific gravity.

STRENGTH-REDUCING CHARACTERISTICS

Knots

Knots are one of the major causes of reduced strength of structural members made from wood. Knots are classified according to form, size, quality, and occurrence (National Lumber Grades Authority 1975). The shape of the knot depends on the direction of the saw cut; a round knot occurs when a limb is cut at a right angle to its length, an oval knot is produced by a diagonal cut, and a spike knot results if the saw cut is made longitudinally through the knot.

The effect of knots on strength depends on their location in a member as well as their size. Knots on the edge are a more serious defect than those on the center line. Extensive tables have been prepared (ASTM Standard D245-70) that relate knot size and location to the percentage reduction in strength for various lumber sizes. In conjunction with the grading rules, which specify the maximum size of knots permitted in a grade, these tables make it possible to determine the anticipated reduction in bending strength for different grades and sizes of lumber.

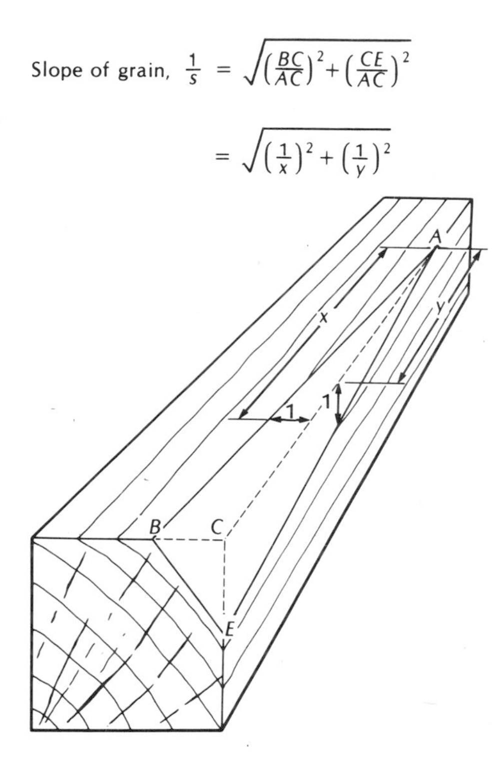

Figure 4.2 Determination of combined slope of grain.

Slope of Grain

Slope of grain refers to the deviation of the line of the fibers from a straight line parallel to the sides of a piece of lumber. It is usually expressed as a ratio, such as a slope of grain of 1 in 12 or 1 in 15. Deviation of the fibers can result either from the pattern of tree growth or from the manner in which the log was sawn. For example, because logs are tapered, sawing the log parallel to the pith, rather than the bark, produces slope of grain in the resulting lumber.

Severe slope of grain is often associated with an unusual growth condition called spiral grain, in which the fibers grow in a spiral manner about the longitudinal axis of the tree. If slope of grain ($1/s$) is evident on two adjoining faces of a member (Figure 4.2), then the true slope of grain is higher than on either of the two faces, and is determined by measuring the respective slopes $1/x$ and $1/y$ and using the equation (Gurfinkel 1973)

$$1/s = \sqrt{(1/x)^2 + (1/y)^2}. \qquad (2)$$

For example, if the slope of grain is 1 in 10 on one face, and 1 in 8 on the adjoining face, then the combined slope of grain is approximately 1 in 6. A slope of grain of 1 in 6 results in a 60% reduction in bending strength, while a slope of 1 in 16 will cause only a 20% reduction. Further information on the influence of slope of grain is available in ASTM Standard D245-70.

Shakes and Checks

Shakes and checks refer to lengthwise separations or splits that occur between or through the annual growth rings. Shakes are thought to originate in the tree, either from lateral growth stresses or from wind and frost action. Checks are defects arising from the seasoning process.

In grading strength, shakes, checks, and splits are treated similarly, and standard reductions are available (ASTM standard D245-70) that permit grading authorities to calculate the percentage of reduction in longitudinal shear strength for different sizes of these defects in various lumber thicknesses. For example, a shake 50 mm (2 in.) long decreases the longitudinal shear strength of a bending member that is 140 mm (6 in. nominal) thick by 30%.

Decay

Decay is not permitted in most structural grades of wood. Frequent inspection of highly stressed wood members is recommended when untreated wood has been mistakenly used in applications where decay may originate, such as in areas with poor ventilation and drainage. Further information on the practical aspects of decay prevention are discussed in Chapter 8, under 'Treatments to Control Stain and Decay.'

Aging

Normal aging of wood used in dry environments has no influence on its structural strength properties. This conclusion has been confirmed by testing both the stiffness and the strength properties of woods in service after periods in excess of 100 years.

However, it has been postulated (Barrett and Foschi 1977) that the effect of loads above a certain threshold that have been carried during the aging period may cumulatively reduce the capacity of a member to continue supporting load indefinitely. This reduced capacity will not necessarily be evident from a load test.

Reaction Wood

Reaction wood is abnormal tissue that generally has an adverse effect on many of the mechanical and physical properties of wood. In softwoods it forms on the lower side of leaning branches and stems and is called compression wood; in hardwoods it forms on the upper side and is called tension wood. For a detailed discussion of the effect of compression and tension wood on strength, see 'Natural Defects and Abnormalities' in Chapter 3.

STRENGTH OF WOOD COMPONENTS IN STRUCTURES

The criteria and equations that an engineer uses in the design of structures are usually independent of the material used. The particular advantage of one material over another is largely determined by its properties and cost. This chapter summarizes the important mechanical properties of wood and how they are measured and includes an explanation of how the strength data became part of the Code for the Engineering Design of Wood (Canadian Standards Association 1976). A corresponding standard for the United States is also available (National Forest Products Association 1977). For further reference, see 'Design of Timber Structures' in Chapter 11.

The mechanical properties of major interest to a designer are as follows (see *Timber Design Manual* [Laminated Timber Institute of Canada 1972] for design values and Tables 4.2 and 4.3 for properties of clear wood):

Figure 4.3 Stresses of wood. (A) Stress at the extreme fiber in bending F_b and longitudinal shear stress in bending F_v. (B) Compression perpendicular to grain $F_{c\perp}$. (C) Compression parallel to grain $F_{c\parallel}$. (D) Tension parallel to grain $F_{t\parallel}$. (E) Modulus of elasticity.

1. Stress at the Extreme Fiber in Bending, F_b

The extreme fibers in bending refer to those fibers along the very top and bottom faces of a member, as shown in Figure 4.3A. Note that the fibers on the face nearest the load are in compression, while those on the opposite face are in tension.

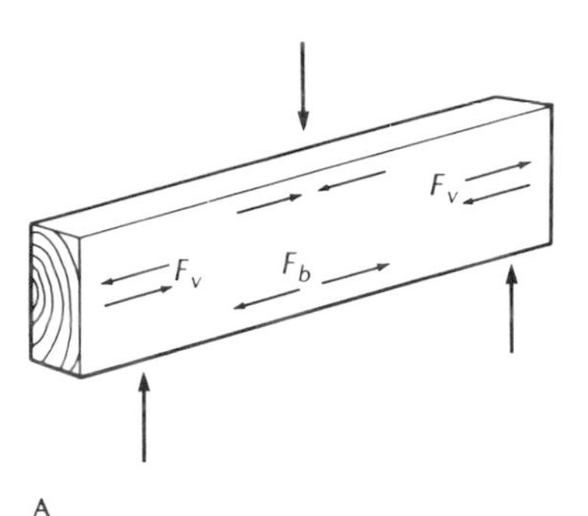

2. Longitudinal Shear Stress in Bending, F_v
Bending members are also subjected to a stress that tends to slide the fibers over each other horizontally. This stress is known as longitudinal, or horizontal, shear, and, if effects at supports are ignored, it reaches a maximum along the center line of the member (Figure 4.3A).

3. Compression Perpendicular to Grain, $F_{c_\perp}$
Joists and beams are usually supported at certain end and intermediate locations, as shown in Figure 4.3B. In this case, the strength of the wood in compression perpendicular to the grain must be sufficiently high to prevent crushing.

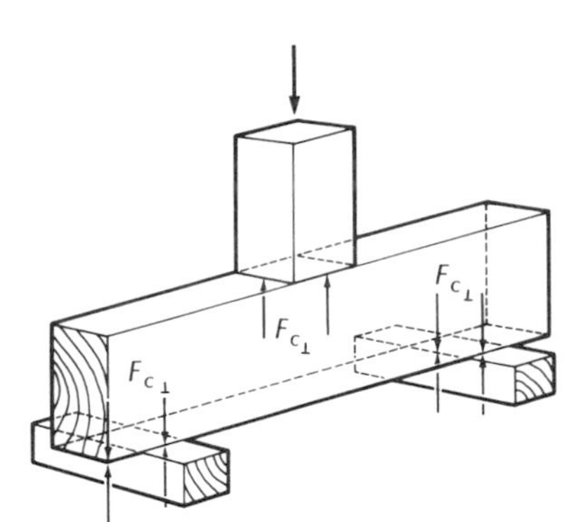

4. Compression Parallel to Grain, $F_{c_\parallel}$
Members such as columns, posts, and struts must resist loads that apply compressive forces parallel to the full length of the piece (Figure 4.3C).

5. Tension Parallel to Grain, $F_{t_\parallel}$
Some members, such as the lower chords of a truss, are subjected to tension loads parallel to the grain (Figure 4.3D). When members are loaded in this manner, stress concentrations produced by knots and slope of grain have a significant influence on the strength.

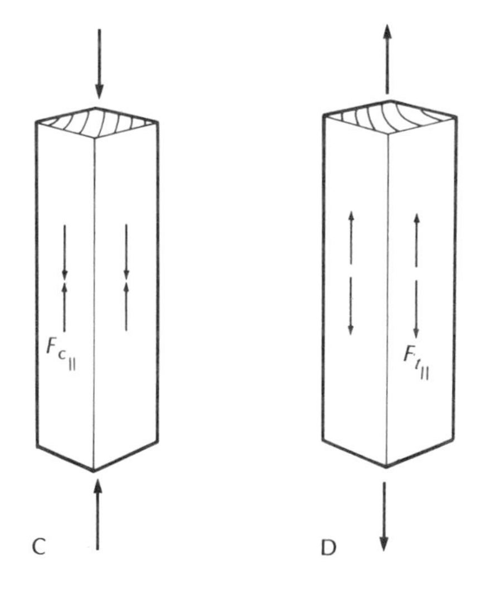

6. Tension Perpendicular to Grain, $F_{t_\perp}$
Wood does not possess high strength in tension perpendicular to the grain, and designers should avoid creating these stresses whenever possible. For example, it is poor design practice to notch the bottom of beams, to join truss webs independently to timber chords rather than to each other, and to attach secondary beam hangers close to the bottom of main girders.

7. Modulus of Elasticity, E
The modulus of elasticity of a member is a measure of its resistance to deflection under load (Figure 4.3E). The most commonly used modulus of elasticity is that measured parallel to the grain, which is designated by E_L. In some applications, such as plywood structures, it is necessary to know the modulus of elasticity perpendicular to the grain, denoted by E_T if the deformation takes place tangentially to the annual rings and E_R if it takes place radially.

Detailed knowledge of E_T and E_R does not exist for most Canadian species. Estimates of E_T and E_R may be made, however, by using the ratios $E_T/E_L \simeq 0.05$ and $E_R/E_L \simeq 0.07$ in association with the known value of E_L(see *Timber Design Manual* [Laminated Timber Institute of Canada 1972]).

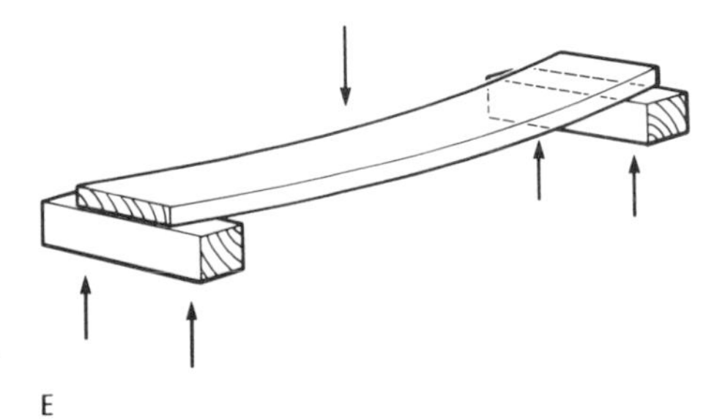

In applications where a body is subjected to shear stresses, the modulus of elasticity in shear – otherwise known as the modulus of rigidity, or G – is sometimes needed. Subscripts are added to this modulus to indicate the plane of deformation. For example, G_{LT} refers to deformation on the LT (longitudinal-tangential) plane resulting from shear stresses in the LR (longitudinal-radial) and RT (radial-tengential) planes.

Information on the modulus of rigidity of the various species is still inadequate, and one must often resort to using approximate values derived from the relations $G_{LT}/E_L \simeq 0.06$, $G_{LR}/E_L \simeq 0.075$, and $G_{RT}/E_L \simeq 0.018$.

Species Groups

In order to simplify design and marketing procedures, each of the 24 commercially important Canadian lumber species have been assigned to one of eight Species Groups, as shown in Table 4.4 (Canadian Standards Association 1976). When certain species grow and are harvested and manufactured together, it is often advantageous to

Table 4.4 *Species combination**

Commercial designation	Stamp identification	Species included in the combination	Species group
Douglas-fir –larch	D. Fir-L(N)	Douglas-fir Western larch	A
Hem-fir	Hem-Fir (N)	Pacific coast hemlock Amabilis fir Grand fir	B
Eastern hemlock –tamarack	Hem-Tam (N)	Eastern hemlock Tamarack	C
Coast species	Coast species	Douglas-fir Western larch Pacific coast hemlock Amabilis fir Grand fir Coast Sitka spruce	
Spruce –pine-fir	S-P-F or Spruce-Pine-Fir	Spruce (all species except coast Sitka spruce) Jack pine Lodgepole pine Ponderosa pine Balsam fir Alpine fir	D
Western cedars	W. Cedar (N)	Western red cedar Pacific coast yellow cedar	
Northern species	North Species	All above species Red pine Western white pine Eastern pine	E
Northern aspen	N. Aspen	Trembling aspen Largetooth aspen Balsam poplar	F

Note: Names of species are standard commercial names. Additional information on botanical and commercial names is given in Chapter 2, 'Commercial Woods.'

*Taken from CSA standard 086-76, *Code for the Engineering Design of Wood* (Canadian Standards Association 1976).

market them together in groups or combinations, particularly when species cannot be readily distinguished from each other in lumber form. Similar properties are desirable, but not essential, for giving two or more species a common designation or grade-stamp identification. The allowable stresses for each property in such a group or combination are determined by the species having the least capacity in that property.

Allowable Unit Stresses

The stresses that a designer of a wooden structure uses in his calculations are known as allowable unit stresses. To obtain these allowable unit stresses, the strength values for small, clear specimens are compiled using the procedures outlined in ASTM standard D143-52. In Canada, data have been systematically obtained from tens of thousands of clear-wood specimens for all the commercially important woods. Much of this information has been published (Jessome 1977), and some of it is summarized in Tables 4.2 and 4.3.

The strength properties of wood, like those of other structural materials, are variable. As a consequence, instead of using average property values to determine allowable unit stresses for bending, shear, and compression parallel to grain, designers establish a level below which no more than 5% of the population is expected to fall. This level of strength is known as the 5% exclusion limit, and methods for determining it are described in ASTM standard D2555-73.

The values obtained for the 5% exclusion limit of these properties and the average values of compression perpendicular to grain and modulus of elasticity are then multiplied by the inverse of the adjustment factors shown in Table 4.5. These values include a correction for normal duration of loading and a factor of safety. The results of this computation are the allowable design stresses for clear, straight-grained wood of a species.

Structural lumber is seldom clear or straight-grained. It is visually graded, however, under a standard set of grading rules that specify the *maximum* allowable defects that a grade may contain (see 'Grading Rules' in Chapter 6).

To convert from the strength of clear wood to the strength of a particular grade of lumber, a concept known as strength ratio, or the hypothetical ratio of the strength of a structural member to the strength it would have if it contained no strength-reducing characteristics, is used. For example, a grade of structural joists and planks assigned a strength ratio of 55% extreme fiber stress in bending would be expected to have at least 55% of the strength of a clear piece.

Tables of strength ratios for various sizes of knots, splits, shakes, and slopes of grain are available in ASTM standard D245-70. Use of these tables permits grading authorities to determine the maximum defects that are acceptable if a grade is to have a predetermined strength ratio.

The allowable unit stresses are obtained by applying the adjustment factors of Table 4.5 to the clear-wood strength and the strength ratios corresponding to a particular grade of lumber from ASTM standard D245-70. Tables of allowable unit stresses, both for sawn lumber and for glued-laminated construction, are presented in CSA standard 086-76. Allowable unit stresses can also be determined from strength tests of

Table 4.5 *Adjustment factors to be applied to the clear-wood properties*

	Softwoods	Hardwoods
Bending strength	2.1	2.3
Modulus of elasticity in bending	0.94	0.94
Tensile strength parallel to grain	2.1	2.3
Compressive strength parallel to grain	1.9	2.1
Horizontal shear strength	4.1	4.5
Proportional limit in compression perpendicular to grain	1.5	1.5

full-size dimension lumber. Such stresses can be expected to be more realistic and reliable than those derived from small, clear specimens. At the time of writing, research is underway to develop and standardize methods for evaluating the different strength properties by means of sampling and testing in-grade, full-size dimension lumber, and to develop corresponding methods for deriving allowable unit stresses or other design information from the data so obtained.

The final step in arriving at the allowable working stresses for glued-laminated and sawn timber is to multiply the allowable unit stresses by a series of stress modification factors for the application in question. These factors take into account such variables as duration of load, service conditions, and whether the wood has been subjected to preservative or fire-retardant treatment (see also 'Design of Timber Structures' in Chapter 11).

Figure 4.4 Testing a wood beam to determine strength in bending.

Machine Stress Rating

Historically, lumber has been graded visually on the basis of the size and distribution of various defects. During the past 10–15 years, a concerted effort has been made to use the full strength potential of every piece of structural lumber. Many laboratories and private companies have attempted to develop nondestructive methods for assessing the strength of each piece of commercial lumber instead of using the strength of small, defect-free specimens as a basis for assigning allowable unit stresses.

As a result of these studies, a useful correlation has been noted between the stiffness of lumber, as measured by E_L, and its ultimate strength in bending, F_b. A number of machines are now available that measure the stiffness of lumber using one of three principles of operation: (1) determining the load to produce a given deflection; (2) applying a known load and measuring the resulting deflection; (3) using an equation relating E to the velocity of sound and density.

Machine stress rating of lumber does not replace visual grading because visual graders are still needed to override machine decisions when the lumber does not meet the allowable visual defect requirements. The primary advantages of machine stress rating are that one species can easily be substituted for another, and material of higher strength can be segregated more readily than with visual grading techniques. A table for machine-stress-rated lumber is included in CSA standard 086-76.

Preservative and Fire-Retardant Treatments

Wood is often treated with preservatives to extend its service life. If the preservation process follows nationally approved procedures, then the currently published allowable stresses may be used without further reduction.

The effect of a fire-retardant treatment on strength depends upon the particular process, species, and product treated. It is difficult to provide one single factor to cover all possible combinations. Nevertheless, when no better information is available, the strength rating of fire-retardant-treated wood should be reduced by not less than 10% for sawn lumber and glued-laminated beams and by a similar amount in applications involving bolts or lag screws. Similarly, a reduction of at

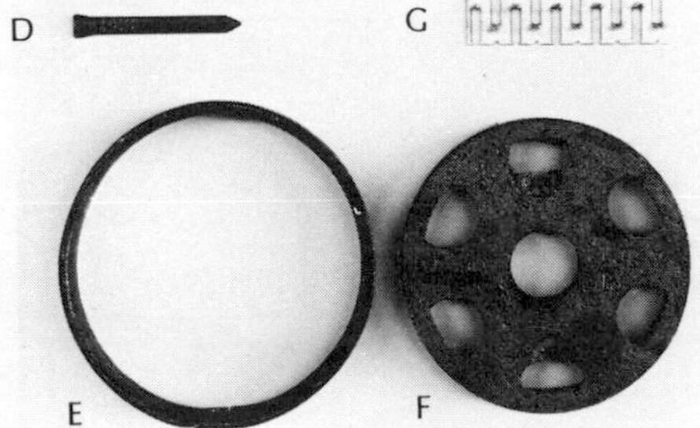

Figure 4.5 Mechanical fasteners for wood: A, helically threaded nails; B, wood screw; C, lag screws; D, glulam rivet; E, split ring; F, shear plate; G, toothed plate connector.

least 20% is recommended for treated plywood construction and connectors using split rings or shear plates.

CREEP AND RELAXATION

Two phenomena characteristic of all building materials are creep and relaxation. A material's mechanical properties are not constant over time. For example, if a load is placed on a beam, deflection, or bending, occurs immediately. What is not so evident is that with the passage of time this deflection increases. This increasing deflection is called creep. Designers using glued-laminated timber often include an allowance for creep in the design of trusses and beams by providing for a certain amount of camber or bend in them to offset the effect of creep.

A constant load on a beam will produce increasing deflection with time, whereas the magnitude of the load required to maintain a constant deflection decreases with the passage of time. This behavior is known as relaxation. Bolted connector joints must be periodically tightened after initial installation because of relaxation.

STRENGTH OF JOINTS

Wooden structures can be used to span distances in excess of 46 m (150 ft) because normal lengths of lumber may be efficiently joined with mechanical and adhesive fasteners. Mechanical fasteners for wood include not only nails, screws, bolts, and lag screws, but also more sophisticated hardware such as toothed plate connectors, split rings, shear plates, and glulam rivets. Examples of these fasteners are shown in Figure 4.5.

Standard procedures have been developed for conducting tests on timber-connector joints, and these are described in ASTM standard D1761-74. The factors that determine the strength of a joint are the type of fasteners, the number of units, the spacing between fasteners, the member thickness and width, the distance of the fasteners to the end and edge of the member, the species of wood, and its moisture content.

In Canada, the allowable load per timber fastener is determined by multiplying the unit capacity of the fastener by a series of load modification factors (Canadian Standards Association 1976).

Nails

Nails are undoubtedly the most familiar form of mechanical fastener for wood (Figure 4.5A). Their effectiveness, however, is dependent on their resistance to withdrawal. This resistance is least when nails are loaded in the direction of withdrawal and greatest when they are subject to lateral loads. Therefore, nails should be loaded laterally, perpendicular to the shank or direction of driving. Tables of unit capacity per nail have been established for all species groups and are available in the *Timber Design Manual* (Laminated Timber Institute of Canada 1972).

Wood Screws

Wood screws are primarily used in furniture construction and are not often used in structural applications, probably because of the higher cost of labor for their installation. Screws have several advantages over nails, including higher withdrawal resistance, insensitivity to vibration, and the capacity to pull two members into close contact for gluing.

Lag Screws

Lag screws come in sizes ranging from 75 mm (3 in.) to 300 mm (12 in.) in length and act as bolts in applications where the main member is too thick for conventional machine bolts or where one face of the member is not accessible for installing nuts and washers. Physically, lag screws differ from bolts in that the screw threads taper to a point, as shown in Figure 4.5C.

Like nails and screws, lag screws are not recommended for use in the end grain of wood. Lag screws should be installed in holes predrilled in accordance with standard specifications (Stern, Reeves, and Griggs 1973). In addition, they should be installed by rotating with a wrench rather than by driving with a hammer.

Graphs illustrating the lateral and withdrawal load capacity of lag screws are available in the *Timber Design Manual* (Laminated Timber Institute of Canada 1972). The allowable load is determined by multiplying the unit capacity by a series of factors to account for such variables as service conditions, depth of penetration, and type of side plates.

Often in engineered wood construction, the load is applied at an angle to the grain rather than parallel or perpendicular to it. In these situations, it is usual to apply Hankinson's formula for determining the allowable load on the connection:

$$N = PQ/(P \sin^2 \theta + Q \cos^2 \theta) \qquad (3)$$

where N is the allowable load at angle θ to the grain direction, P is the allowable load parallel to the grain, Q is the allowable load perpendicular to the grain, and θ is the angle between the direction of the grain and the direction of the load. This formula is used not only for lag screws but also for bolts, truss plates, and glulam rivets.

In a joint involving multiple lag screws, split rings, shear plates, or bolts, the load on the joint is not borne equally by all fasteners; the load-carrying capacity of N fasteners in a row is not N times the allowable load per fastener. With multiple fasteners, therefore, the specific spacing rules must be followed, and the designer must be cognizant of the uneven loads placed on the fasteners forming the joint.

Toothed Plate Connectors

Prefabricated trusses, in which the members are held together with toothed plate connectors, are being used in an increasing number of homes. These connectors are 1.6–1.00 mm (16–20 gauge) galvanized steel plates with teeth punched out at right angles to the plate (Figure 4.5G). They are usually available only to franchised distributors who have received approval to use them from local building code authorities. A study (Aplin 1973) has shown that using partially seasoned lumber for trussed rafters, and allowing the lumber to season in service has no influence on the strength of the truss.

Split Rings and Shear Plates

Split rings and shear plates are more efficient than either lag screws or bolts and are commonly used in glued-laminated construction. Split rings are either 63 or 101 mm (2½ or 4 in.) in diameter and are used solely for wood-to-wood connections (Figure 4.5E). They are installed in a precut circular groove. The ring has a special wedge-shaped cross-section, which provides a tight-fitting joint when the bolt and nut are drawn tight. The function of the bolt is merely to hold the members together – the load is transferred by the inner and outer surfaces of the ring acting on the adjoining wood surfaces.

Shear plates (Figure 4.5F) are intended primarily as connectors between wood and metal members, although they are sometimes used in wood-to-wood applications. They are available in two diameters, 66 and 101 mm (2⅝ and 4 in.), and are placed flush with, or fully embedded in, the wood member. Shear plates also differ from split rings in that the bolt transfers the shear stress from one shear plate to a second shear plate or from one shear plate to an adjoining steel side member.

The load capacity of a 66 mm (2⅝ in.) shear plate connector is approximately the same as that of a 63 mm (2½ in.) split ring connector. Similarly, the load ratings for a 101 mm (4 in.) split ring correspond closely to those for a 101 mm (4 in.) shear plate. Charts providing the unit load capacity for both split rings and shear plates are published in CSA standard 086-76 for the Canadian species groups. The allowable design loads are subject to a variety of adjustments for such factors as service conditions, load duration, the location of the fastener within the member, and its position relative to other fasteners.

Glulam Rivets

In a continuing search for a better fastener for glued-laminated construction, the glulam rivet was developed (Foschi 1973a,b) to provide an extremely efficient fastener that gives a very high load transfer per contact area. Physically, the glulam rivet resembles a flattened nail and is installed in conjunction with predrilled steel plates, using a rectangular hole pattern (Figure 4.5D).

Two major advantages are claimed for glulam rivets. One is easy field assembly, and the other is that the member may be designed on the basis of its gross cross-sectional area rather than the area remaining after deductions for grooving and drilling, as when split rings and shear plates are used. Current design methods for glulam rivets are described by Foschi (1973a,b) and included in CSA standard 086-76.

Bolts

Bolts and bolts in combination with split rings or shear plates are the two most common forms of timber connectors. The allowable load per bolt is significantly less than that for split rings and shear plates, and therefore the latter are usually preferred for prefabricated structures. In some applications, however, such as assembly at the job site, the ease of merely boring a hole – usually 0.8–1.6 mm (3/32–1/16 in.) oversize – in the members and inserting a bolt has much to recommend it.

When bolts are loaded perpendicular to their axis, the bearing pressure developed on the wood is largely dependent on the relative dimensions of the bolt, that is, on its length/diameter (L/D) ratio. For low L/D ratios, the bearing pressure is nearly uniform. If the bolt is slender, the bearing pressure is nonuniform and is concentrated near the surfaces of the member. As a result, bolts with large diameters have higher efficiencies. The factors that influence allowable load per bolt are identical with those for other connectors, except for the addition of a bolt size parameter to take into account the influence of the L/D ratio.

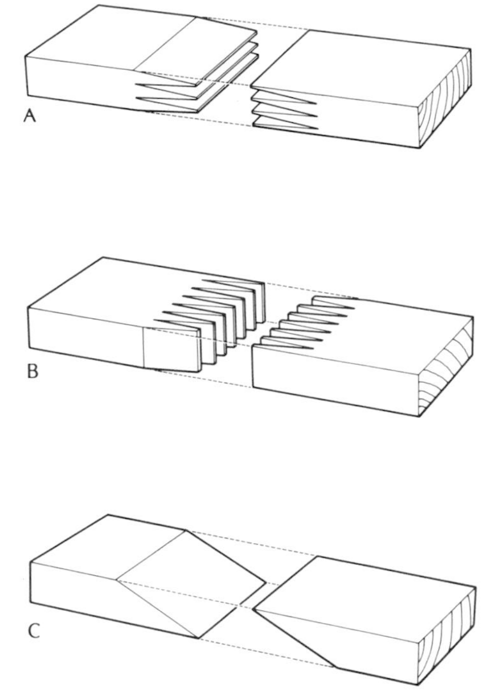

Figure 4.6 Finger joints – A, fingers cut parallel to wide face; B, fingers cut perpendicular to wide face – and a scarf joint (C).

Adhesive Joints

The glued-laminated timber industry resulted largely from the development of improved glues that could meet severe service conditions. By end jointing lumber, using either plain-scarf or finger joints, large structural elements in excess of 30 m (100 ft) in length can be fabricated from readily available lumber sizes. In this type of connection, 645 mm² (1 sq. in.) of glue line has more lateral strength than almost any type of nail (Hoyle 1972).

Typical scarf and finger joints are illustrated in Figure 4.6. Well-made scarf joints can have strength properties varying between 75% and 85% of those for the clear wood. The most common slopes are in the range of 1 in 8 to 1 in 12, with the flatter slope providing the stronger joint.

Finger joints have the advantage of consuming less material in the joint. They are made using a variety of finger shapes, the fingers ranging in length from 8 to 38 mm ($\frac{5}{16}$ to $1\frac{1}{2}$ in.). The strength of finger joints is highly dependent on the geometry of the joint, with sharp tips (1 mm [0.045 in.]) and low slopes (1 in 14) giving higher strengths (Selbo 1963). The important point for laminators planning to use a particular

Figure 4.7 The attractive all-wood roof, featuring massive glued-laminated beams, adds to the appeal of this arena at Scarborough, Ontario. (Photo: Koppers International Canada Ltd.)

scarf or finger joint is that the joint must exceed the *minimum* strength requirements as specified in the current CSA standard 0268-74.

A reliable butt joint has not yet been developed for lumber. Tests using adhesives that are now available indicate that butt joints have quite variable strength characteristics. Thus, their use is not recommended except under special circumstances.

Glued-laminated beams can be designed on the basis of both the strength requirement and their aesthetic appeal (Figure 4.7). If lumber is seasoned before beam fabrication, it is possible to produce members, often as large as 300 × 1825 mm (1 × 6 ft) in cross-section, that are virtually free from seasoning checks.

Glued-laminated beams may be either horizontally or vertically laminated, but the former are by far the more common. Standard sizes range from 75 to 360 mm (3 to 14½ in.) in width and from 114 to 2133 mm (4½ to 84 in.) in depth. Because the maximum lumber width is 286 mm (12 in. nominal), beams wider than about 273 mm (10¾ in.) normally require edge joining of two boards side by side with the joints staggered throughout the depth of the member. Two standard thicknesses of laminations are used in Canada – 38 mm (1½ in.) and 19 mm (¾ in.). The thicker material is used for straight members or for members having a radius of curvature greater than 8 m (27½ ft). (See Chapter 9 for further discussion of glues and gluing; also 'Glued-Laminated Lumber' in Chapter 11.)

POLES AND PILING

Because of their inherent decay resistance, western red cedar and eastern white cedar were originally the preferred species for use as poles. Now that modern preservative treatments are available, Douglas-fir, jack pine, red pine, and lodgepole pine are widely used.

The important characteristics of poles are high bending strength, moderate weight, straightness, evenness of taper, durability, and acceptance of preservative treatment. The strength properties of various species of poles are outlined in several publications: Forest Products Laboratory (1947), ASTM (1960), McGowan (1960, 1962), McGowan and Smith (1965), and Aplin (1967).

For design purposes, poles are considered members subject to combined axial load and bending moment. The allowable unit stresses for all species, except eastern white cedar, are taken as 80% of those given in CSA standard 086-76 for select structural beams and stringers of the appropriate species. The corresponding figure for eastern white cedar is 50% for select structural grade beams and stringers of the species combination North Species.

Several species are used for piles, including Douglas-fir, western larch, spruce, pine, western red cedar, oak, maple, and tamarack. The most popular species for pressure treating are Douglas-fir, jack pine, lodgepole pine, and red pine.

Piles are driven vertically or near vertically into the earth to support buildings and other structures and are classified according to the manner in which they develop their load-carrying capacity. Tip-bearing piles are those that transfer the load directly through soft upper strata to a hard substratum. The load capacity of such piles is usually determined by the unit bearing pressure of the stratum on which the tips of the piles are seated. Friction piles depend on the friction forces developed between the pile surface and the surrounding soil layer. The capacity of friction piles depends primarily on the type of soil, and only secondly on the shape and type of pile itself. The use of pressure-treated wood piling is now well established, and conservative estimates for the life of such foundations are in excess of 100 years (Chellis 1961; Canadian Institute of Timber Construction 1962). Pressure-treated timber piles have particular advantages for permafrost areas. Generally, they are more economical and are far less likely to disturb the thermal equilibrium of the permafrost layer than a conventional concrete foundation is. Concrete piles are difficult to cure at low temperatures, and steel piles are heavy and expensive.

PHYSICAL PROPERTIES OF WOOD

Wood has many unique physical properties that give it a versatility unmatched by any other material. This section briefly describes the acoustical, electrical, and thermal properties of wood, which are important in construction and structural applications.

Acoustical Properties

Wood has valuable acoustical properties. Those of primary interest are the velocity of sound and its capacity to dampen vibrations.

The velocity of sound in wood parallel to the grain, V (mm/sec), may be determined from the equation

$$V = \sqrt{E/\rho} \tag{4}$$

where E is the modulus of elasticity parallel to the grain (MPa) and ρ is the average relative density. Using equation (4) and air-dry Douglas-fir data for E and ρ from Table 4.3, we obtain a value for V of 5.25×10^6 mm/sec, which is approximately the same as the velocity of sound in steel.[2]

An interesting commercial application of equation (4) is a method of machine stress rating lumber by determining its stiffness, E, based on the measurement of the average velocity of sound and the density. This is done by rearranging the equation so that

$$E = V^2\rho. \tag{5}$$

The velocity of sound across the grain can also be deduced from this equation if one realizes that E across the grain is often only $\frac{1}{16}$ E parallel to the grain.

An important acoustical property of wood is its great ability to dampen vibrations. Wood has high internal friction, which gives it much more damping capacity than most other structural materials, especially metals. This capacity to dampen out vibration is expressed by the term logarithmic decrement, which is defined by the equation

$$\text{logarithmic decrement} = \ln (A_1/A_2) \tag{6}$$

2 In imperial units, using data for E (in psi) and ρ from Table 4.3A (see Appendix), equation (4) leads to the value $V = 2.1 \times 10^5$ in./sec for the velocity of sound in air-dry Douglas-fir.

where A_1 and A_2 are the amplitudes of any two successive oscillations.

Logarithmic decrement values for wood range from 0.02 to 0.10, depending on moisture content and temperature. Similar values for steel and aluminum, for example, are much lower; both have values much less than 0.01. James (1961) has made an extensive study of variations in the velocity of sound and logarithmic decrement in Douglas-fir over a wide range of temperature and moisture content and has concluded that the speed of sound and modulus of elasticity both decrease with increasing moisture content and temperature.

The problems of vibration and noise have traditionally been approached using the techniques of absorption and isolation. The sound absorption coefficient of a material, α, is defined as the ratio of the amount of sound energy absorbed to the total energy of impact. For unfinished wood the sound absorption coefficient is 0.06; comparative values are 0.25 for heavily lined carpet and 0.03 for brick, glass, and varnished wood. Increasing the surface roughness of wood with striations or sandblasting will improve its absorption properties.

The sound insulating qualities of a wall depend on the mass of the wall, its stiffness, and the nature of the fasteners used. Because the transmittance decreases as the logarithm of the weight, adding mass is not particularly effective in increasing the insulation provided by a wall. A wall weighing 16 018.46 kg/m^3 (1000 lb/cu ft) is only 50% more effective than one weighing 1601.85 kg/m^3 (100 lb/cu ft). A far more effective solution is to sandwich a thin elastomer between an outside constraining layer and the structure itself. This technique is called constrained layer damping. Solutions to specific problems in the wood industry are described in several publications: Timber Engineering Company (1956); Smith (1971); and Kugler, Niemic, and Pope (1973).

Electrical Properties

DC Properties

Wood in the oven-dry condition is an excellent insulator, but it loses its insulating qualities as its moisture content increases. In the oven-dry condition and at room temperature, the electrical resistivity of wood is approximately 10^{16} ohm-metres, but it decreases to 10^2 ohm-metres when the wood is green. This dramatic change in the electrical resistance of wood when moisture content goes from 0% to 30% led to the development of moisture meters based on DC resistivity. As discussed in Chapter 7, errors can result in the use of these instruments unless the wood has a uniform moisture content throughout its cross-section.

Figure 4.8 The versatility of wood makes it the natural choice for building homes.

AC Properties

High-frequency alternating currents are used in the wood industry primarily for the curing of resin adhesives and to a lesser degree for radio-frequency moisture meters and for drying wood rapidly. The electrical properties of wood of primary interest are the dielectric constant (ε), the power factor ($\cos\theta$), and the resistivity (r). These properties vary with moisture content, density, grain direction, temperature, and frequency. For a full discussion of the complex interactions, the reader is referred to Skaar (1948); Brown, Panshin, and Forsaith (1952); and Lin (1967a, 1967b).

The dielectric constant, ε, can be defined as the ratio of the capacitance of a capacitor with some chosen material as a dielectric to its capacitance using a vacuum as the dielectric. It was initially thought that the dielectric constant of wood would lie between the low value of 2 for oven-dry wood and the high value of 81 for water. More recent work (James 1974) indicates that this is true only for relatively high radio frequencies, such as those used in dielectric heating applications. For frequencies of the order of 1 kilohertz (kHz) and lower, the dielectric constant of wood can range up to 100 000 and even 1 million. This wide variation is due to various forms of polarization – interfacial, dipolar, and ionic – occurring at low frequencies.

The power factor of a dielectric is defined as the ratio of the electrical energy dissipated per cycle of oscillation to the total energy stored per cycle. This energy is dissipated in the form of heat as the molecules rotate under the influence of the alternating electric field. It is this heating effect which forms the basis for dielectric heating in drying and adhesive curing applications.

As a rough approximation, the power factor at 2 megahertz (MHz) may be taken as 0.02 for oven-dry wood and 0.10 for wood at 20% moisture content. The following equation is useful in dielectric heating applications, as it expresses the relation between the key variables of the power factor ($\cos \theta$) – frequency (f), resistivity (r), and dielectric constant (ε):

$$\cos \theta = 1.8 \times 10^{12}/(fr\varepsilon). \quad (7)$$

Thermal Properties

Thermal Expansion

A comprehensive study of the thermal expansion of wood (Weatherwax and Stamm 1946) indicates that the coefficient of thermal expansion parallel to the grain is approximately $4 \pm 1 \times 10^{-6}$ per degree C for the nine species studied. The thermal expansion perpendicular to the grain is much higher, ranging between 15 and 45×10^{-6} per degree C and averaging 30×10^{-6} per degree C.

Dimensional changes occurring as a result of temperature variation are small in comparison with those arising from changes in moisture content. The dimensional change associated with a 100°C change in temperature, for example, is $100 \times 30 \times 10^{-6} = 0.003$ mm per mm of length. Douglas-fir undergoes about a 6% shrinkage in drying from 30% moisture content to oven-dry, or a dimensional change of approximately 0.002 mm per mm of length for a 1% change in moisture content. In other words, where dimensional stability is concerned, a 1% change in moisture content is nearly equivalent to a 100°C temperature change.

Thermal Conductivity

Because wood is one of the poorest conductors of heat, it makes an excellent thermal insulator. As a result, wood is often used in buildings to lower heating costs. It is also used in the handles of cooking utensils and tools in order to prevent the passage of excessive heat.

The conduction of heat in wood depends on its thermal conductivity and specific heat. Thermal conductivity is directly related to the density of wood. The heaviest woods have the least insulating effect. This is due to the cellular nature of wood because the lighter woods contain the larger number of cell cavities in relation to the solid substance of the cell walls. The cell cavities in dry wood are filled with air, which is one of the poorest conductors known, thus the lighter porous woods conduct less heat than the heavier woods. Thermal conductivity is also affected by the moisture content of wood; material that is kiln-dried to a moisture content of less than 12% has lower conductivity than material containing more moisture.

The thermal conductivity of wood is expressed in watts per square metre per degree Celsius for one millimetre thickness, or, in imperial units, in BTU per hour per square foot per degree Fahrenheit for one inch thickness. A typical value for thermal conductivity in wood is that for Douglas-fir at 12% moisture content: 0.1956 $W/(m^2.°C)$ for one millimetre thickness, or 0.875 BTU per hour per square foot per degree F for one inch thickness, which is approximately 5 times lower than that for building brick, 14 times lower than that for concrete, and 370 times lower than that for steel.

Thermal conductivity is measured across the grain of wood because in sheathing and partitions this is the general direction of the flow of heat. A general equation for determining the thermal conductivity across the grain of wood at a moisture content below 40% is

$$K = 0.2236G(1.39 + 0.028M) + 0.165 \tag{8}$$

where K is the thermal conductivity in $W/(m^2.°C.mm)$, G is the basic relative density (Table 4.2), and M is the percent moisture content.[3] For wood at a moisture content of 40% or greater, the following equation is recommended:

$$K = 0.2236G(1.39 + 0.038M) + 0.165. \tag{9}$$

Thermal conductivity is two to three times higher along the grain than across it.

The other factor in the conduction of heat in wood is specific heat. This is the amount of heat required to raise the temperature of one gram of wood one degree Celsius. Studies have shown that the specific heat of wood is not influenced by species or relative density, but it does vary with temperature. The following equation may be used to calculate the specific heat of dry wood at a given temperature, T (°C):

$$\text{specific heat} = 0.226 + 0.00116T. \tag{10}$$

The specific heat of wood is about 50% higher than the specific heat of air and four times higher than that of copper. The specific heat is increased when wood contains water, because the specific heat of the water present is higher than that of dry wood.

3 In imperial units equations (8) and (9) are written as $K = G(1.39 + 0.028M) + 0.165$ and $K = G(1.39 + 0.038M) + 0.165$, respectively, where K is the thermal conductivity in BTU per hour per square foot per degree F for one inch thickness, G is the basic relative density (Table 4.2A in the Appendix), and M is the percent moisture content.

Knowledge of the specific heat of wood is useful in determining the amount of heat required for the heat treatment or conditioning of veneer bolts during the manufacture of veneer. If the bolts are softened by heat before the veneer is debarked and cut, the bark can be removed more readily and the knife will cut smoothly with minimum power consumption. The heat requirement for softening veneer bolts can be calculated by the equation

$$Q = W(T_F - T_I)(C + M/100) \times 0.004184 \quad (11)$$

where Q is the heat requirement in KJ, W is the oven-dry weight of wood in kg, T_F and T_I are the final and initial temperatures in °C, C is the average specific heat for the temperature range, and M is the moisture content (%).[4]

The combined effect of thermal conductivity and specific heat is known as the thermal diffusivity of wood. This is a measure of how quickly wood can absorb heat from its surroundings; it is the ratio of the thermal conductivity to the product of the density and specific heat (U.S. Forest Products Laboratory 1974).

Because of its relatively high specific heat and poor conductivity, wood absorbs heat very slowly. As a result, its thermal diffusivity is much lower than that of other structural materials, such as metals, brick, and stone. This explains why wood does not seem extremely hot or cold to the touch as many other materials do. The high resistance offered by wood to the passage of heat through it not only makes it an excellent thermal insulator but also has an important bearing on its suitability as a fire-resistant material.

REFERENCES

Aplin, E.N. 1967. *The strength of red pine poles* (An evaluation of plantation-grown and artificially-seasoned poles). East. For. Prod. Lab. Inf. Rep. OP-X-10. Ottawa

– 1973. *Factors affecting the stiffness and strength of metal plate connector joints*. East. For. Prod. Lab. Inf. Rep. OP-X-57. Ottawa

ASTM. 1960. *Strength and related properties of wood poles*. Wood Pole Research Program Final Rep. Philadelphia

ASTM standard D143–52. 1952. *Standard methods of testing small clear specimens in timber*. Philadelphia: ASTM

ASTM standard D245–70. 1970. *Standard methods for establishing structural grades and related allowable properties for visually graded lumber*. Philadelphia: ASTM

ASTM standard D2555–73. 1973. *Standard methods for establishing clear wood strength values*. Philadelphia: ASTM

ASTM standard D1761–74. 1974. *Standard methods for testing metal fasteners in wood*. Philadelphia: ASTM

ASTM standard D2915–74. 1974. *Standard methods for evaluating allowable properties for grades of structural lumber*. Philadelphia: ASTM

Barrett, J.D., and Foschi, R.O. 1978. Duration of load and failure probability in wood: Cyclic loads of random amplitude. *Can. J. Civ. Eng.* 5: 515–32.

Brown, H.P.; Panshin, A.J.; and Forsaith, C.C. 1952. *The physical, mechanical and chemical properties of the commerical woods of the United States*. Textbook of wood technology, Vol. II. New York: McGraw-Hill

4 The heat requirement for softening veneer bolts may be expressed in imperial units by the equation $Q = W(T_F - T_I)(C + M/100)$, where Q is the heat requirement in BTU, W is the oven-dry weight of wood in lb, T_F and T_I are the final and initial temperatures in °F, C is the average specific heat for the temperature range, and M is the moisture content (%).

Canadian Institute of Timber Construction. 1962. *Pressure treated timber piles.* Ottawa
Canadian Standards Association. 1974. *Qualification code for manufacturers of glued end-jointed structural lumber.* CSA standard 0268–74. Rexdale, Ontario
– 1976. *Code for the engineering design of wood.* CSA standard 086–76. Rexdale, Ontario
Chellis, R.D. 1961. *Pile foundations.* New York: McGraw-Hill
Forest Products Laboratory. 1947. *The strength of telephone poles. Eastern cedar, red pine and jack pine.* Circ. 31 (rev.). Ottawa
Foschi, R. O. 1973a. *Stress analysis and design of glulam rivet connections for parallel-to-grain loading of wood.* West. For. Prod. Lab. Inf. Rep. VP-X-116. Vancouver
– 1973b. *Stress analysis and design of glulam rivet connections for perpendicular-to-grain loading of wood.* West. For. Prod. Lab. Inf. Rep. VP-X-117. Vancouver
Gurfinkel, G. 1973. *Wood engineering.* New Orleans: South. For. Prod. Assoc.
Hoyle, R.J. 1972. *Wood technology in the design of structures.* Missoula, Mont.: Mountain Press
James, W.L. 1961. Effect of temperature and moisture content on internal friction and speed of sound in Douglas-fir. *For. Prod. J.* 11 (9): 383–90
– 1974. *Dielectric properties of wood and hardboard; variation with temperature, frequency, moisture content.* U.S. Dep. Agric., For. Serv. Res. Pap. FPL-245. Madison, Wis.
Jessome, A.P. 1977. *Strength and related properties of woods grown in Canada.* Dep. Fish. Environ. For. Tech. Rep. 21. Ottawa
Kugler, B.A.; Niemic, K; and Pope, L.D. 1973. *Noise control design guide for moulding and millwork plants.* Portland, Oreg.: West. Wood Moulding Millwork Prod.
Laminated Timber Institute of Canada. 1972. *Timber design manual.* Ottawa: LTIC
Lin, R.T. 1967a. Review of the electrical properties of wood and cellulose. *For. Prod. J.* 17 (7): 54–61
– 1967b. Review of the dielectric properties of wood and cellulose. *For. Prod. J.* 17 (7): 61–6
McGowan;, W.M. 1960. *The strength of Douglas-fir telephone poles.* For. Prod. Lab. Tech. Note 15. Vancouver
– 1962. *The strength of western hemlock power and communication poles.* For. Prod. Res. Branch Tech. Note 27. Vancouver
McGowan, W.M., and Smith, W.J. 1965. *Strength and related properties of western red cedar poles.* Dep. For. Publ. 1108. Ottawa
National Forest Products Asociation. 1973. *National design specification for stress-grade lumber and its fastenings.* Washington, DC
– 1977. *National design specification for stress-grade lumber and its fastenings.* Washington, DC
National Lumber Grades Authority. 1975. *Standard grading rules for Canadian lumber* (rev. Jan. 1, 1975). Vancouver
Selbo, M.L. 1963. Effect of joint geometry on tensile strength of finger joints. *For. Prod. J.* 13 (9): 390–400
Skaar, C. 1948. *The dielectric properties of wood at several radio frequencies.* NY State Coll. For. Tech. Publ. 69. Syracuse, NY
Smith, J.H. 1971. Noise in the woodworking industry – A review of the literature. *For. Prod. J.* 21 (9): 82–3
Stern, G.E.; Reeves, J.R.; and Griggs, W.C. 1973. *Mechanical fastening of wood – A review of the state of the art.* Madison, Wis.: For. Prod. Res. Soc.
Timber Engineering Company. 1966. *Timber design and construction handbook.* New York: McGraw-Hill
U.S. Forest Products Laboratory. 1974. *Wood handbook: Wood as an engineering material.* Agric. Handb. No. 72 (rev.). Washington: U.S. Dep. Agric.
Weatherwax, R.C., and Stamm, A.J. 1946. *The coefficients of thermal expansion of wood and wood products.* U.S. For. Prod. Lab. Rep. No. R1487. Madison, Wis.

APPENDIX

Tables 4.1, 4.2, and 4.3, which give data in SI metric units, are repeated in this appendix in imperial units of measurement as Tables 4.1A, 4.2A, and 4.3A.

Table 4.1A *Functions relating mechanical properties* to relative density† of clear, straight-grained wood*

Property	Relative density–strength relation‡	
	Green wood	Air-dry wood (12% moisture content)
Static bending		
Fiber stress at proportional limit, psi	$10\,200G^{1.25}$	$16\,700G^{1.25}$
Modulus of elasticity, million psi	$2.36G$	$2.80G$
Modulus of rupture, psi	$17\,600G^{1.25}$	$25\,700G^{1.25}$
Work to maximum load, in.-lb per cu in.	$35.6G^{1.75}$	$32.4G^{1.75}$
Total work, in.-lb per cu in.	$103G^{2}$	$72.7G^{2}$
Impact bending, height of drop causing complete failure, in.	$114G^{1.75}$	$94.6G^{1.75}$
Compression parallel to grain		
Fiber stress at proportional limit, psi	$5\,250G$	$8\,750G$
Modulus of elasticity, million psi	$2.91G$	$3.38G$
Maximum crushing strength, psi	$6\,730G$	$12\,200G$
Compression perpendicular to grain, fiber stress at proportional limit, psi	$3\,000G^{2.25}$	$4\,630G^{2.25}$
Hardness		
End, lb	$3\,740G^{2.25}$	$4\,800G^{2.25}$
Side, lb	$3\,420G^{2.25}$	$3\,770G^{2.25}$

*Data from U.S. Forest Products Laboratory (1974).
†Formerly specific gravity.
‡The properties and values should be read as equations; for example, modulus of rupture for green wood = $17\,600G^{1.25}$, where G represents the relative density of oven-dry wood, based on the volume at the moisture condition indicated.

Table 4.2A *Average clear-wood strength values for commercial species in green condition*

	Property									
	Relative density,*	Shrinkage, green to oven-dry based on dimensions when green (%)			Modulus of rupture	Modulus of elasticity	Compression parallel to grain, crushing strength	Shear strength	Compression perpendicular to grain, fiber stress at proportional limit	Tension perpendicular to grain
Species	basic	Radial	Tangential	Volumetric	(psi)	(1 000 psi)	max. (psi)	(psi)	(psi)	(psi)
SOFTWOODS										
Cedar										
Eastern white	0.30	1.7	3.6	6.4	3 900	520	1 890	660	200	330
Western red	0.31	2.1	4.5	7.8	5 300	1 050	2 780	700	280	240
Yellow	0.42	3.7	6.0	9.4	6 600	1 340	3 240	880	350	390
Douglas-fir	0.45	4.8	7.4	11.9	7 500	1 610	3 610	920	460	410
Fir										
Amabilis (Pacific silver)	0.36	4.2	8.9	12.5	5 500	1 350	2 770	710	230	270
Balsam	0.34	2.7	7.5	10.7	5 300	1 130	2 440	680	240	290
Hemlock										
Eastern	0.40	3.5	6.7	11.2	6 800	1 270	3 430	910	400	340
Western	0.41	5.4	8.5	13.0	7 000	1 480	3 580	750	370	390
Tamarack	0.48	2.8	6.2	11.2	6 800	1 240	3 130	920	410	400
Larch, western	0.55	5.1	8.9	14.0	8 700	1 650	4 420	920	520	420
Pine										
Eastern white	0.36	2.5	6.3	8.2	5 100	1 180	2 590	640	240	320
Jack	0.42	4.0	5.9	9.6	6 300	1 170	2 950	820	340	350
Lodgepole	0.40	4.7	6.8	11.4	5 700	1 270	2 860	720	280	330
Ponderosa	0.44	4.6	5.9	10.5	5 700	1 130	2 840	720	350	390
Red	0.39	3.7	6.3	9.6	5 000	1 070	2 370	710	280	350
Western white	0.36	3.7	6.8	10.7	4 800	1 190	2 520	650	240	230
Spruce										
Black	0.41	3.8	7.5	11.1	5 900	1 320	2 760	800	300	340
Engelmann	0.38	4.2	8.2	11.6	5 700	1 250	2 810	700	270	320
Red	0.38	4.0	7.9	11.7	5 900	1 320	2 810	810	270	350
Sitka	0.35	4.6	7.8	11.7	5 400	1 370	2 560	630	290	310
White	0.35	3.2	6.9	11.3	5 100	1 150	2 470	670	240	310
HARDWOODS										
Aspen, trembling	0.37	3.6	6.6	11.8	5 500	1 310	2 350	720	200	440
Birch, yellow	0.56	5.8	7.1	15.1	8 200	1 540	3 390	1 140	490	760
Maple, sugar	0.60	4.6	8.8	15.7	10 200	1 700	4 560	1 620	850	1 040
Oak, red	0.58	3.6	6.7	12.0	9 400	1 560	3 940	1 360	790	950

*Formerly specific gravity.

Table 4.3A *Average clear-wood strength values for commercial species in air-dry condition*

	Property										
	Relative density*		Shrinkage, green to air-dry based on dimensions when green (%)			Modulus of rupture (psi)	Modulus of elasticity (1 000 psi)	Compression parallel to grain, crushing strength max. (psi)	Shear strength (psi)	Compression perpendicular to grain, fiber stress at proportional limit (psi)	Tension perpendicular to grain (psi)
Species	Nominal	Oven-dry	Radial	Tangential	Volumetric						
SOFTWOODS											
Cedar											
Eastern white	0.30	0.31	–	–	3.8	6 100	640	3 600	1 000	390	280
Western red	0.34	0.34	–	–	4.8	7 800	1 200	4 910	810	500	210
Yellow	0.43	0.46	–	–	5.0	11 600	1 600	6 650	1 340	690	510
Douglas-fir	0.49	0.51	–	–	7.0	12 800	1 960	7 270	1 380	870	440
Fir											
Amabilis (Pacific silver)	0.39	0.41	–	–	7.5	10 000	1 650	5 920	1 090	520	440
Balsam	0.35	0.37	1.2	4.3	5.7	8 500	1 400	4 980	910	460	300
Hemlock											
Eastern	0.43	0.45	2.4	4.7	6.2	9 700	1 400	5 950	1 270	620	300
Western	0.43	0.47	–	–	8.1	11 800	1 790	6 780	940	660	420
Tamarack	0.51	0.54	–	–	7.1	11 000	1 360	6 500	1 310	890	500
Larch, western	0.58	0.64	–	–	8.0	15 500	2 080	8 840	1 340	1 060	520
Pine											
Eastern white	0.37	0.38	–	–	4.5	9 400	1 360	5 250	880	490	380
Jack	0.44	0.45	2.1	3.8	5.7	11 300	1 480	5 880	1 190	830	530
Lodgepole	0.41	0.46	–	–	6.6	11 000	1 580	6 270	1 240	530	550
Ponderosa	0.46	0.49	–	–	6.1	10 600	1 380	6 140	1 020	760	500
Red	0.40	0.42	1.9	4.1	6.5	10 100	1 370	5 490	1 090	720	510
Western white	0.37	0.40	–	–	6.0	9 300	1 460	5 240	920	470	380
Spruce											
Black	0.43	0.44	1.7	4.0	6.5	11 400	1 510	6 020	1 250	620	500
Engelmann	0.40	0.42	–	–	6.8	10 100	1 550	6 150	1 100	540	400
Red	0.40	0.42	–	–	6.2	10 400	1 600	5 590	1 340	550	540
Sitka	0.39	0.39	–	–	6.0	10 100	1 630	5 480	980	590	360
White	0.37	0.39	1.4	4.0	6.8	9 100	1 440	5 350	980	500	480
HARDWOODS											
Aspen, trembling	0.41	–	2.7	5.7	8.3	9 800	1 630	5 270	980	510	610
Birch, yellow	0.61	–	–	–	9.9	15 400	2 040	7 560	2 130	1 050	1 090
Maple, sugar	0.66	–	2.9	6.4	9.3	16 700	2 040	8 180	2 420	1 410	1 340
Oak, red	0.61	–	–	–	6.9	14 300	1 730	7 230	2 080	1 290	950

*Formerly specific gravity.

5

G.M. BARTON
Western Forest Products Laboratory, Vancouver
H.H. BROWNELL
Eastern Forest Products Laboratory, Ottawa

The Chemistry of Wood

THE COMPONENTS OF WOOD

In contrast to petroleum and other nonrenewable raw materials, wood is a constantly renewable resource. All wood is formed from carbon dioxide, which is taken from the air, and from water, which is taken from the soil along with small amounts of dissolved minerals. The element composition of dry wood is about 50% carbon, 6% hydrogen, 44% oxygen, and less than 0.1% nitrogen. There is little variation in these figures from one species of wood to another. During photosynthesis, cells containing chlorophyll in the leaves and needles absorb radiant energy from sunlight and use it to convert these simple compounds into more complex substances, which eventually form the different components of wood.

Figure 5.1 The chemical composition of wood.

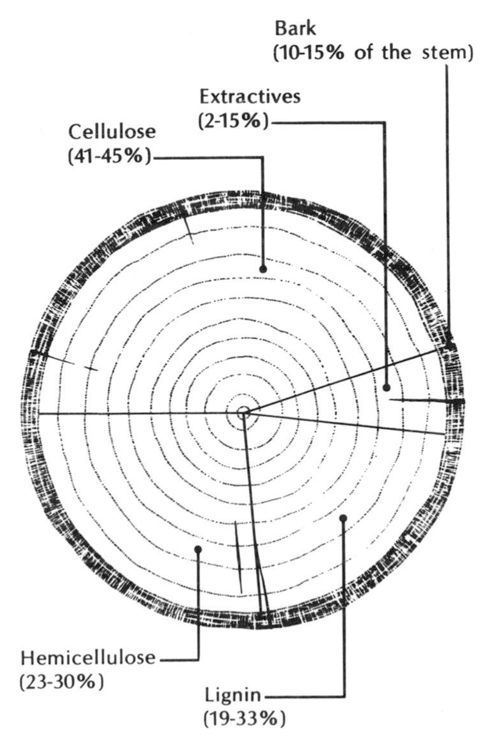

As indicated in Figure 5.1, wood is composed of several different types of organic material, the principal components being cellulose, hemicelluloses, and lignin. Other components present in lesser amounts are extractives (approximately 5%) and inorganic material (approximately 0.3%), consisting mainly of calcium, magnesium, sodium, potassium, and manganese, and including a wide variety of trace elements. Bark makes up 10–15% of the bole or main stem of the tree; foliage, 6–9% of the bole.

Wood is stronger than steel per unit of weight. Its great strength is due to the cellulose and in part to the hemicelluloses. Lignin acts as a glue, binding the cellulose molecules together to form the basic wood structure. Wood can thus be compared with glass-reinforced resin, or fiberglass, in which the strong glass fibers are held together by the synthetic resin. Wood differs from fiberglass, however, in having a cellular structure.

A wide variety of substances, commonly called extractives, are nonuniformly distributed within the walls and cavities of the cells. Many of these extractives are of relatively low molecular weight and tend to make one wood species superior or inferior to another for certain uses. Western red cedar, for example, contains fungicides, which make it a preferred species for long-lasting poles and posts. Another is Douglas-fir; because of some of its extractives, it cannot be pulped by the sulfite process. The chemical composition of some Canadian wood species is given in Tables 5.1 and 5.2.

Table 5.1 *Chemical composition* (as percentages) *of some Canadian wood species* (extractive-free wood basis)

	Analysis								
Species	Ash	Lignin	Acetyl	Uronic acid anhydride	Glucan	Galactan	Mannan	Araban	Xylan
Trembling aspen	0.2	16.3	3.4	3.3	57.3	0.8	2.3	0.4	16.0
White elm	0.3	23.6	3.9	3.6	53.2	0.9	2.4	0.6	11.5
Beech	0.4	22.1	3.9	4.8	47.5	1.2	2.1	0.5	17.5
White birch	0.2	18.9	4.4	4.6	44.7	0.6	1.5	0.5	24.6
Yellow birch	0.3	21.3	3.3	4.2	46.7	0.9	3.6	0.6	20.1
Red maple	0.2	24.0	3.8	3.5	46.6	0.6	3.5	0.5	17.3
Sugar maple	0.3	22.7	2.9	4.4	51.7	–	2.3	0.8	14.8
Balsam fir	0.2	29.4	1.5	3.4	46.8	1.0	12.4	0.5	4.8
Eastern white cedar	0.2	30.7	1.1	4.2	45.2	1.5	8.3	1.3	7.5
Eastern hemlock	0.2	32.5	1.7	3.3	45.3	1.2	11.2	0.6	4.0
Jack pine	0.2	28.6	1.2	3.9	45.6	1.4	10.6	1.4	7.1
White spruce	0.3	27.1	1.3	3.6	46.5	1.2	11.6	1.6	6.8
Tamarack	0.2	28.6	1.5	2.9	46.1	2.3	13.1	1.0	4.3

	Estimation					Gross composition		
Species	Cellulose	Non-cellulosic glucan	Gluco-mannan (acetate)	Arabino-galactan	4-*O*-methyl glucurono (arabino) xylan (acetate)	Cellulose	Hemicelluloses	Lignin
Trembling aspen	53	3	4	1	23	53	31	16
White elm	49	2	4	2	19	49	27	24
Beech	42	4	4	2	25	42	36	22
White birch	41	2	3	1	34	41	40	19
Yellow birch	40	3	7	1	28	40	39	21
Red maple	41	2	7	1	25	41	35	24
Sugar maple	–	–	4	1	22	–	–	–
Hardwoods, average	45	3	5	1	25	45	34	21
Balsam fir	44	0	18	1	8	44	27	29
Eastern white cedar	44	0	11	2	12	44	25	31
Eastern hemlock	42	0	17	1	7	42	26	33
Jack pine	41	0	16	2	12	41	30	29
White spruce	44	0	17	2	10	44	29	27
Tamarack	43	0	18	3	7	43	28	29
Softwoods, average	43	0	16	2	9	43	28	29

Sources: Data from Rydholm (1965) and Timell (1957a, b)

Table 5.2 *Chemical analysis of some Canadian wood species* (unextracted, dry-wood basis)

Species	Ash content, %	Extractives, % in Alcohol-benzene	Ethyl ether	1% NaOH	Hot water	Lignin content, %	Pentosans, %
Western white pine	0.3	8.3	5.6	15.6	3.7	25.4	7.9
Red pine	–	3.5	2.5	13.4	4.4	26.2	10.0
Jack pine	0.3	3.7	2.3	11.2	2.9	26.7	9.7
Lodgepole pine	0.2	3.5	2.3	12.6	3.6	25.0	9.2
Tamarack	0.3	2.0	0.7	11.5	4.8	26.4	8.3
Western larch	0.4	1.4	0.4	13.4	4.9	26.8	7.8
Engelmann spruce	0.2	2.8	1.4	12.2	3.7	26.3	9.2
Douglas-fir	0.2	4.4	1.2	15.1	5.6	27.2	6.8
Western hemlock	0.3	1.6	0.8	9.2	0.4	27.8	9.2
Mountain hemlock	0.5	4.6	1.0	11.6	4.8	27.0	7.0
Western red cedar	0.3	14.1	2.5	21.0	11.0	31.8	9.0
Trembling aspen	–	2.9	1.1	18.0	2.1	19.3	18.8
Yellow birch	0.8	2.6	0.8	15.4	2.7	22.7	22.6
White birch	0.4	2.8	1.3	14.1	1.5	20.0	22.6
Beech	0.5	1.8	0.7	14.7	1.5	21.0	20.2
Chestnut oak	0.4	4.7	0.6	21.1	7.2	24.3	19.2
White elm	0.4	2.0	0.5	14.3	1.6	20.5	16.2
Red maple	0.7	2.5	0.8	17.9	4.4	22.8	17.1
Basswood	0.7	4.1	2.1	19.9	2.4	20.0	16.6

Source: Data from Wise and Jahn (1952) and Bray and Martin (1947)

Figure 5.2 The potential of forest chemicals and other uses for wood chemistry are being developed through continuing research.

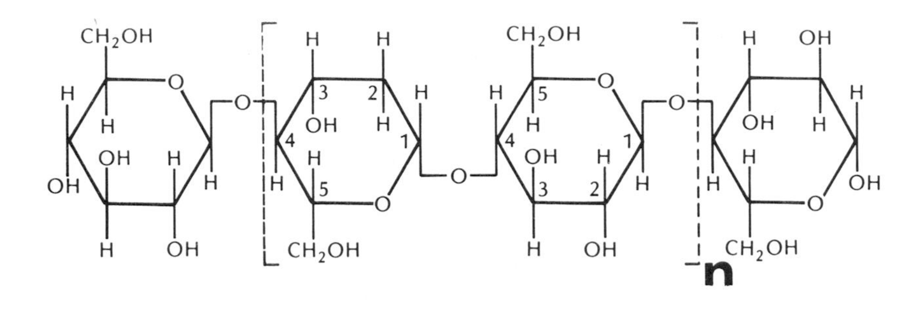

Formula 1 Cellulose.

Cellulose

The most important component of both hardwoods and softwoods is cellulose, which generally amounts to 41–45% of the dry wood by weight. It is the main structural element of tree fibers (Chapter 3) and the major constituent of pulp (Chapter 13).

—Glucose—Glucose—Glucose—

Formula 2 Cellulose (abbreviated formula).

Cellulose is a linear polymer composed of several thousand glucose units (Stamm 1964) linked end to end, as shown in Formula 1. It can also be represented in an abbreviated form, as shown in Formula 2. The chemical linkages between the glucose units can be broken by mineral acid, reverting the cellulose polymer to the glucose molecules from which it was built.

Each glucose unit of cellulose contains three hydroxyl (—OH) groups, except the two end units, which contain four. The hydroxyl groups give cellulose its principal chemical properties and are the reactive sites to which other chemical groups can be attached in the preparation of derivatives, such as cellulose acetate. The hydroxyl groups strongly attract water molecules and thus are the major cause of the swelling and shrinking of wood. They also attract the hydroxyl groups of adjacent cellulose molecules, creating microfibrils, which are threadlike bundles of cellulose molecules that lie approximately parallel to each other (Mühlethaler 1965).

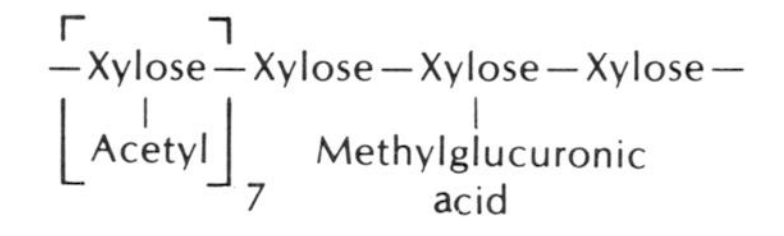

Formula 3 Hardwood xylan.

Microfibrils contain regions where the cellulose molecules are sufficiently rigid and regular to be distinguished by x-ray diffraction as crystalline. These crystalline regions, or crystallites, which form up to 70% of natural wood cellulose, make much of it relatively inaccessible to chemical reactions. Water, for example, readily penetrates the amorphous regions of cellulose but cannot enter the crystallites (Howsmon and Sisson 1954).

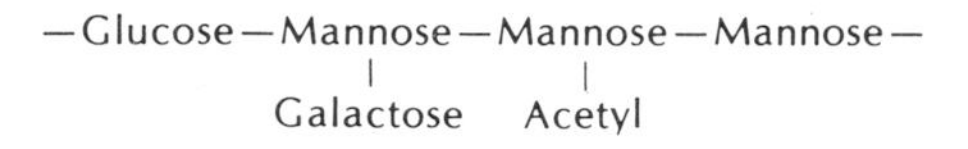

Formula 4 Softwood galactoglucomannan.

Hemicelluloses

Hemicelluloses are the noncellulosic polysaccharides of the cell wall. Polysaccharides are polymers formed from simple sugars. Cellulose is a polysaccharide formed from glucose sugar only. The hemicelluloses, in contrast, are formed from a number of sugars, the most important of which are glucose, galactose, mannose, xylose, and arabinose. Hemicelluloses usually constitute 23–30% of both hardwoods and softwoods. They are structurally more complex than cellulose, lower in molecular weight (usually between 100 and 200 sugar units per molecule), and generally amorphous before isolation.

All hardwoods are believed to contain only one major hemicellulose, commonly called hardwood xylan. This material usually constitutes

Formula 5 Softwood xylan.

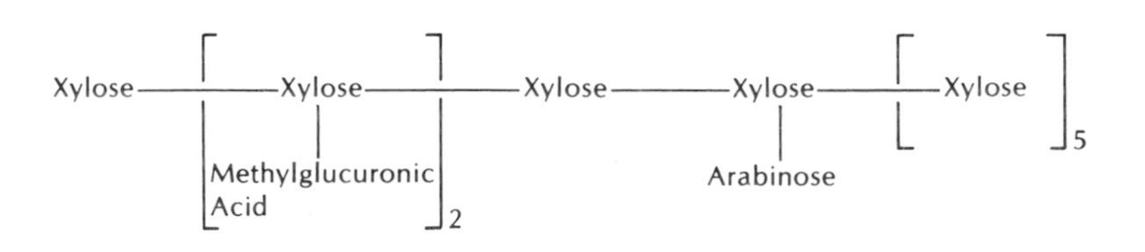

19–26% of the wood, although white birch contains an exceptional 35%. Hardwood xylan consists of a backbone of xylose sugar units linked end to end, much like the glucose units in cellulose. As Formula 3 illustrates, however, methylglucuronic acid groups (which are acidic derivatives of glucose sugar) are attached, apparently randomly, to one out of every ten xylose units (Timell 1967). Seven out of every ten xylose units carry an acetyl group, which is the source of much of the acetic acid liberated during wood pyrolysis.

Hardwoods also contain 3–5% of hardwood glucomannan, which consists of glucose and mannose sugar units usually in a 1:2 ratio, linked in a chain much like the chain of glucose units in cellulose.

Softwoods contain two major hemicelluloses, softwood galactoglucomannan (approximately 12–18% of the wood) and softwood xylan (approximately 7–14% of the wood). The galactoglucomannan consists of a backbone of glucose and mannose sugar units, distributed apparently at random, in a 1:3 ratio, and linked together much like the glucose units in cellulose. Single galactose sugar units are attached to some of the sugar units of the backbone, as indicated in Formula 4 (Timell 1967). The more soluble fractions contain one galactose unit per four units of the backbone; the less soluble fractions, which are sometimes called glucomannan, contain only 0.1 galactose unit per four units. In both fractions, acetyl groups are attached to some of the mannose units.

Softwood xylan, like hardwood xylan, consists of a backbone of xylose units carrying methylglucuronic acid groups (Formula 5). Softwood xylan is more acidic, however, having approximately one such group for every five xylose units (Timell 1967). Unlike hardwood xylan, it contains no acetyl groups but has an average of one arabinose sugar unit for every eight or nine xylose units.

Other hemicelluloses are usually present only in minor amounts. However, a polymer of galactose and galacturonic acid (an acidic derivative of galactose sugar) accounts for about 10% of reaction wood (both tension wood and compression wood) of several species. In addition, larch wood contains 10–20% of a highly branched polymer of arabinose and galactose sugars. This polymer, which is present largely in the cell lumen and can be extracted easily with water, is sometimes classified as an extractive rather than a hemicellulose.

The hemicelluloses are isolated either directly from wood or, more readily, from wood that has first been freed of its lignin. In both cases, the hemicelluloses are extracted with strongly alkaline solutions. The hemicelluloses are related to cellulose and therefore have similar chemical reactions. Because the hemicelluloses are degraded at different rates by both acids and alkalis, however, different fractions survive different pulping processes.

The hemicelluloses play an important role in developing the

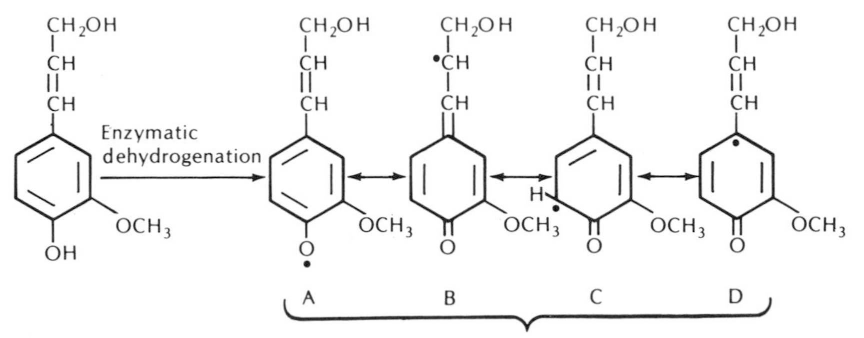

Formula 6 Coniferyl alcohol.

strength of paper. They function as a binder through the mutual bonding of their hydroxyl groups when dried in contact with adjacent fiber surfaces. Cellulose does not bond as readily, despite its many hydroxyl groups, because many of these groups are held rigidly in the crystalline regions of the microfibrils and are not accessible for bonding.

Lignin

Lignin is an amorphous three-dimensional polymer. It permeates the matrix of the cellulose microfibrils in the cell walls and largely fills the spaces between the wood cells. Generally, lignin constitutes 24–33% of softwoods and 19–28% of hardwoods.

Each lignin molecule in wood is built up from probably several hundred phenyl propane units, which are linked in various ways. One such phenyl propane unit, coniferyl alcohol, is shown in Formula 6. Although several different types of phenyl propane units are involved, all consist of a benzene ring, which is represented by the hexagon in Formula 6, and a 3-carbon side chain. The benzene ring carries one weakly acidic hydroxyl group (OH), known as a phenolic hydroxyl group, and usually one or two methoxyl groups ($—OCH_3$), which are responsible for part of the methyl alcohol produced during wood pyrolysis.

Polymerization of the phenyl propane units occurs after enzymes remove the hydrogen atom from the hydroxyl group (OH), leaving an unpaired electron (represented by the dots in Formula 6). This electron wanders about in the resulting structure and is most likely to be found in one of the four positions shown in structures A, B, C, and D of Formula 6.

Polymerization is believed to begin with bond formation between any two of these structures which occurs when the unpaired electron in one structure forms a pair, or bonds, with the electron of another structure. A continuation of this process, together with additions to the double bonds of structures like B, is believed, ultimately, to give the lignin polymer depicted in Formula 7. For example, bond formation between structures A and B, from Formula 6, gives the linkage between units 1 and 4 in Formula 7. Similarly, a pair of B structures gives the carbon-carbon bond joining units 8 and 9. These stable bonds make lignin resistant to degradation into simple units by any method analogous to the breaking down of cellulose, or the hemicelluloses, into simple sugars.

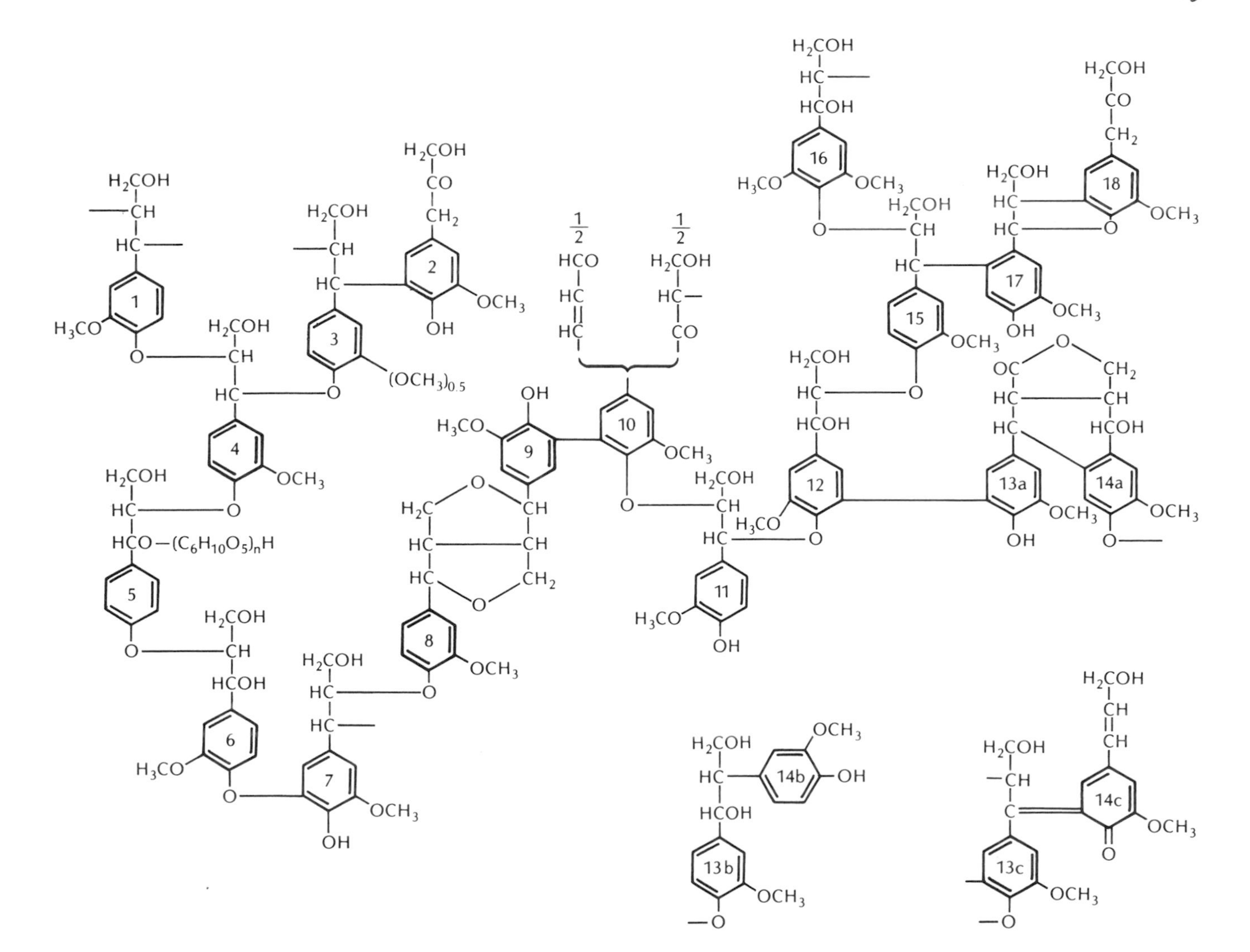

Formula 7 Freudenberg's diagrammatic representation of sprucewood lignin.

The resulting lignin polymer does not have a structure built up from regularly repeating units, and, consequently, no exact formula can be written. Nevertheless, Freudenberg's diagrammatic representation of spruce wood lignin (Formula 7) does illustrate the types and relative frequencies of the different linkages and functional groups (Freudenberg and Neish 1968).

The lignins of most softwoods are similar, if not identical. Most of the units in their structure are derived from phenyl propane units carrying one methoxyl group ($—OCH_3$), as in coniferyl alcohol. Approximately 70% of the original phenolic hyydroxyl groups (OH) are blocked by bond formation to adjacent units. The remaining 30%, which are free, help make lignin soluble in alkali and give greater reactivity to the phenyl propane units during pulping.

Hardwood lignin differs from softwood lignin. Phenyl propane units carrying two methoxyl groups (e.g., unit 16 in Formula 7) represent only a minor part of softwood lignin, but constitute 24–59% of the units of hardwood lignin (Sarkanen and Hergert 1971).

During sulfite pulping, lignin is dissolved by the addition of water-solubilizing sulfonic acid groups ($—SO_3H$) to about half the carbon atoms directly attached to the benzene rings. This addition occurs either by a displacement of the hydroxyl groups originally present (e.g., in units 6, 13b, and 16 of Formula 7) or by cleavage of the carbon-oxygen-carbon linkage (e.g., between units 3 and 4 and between units 11 and 12 of Formula 7). During kraft pulping, one of the main solubilizing reactions is the cleavage of the carbon-oxygen-carbon linkages joining adjacent phenyl propane units (e.g., those joining units 1 and 4, 5 and 6, and 12 and 15 of Formula 7). This cleavage breaks the lignin into smaller fragments and liberates the phenolic hydroxyl groups. The fragments are then soluble in the alkaline pulping liquor.

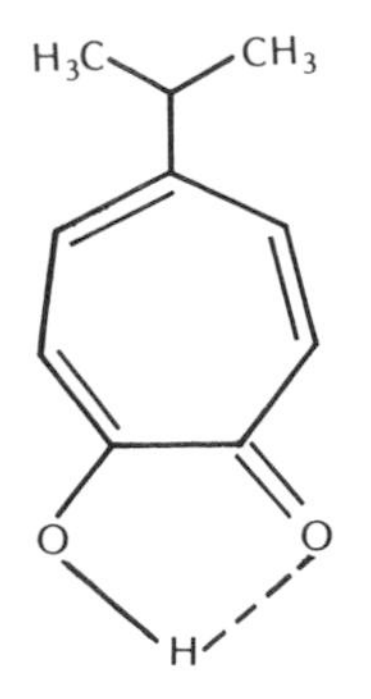

Formula 8 γ-thujaplicin, 0.2% (from western red cedar, percentage for moisture free wood).

Extractives

Extractives may be defined as those substances that can be removed from wood with neutral solvents such as ether, alcohol, and water. Typical classes of extractives soluble in ether are terpenes, resin acids, fatty acids, and tropolones; those soluble in alcohol are phenols, lignans, and polyphenols; and those soluble in water are sugars, acids, and tannins. Although both softwoods and hardwoods contain all three classes of extractives, softwoods generally have a higher content of ether solubles. Resin acids, for example, occur only in softwoods, where they account for 25–50% of the ether solubles. Terpenes occur mainly in softwoods, and the pinenes (especially α-pinene) usually constitute 50–100% of the terpenes. Fatty acids, mainly esterified to fats, dominate the hardwood extractives, amounting to 50–90% of the hardwood ether solubles. They also constitute 25–50% of the ether solubles of softwoods.

Extractives do not form an integral part of a cell wall and are never found uniformly distributed in a tree. Variation occurs not only vertically in the stem but radially as well. A typical distribution pattern can be seen in Figure 5.3, which shows the radial variation of dihydroquercetin, the major extractive of Douglas-fir (Gardner and Barton 1960).

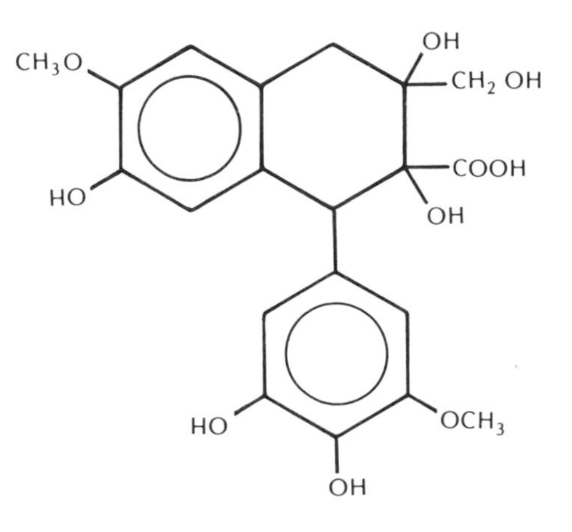

Formula 9 Plicatic acid, 1–5% (from western red cedar, percentages for moisture-free wood).

Extractives vary far more in composition and quantity from one species to another than do the structural components of the cell wall, hemicellulose and lignin. While some extractives, such as α-pinene and glucose, are common to many species, a significant number occur only in one species. In spite of their low concentrations (usually less than 5% of the wood), they are the 'personality' chemicals and are responsible for many variations in the properties of wood from different species and from individual trees within the same species (Barton and MacDonald 1971; Rogers and Manville 1972). Thus, the resistance to decay of western red cedar and the occurrence of chemical stains on moist cedar in contact with iron are caused by the extractives present in cedar. The chemical structure of wood extractives varies greatly, as can be seen in Formulas 8–13.

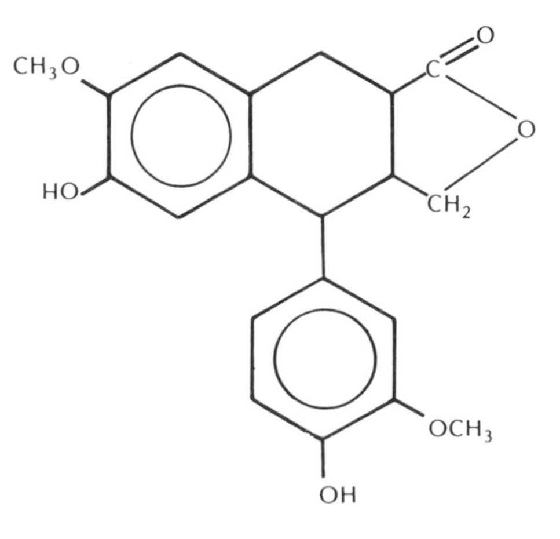

Formula 10 α-conidendrin, 0.15% (from western hemlock; percentage for moisture-free wood).

Bark

Because bark is more heterogeneous than wood and its cellulose and lignin are difficult to isolate in clearly defined fractions, bark chemistry presents special problems. Thus, a summative analysis of the type shown for wood in Tables 5.1 and 5.2 is rarely attempted. A range of values for conifer bark might be holocellulose, 30–40%; lignin, 17–42% (extractive-free basis). This range is much greater than that for wood.

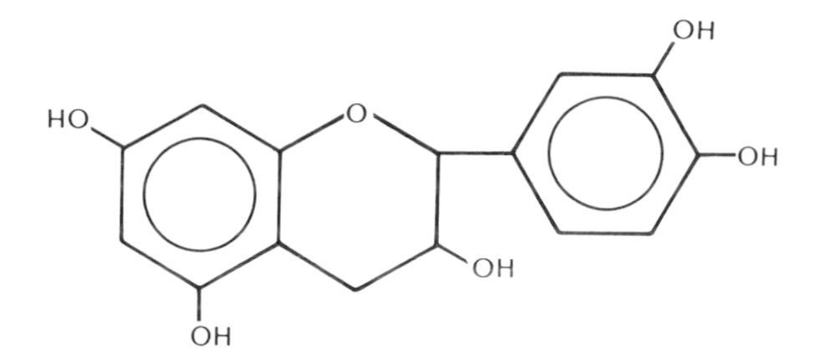

Formula 11 Catechin, 0.05% (from western hemlock; percentage for moisture free-wood).

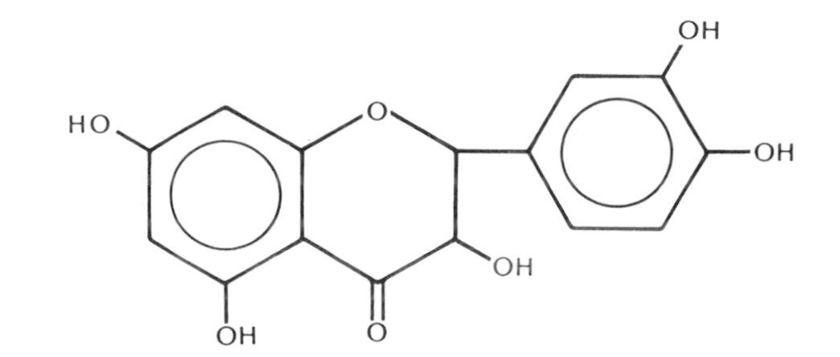

Formula 12 Dihydroquercetin, 0.2–1.5% (from Douglas-fir; percentages for moisture-free wood).

$CH_3(CH_2)_4CH{=}CHCH_2CH{=}CH(CH_2)_7COOH$

Formula 13 Linoleic acid, 0.7–1.6% (from white birch; percentages for moisture-free wood).

In general, bark contains a much higher proportion of extractives than wood does. For example, Douglas-fir bark is reported to contain 5% dihydroquercetin (Hall 1971), whereas the wood seldom contains more than 1% (Gardner and Barton 1960). Because many extractives are toxic to fungi and insects, it is reasonable to assume that bark extractives help protect the tree from these pests. The chemical content of the barks of such species as aspen, Douglas-fir, western hemlock, and western red cedar has been examined and found to yield many new constituents of potential commercial value.

In view of their common origin in the living cambium, it might be expected that bark (phloem) and wood (xylem) from the same species would contain the same extractives. Although they usually do, there are notable exceptions. For example, tropolones are found in the heartwood of western red cedar but are completely absent from the bark (Quon and Swan 1972). These tropolones, such as γ-thujaplicin (Formula 8), are highly toxic to many wood-destroying fungi; they have the same order of activity as pentachlorophenol, which is a well-known and extensively used wood preservative.

Figure 5.3 Radial analyses of dihydroquercetin in Douglas-fir.

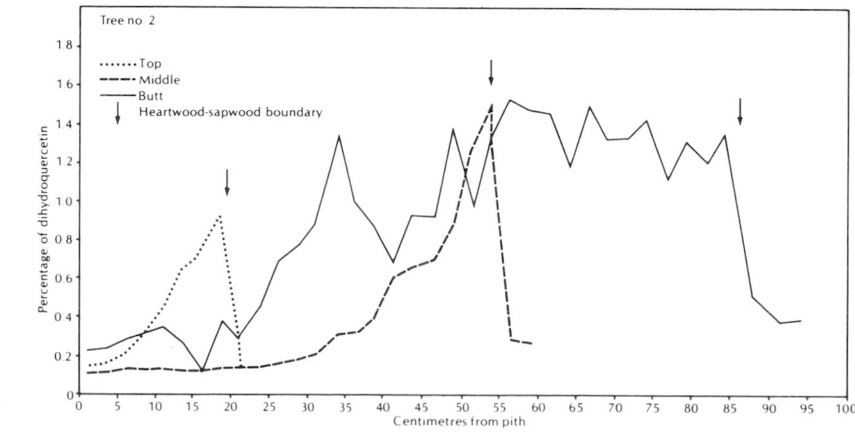

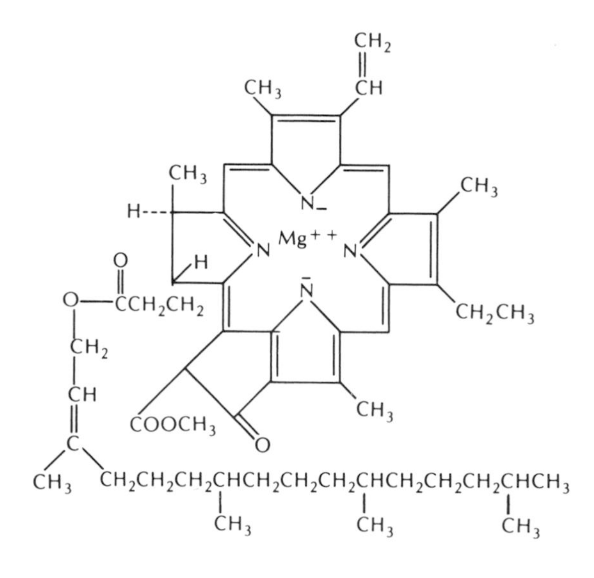

Formula 14 Structure of chlorophyll α. (Source: Robert F. Steiner, *The Chemical Foundations of Molecular Biology* [New York: Van Nostrand Reinhold Co., Inc., 1965])

Foliage

Depending on the content of small branches, dry foliage has the following composition: cellulose, 32–36%; lignin, 33–37%; protein, 8–11%; extractives, 9–12%; and ash, 6–7%. Foliage contains chemicals in the living cells directly or indirectly related to photosynthesis. Such chemicals are generally not present in wood or bark. Nitrogen, carbon, hydrogen, and oxygen, which are essential to life processes, are combined in structures such as chlorophyll (Formula 14). In addition, green foliage contains other related chemicals, such as xanthophyll and carotene. Carotene yields vitamin A, which is easily obtained from it by a simple hydrolytic reaction.

Foliage has a greater quantity of odoriferous oils than wood does. Foliage oils belong to a large group of plant substances called essential oils (Bender 1963). The term essential oil is derived from a traditional belief that the fragrance of a substance contains the very essence of it, the vitally important ingredient. Essential oils, which are usually pleasant smelling, are isolated simply by passing steam through finely divided twigs and foliage. The foliage of several varieties of North American species, including white, red, western red, and yellow cedar, as well as Douglas-fir and western hemlock, yields essential oils.

New logging practices, in which the whole tree is brought to a central location where tops and branches are removed, may make the recovery of these chemicals from foliage economically feasible. The structures of a few foliage chemicals are shown in Formulas 15–17.

Formula 15 α-pinene.

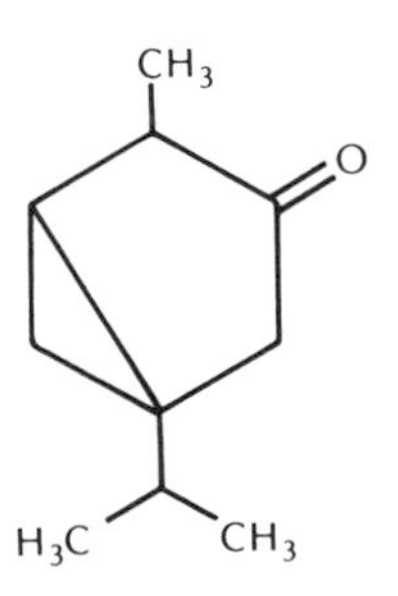

Formula 16 Thujone.

APPLICATIONS OF WOOD CHEMISTRY

Wood chemistry gives part of the basic information necessary for identifying and solving many problems encountered by manufacturers and users of wood products. Following are descriptions of a few problems that have been solved with the help of chemical studies.

Pulping

Both the sulfite and kraft pulping processes, which are described in

Formula 17 β-carotene.

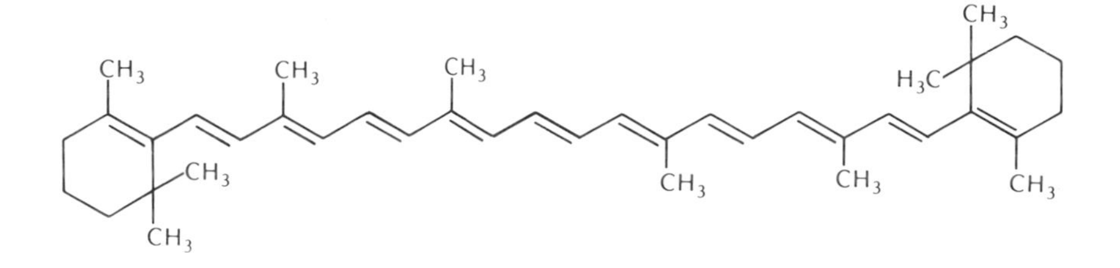

Figure 5.4 Macro gas-liquid chromatographic separation of wood and bark extractives in one of many techniques used in wood chemistry.

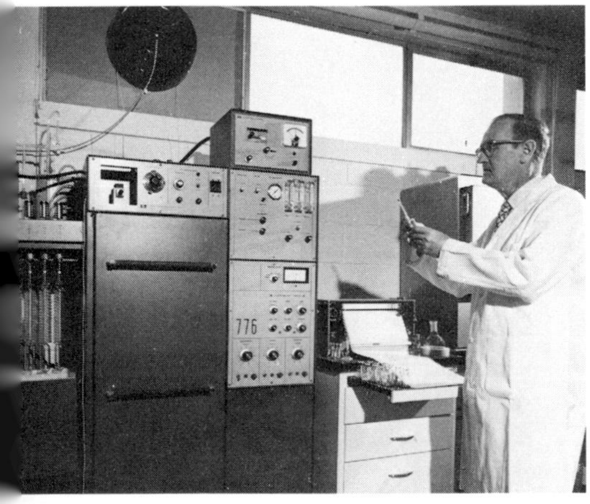

Chapter 13, were invented during the nineteenth century, before much knowledge of wood chemistry existed. Many problems were not understood but were avoided or minimized by strict adherence to inflexible pulping schedules. An extensive literature now exists on the chemistry of these processes (Chapter 13). Some examples of the contribution of wood chemistry to problem solving in the pulp industry follow.

Burnt Cook

During the early use of sulfite pulping, serious financial losses resulted when the wood chips turned dark brown without being converted into pulp. This phenomenon was known as burnt cook (Rydholm 1965). Burnt cooks occurred even when the cooking temperature was not exceeded, and whole digester charges of wood often had to be discarded.

Burnt cook was found to be caused by an acid-catalyzed condensation reaction of lignin, in which lignin molecules reacted with each other to form larger insoluble molecules with stable carbon-carbon bonds; thus, these lignin condensation products could not be dissolved by further pulp cooking. The brown polymerized residue gave the partially pulped wood its burnt appearance.

This problem could not be solved by reducing the acidity because sulfonation, a reaction of major importance in lignin solubilization, is promoted by acidity. By increasing the bisulfite concentration in the cooking liquor, however, the sulfonation reaction could be accelerated, giving the lignin condensation less time to occur. This change in procedure overcame the burnt cook problem, but it was found that acid diffused more rapidly into the wood chips than the bisulfite ion did. As a result, the lignin condensation reaction took place in the inner part of the chips before sulfonation could get underway. To avoid this effect, complete penetration of the cooking liquor had to be ensured before the temperature of the digester was raised.

The Influence of Extractives

Much study of the influences of extractives on pulping has been devoted to overcoming the deleterious effect of wood resins on pulp and to understanding the factors that influence production of tall oil (Hillis 1962). Studies of the chemistry of wood extractives have identified causes of a number of other problems, however, and have facilitated remedial actions.

It was noted many years ago that the wood of Douglas-fir did not pulp well by the calcium-based sulfite process. Hoge (1954) showed that pulping was inhibited by the dihydroquercetin (Formula 12), which is generally present in concentrations of less than 1% in this species. Further, dihydroquercetin was found to be corrosive to steel digesters in alkaline pulping.

Corrosiveness was also a major problem in pulping western red cedar because of two different types of extractives, the steam-volatile thujaplicins, which cause vapor-space corrosion, and the polyphenols, which are responsible for liquor-space corrosion. Many polyphenols are catechol derivatives (MacLean 1970). The adoption of stainless steel liners has largely solved the corrosion problems.

Another factor to contend with in pulping western red cedar is the higher concentration of extractives (up to 10%) and their acidic nature (Barton and MacDonald 1971). Acidic extractives rapidly neutralize some of the alkali in the cooking liquor, making it unavailable for delignification reactions. As a result, chemical requirements per unit of delignification of western red cedar are high compared with those of hemlock and other conifers.

Extractives greatly affect brightness and color reversion in groundwood pulp. Western hemlock has extractives that contain color-producing chromophores or that develop them during the pulping process. It is estimated that the brightness of western hemlock groundwood pulp is five brightness points lower than of eastern spruce pulp. Undoubtedly, western hemlock's extractives, especially its flavonoids with adjacent phenolic hydroxyl groups, contribute to much of this loss of brightness (Polcin and Rapson 1971).

Stains on Wood Products

Stains on finished and unfinished lumber that are not caused by bacteria or fungi are often caused by extractives. Knowledge of the chemical and physical properties of the extractives is often useful in explaining the cause of the stain and suggesting remedial action.

Causes of Hemlock Brown Stain

Both air- and kiln-dried western hemlock lumber may develop an objectionable color, known as hemlock brown stain. The stain, which ranges from orange to almost black, is confined to the surface of the ends and, occasionally, to the sides of the lumber. It is normally concentrated in two main zones: the heartwood-sapwood boundary and the sapwood-cambium boundary. Although the stain mars the appearance of the lumber, it does not lower the strength because it does not penetrate beyond the surface.

The cause of the stain is catechin (Formula 11) (Barton and Gardner 1966), a naturally occurring water-soluble extractive that migrates to the surface. It reacts enzymatically with oxygen to give a brown, innocuous, tanninlike deposit that is not connected with decay, sap stain, or mold. Because the stain can be permanently removed by planing after the lumber is dry, the lumber may be used except where light color on the rough lumber is a necessity.

Today the ends of all western hemlock boards, both stained and unstained, are painted with a reddish paint before they are marketed. The paint is applied to indicate that the lumber has been sufficiently pretreated with fungicides to prevent formation of sap stain and mold.

Causes of Colored Stains on Western Red Cedar and Yellow Cedar

The most interesting extractives in western red cedar are the water-soluble tropolones, such as thujaplicin (Formula 8), which are largely responsible for cedar's high natural durability. Tropolones are very reactive organic chemicals and have the ability to take iron directly into solution by a process known as chelation, forming stable chelates. These chelates are red. Accordingly, if moist cedar wood comes into direct contact with iron (e.g., steel rollers, iron bands) a pink to red stain forms on the wood surface. Since yellow cedar also contains tropolones, although in lesser amounts, it also stains. The color of the stain, however, is usually brown rather than red. In addition to tropolones, western red cedar contains a variety of phenolic compounds that react with iron to produce black stains. These stains are particularly noticeable around holes made by ordinary steel nails.

The use of high-quality galvanized iron or aluminum nails will prevent nail stains. The other stains may be prevented by keeping wet cedar wood away from contact with iron or steel.

Causes of Black Surface Stains

Most wood species react with iron under moist conditions. During railway shipment from the interior of British Columbia to the coast, white spruce and lodgepole pine often develop an unsightly black surface stain. It has been found that iron particles from wear of the flatcars' brake shoes settle on uncovered lumber and react with the natural polyphenols of the wood under moist conditions to produce the stain. As long as a shipment does not get wet, the lumber retains its clean appearance. Thus, covering the lumber during shipment solves the problem.

Causes of Stain on Oak

Dark blue or black stains occasionally appear on oak wood. Oak flooring, for example, that has been rubbed with steel wool may develop such stains when it gets wet. Tannin in the wood reacts with traces of iron to produce the stains. The stains can be removed by treatment with oxalic acid.

Causes of Mineral-Stained Maple Wood

Mineral stain occurs in maple wood and, to some extent, in other hardwoods. Mineral-stained maple wood contains relatively large amounts of metal salts and other colorless compounds not found in unstained maple. These compounds, which are probably associated with the formation of the stain, have been identified as substituted coumarins (Levitin 1972).

The dark color in mineral-stained maple is due to polymerized phenolic material, which is present in both the lumen and the walls of parenchyma cells. Much of the highly colored material in the lumen resembles soluble lignin and can be extracted with organic solvents. The bulk of it, however, in the cell walls, remains insoluble. A surface bleaching process has been developed that selectively removes the most objectionable dark coloring.

Chemical Identification of Species

In general, species identification is relatively easy in a living tree, where foliage and bark provide clues. In lumber or veneer products, however, identification is often difficult or impossible (Chapter 3). For example, it is difficult to distinguish alpine fir from spruce, pine, and other firs of the spruce-pine-fir group. Identification of alpine fir is required for efficient kiln drying because it has different drying characteristics from other species in this group. In addition, alpine fir has different strength properties from those of amabilis fir of the coastal hem-fir group, and, since there is an occasional overlap in the growth ranges, it has become increasingly important that a reliable means be developed for identifying the two species.

Alpine fir contains extractives not found in spruce, pine, or amabilis fir. Certain chemicals typically react with these extractives to provide color differences between alpine fir and the other species. Of the many chemicals investigated, the most positive identification is provided by Ehrlich's reagent, which produces purple streaks in the wood of alpine fir but does not cause such streaks in amabilis fir, white spruce, or lodgepole pine. This relatively quick and simple test (Fraser and Swan 1972) makes it possible to identify alpine fir lumber.

Another chemical identification system is useful for separating white spruce and lodgepole pine veneers. Separation of these veneers is necessary because they have different gluing characteristics. During high-temperature drying of these veneers, the volatile oils evaporate, making it impossible to identify them by odor or visual scanning. A rapid detecting reagent has been developed that permits their identification based on differences in the nonvolatile extractives (Barton 1973). Several chemical color tests (Barton 1973) have also been developed to differentiate sapwood from heartwood in species such as western hemlock when visual identification is difficult.

Bleed-through

Water-soluble extractives from western red cedar, and oil-soluble (or ether-soluble) extractives from red and white pine, can bleed through paint and varnish and affect finishes.

Among its extractives, western red cedar contains highly water-soluble plicatic acid (Formula 9). Moisture condensing in the exterior walls of houses can penetrate western red cedar siding from the back and transport the plicatic acid to the paint surface, where it accumulates as a dark brown blister or spot. One effective cure for this problem is to place an efficient water-vapor barrier on the inner surface of the wall. Another is to vent the moisture from behind the siding.

Western red cedar is one of the best woods for exterior use because of its high natural resistance to weathering and decay and its good dimensional stability or resistance to swelling and shrinkage caused by changes in its moisture content (Gardner 1965). These properties are dependent to a large extent on the quantity and nature of the extractives found in this cedar. Potent fungicides, for example, are responsible for its high resistance to decay. Because these extractives are largely polyphenolic, they may be considered natural antioxidants and absorbers of ultraviolet light, and thus contribute to the photochemical stability of the substrate surface. Photochemical stability is of particular importance in exterior finishes with low pigmentation because it helps protect the finish against discoloration and deterioration caused by ultraviolet radiation. The extractives of western red cedar also act as natural bulking agents in the cell walls of the wood. As a result, where variations in moisture content occur, this species has better dimensional stability than would normally be expected from its low density.

Resins occasionally bleed through painted red and white pine. This problem is related to the extractives found in pine wood.

The extractives of pine have a high content of ether solubles, which consist mainly of volatile terpenes, fatty acids, fats, resin acids, and sterols. These ether solubles have been studied extensively, partly because of their influence on wood pulping and paper manufacture and partly because of their importance as commercial products such as naval stores.

The volatile terpenes or turpentines of pine woods are the traditional solvents for oil-base paints. An examination of the extractives of red pine and eastern white pine revealed the following. Normal wood of red pine contains 3–6% of ether solubles, of which only 1% are volatile terpenes; resinous and knot wood of the species, however, are 30–40% ether soluble. About 10% of the ether solubles, or, in extreme cases, as much as 20% (Levitin 1962), are volatile terpenes. Eastern white pine generally contains fewer ether solubles, with a maximum of 18% in highly resinous areas. Most bleed-through in red and white pine can be attributed to the solvent power of the terpenes, which weakens the paint film and retains fluidity in the resin. Consequently the resin does not harden and will exude through surface coatings of paint or varnish. An effective solution to this problem is to use lumber kiln-dried at 116°C (240°F).

Utilization of Sawdust and Bark

The use of sawdust and ground bark for mulching and soil conditioning is rapidly increasing (Bollen and Glennie 1961).The chief benefits derived from these mulches are increased moisture retention and greater soil aeration. They also retard erosion, hinder weed growth, reduce evaporation of water, and retain warmth. The sawdust and bark of most wood species can be used for mulching without treatment. Exceptions are the residues of western red cedar and yellow cedar, which contain water-soluble tropolones that are toxic to young roots. Cedar sawdust and bark have been used successfully, however, to raise various shrubs once the tropolones are removed from them by leaching.

Sawdust and bark have no fertilizer value and, in fact, deplete the soil of nitrogen as they decompose. Therefore, if they are used, nitrogen must be added at a rate of 1 kg/100 kg (20 lb/ton) of wood waste to prevent nitrogen starvation of the soil. As the sawdust or bark decays, the nitrogen again becomes available to growing plants. Wood residues are also used in animal litter and bedding (Scroggins and Currier 1971).

WOOD CHEMICALS AND CHEMICAL BYPRODUCTS

Cellulose and Its Derivatives

The most valuable constituent of wood is cellulose, which finds its greatest use in basically unchanged chemical form in pulp and paper (Chapter 13). Of secondary importance are cellulose derivatives, which are formed from chemically altered cellulose and are the source of the principal chemical products derived from wood. All are prepared from a special grade of pulp called dissolving pulp, which is essentially pure cellulose. These products include xanthate rayon, acetate rayon, cellulose nitrate, and cellulose ethers such as methyl cellulose, ethyl cellulose, hydroxyethyl cellulose, and carboxymethyl cellulose.

Viscose Rayon and Cellophane

Viscose rayon and cellophane are prepared from cellulose that has been chemically dissolved and then regenerated as cellulose in continuous filament or sheet form (Jahn 1952; Kline 1954). Cellulose pulp is first converted to alkali cellulose by being steeped in caustic soda solution. The excess caustic solution, which dissolves much of the residual hemicellulose of the pulp, is squeezed out, leaving a swollen fibrous material, which is then mixed with carbon disulfide. The resulting cellulose xanthate is dissolved in a dilute caustic soda solution and forced through very small holes (in the case of viscose rayon), or through a long, narrow slot (in the case of cellophane), into an acidic bath. The acid removes the solubilizing xanthate groups, producing regenerated cellulose in the form of filaments (viscose rayon) or sheets (cellophane).

$$\underset{\text{Cellulose}}{\text{Cell—OH}} \xrightarrow{\text{NaOH}} \underset{\text{Alkali cellulose}}{\text{Cell—ONa}} \xrightarrow{CS_2} \underset{\text{Cellulose xanthate}}{\text{Cell—O—}\overset{\overset{\large S}{\|}}{C}\text{—SNa}} \xrightarrow{\text{acid}} \underset{\text{Regenerated cellulose}}{\text{Cell—OH}}$$

Because of their hydroxyl groups, viscose rayon and cellophane have a considerable affinity for water. This property is desirable in rayon clothing because it permits absorption of moisture from the skin. In food packaging, however, absorption of water is undesirable. Thus, cellophane is usually coated with a thin film of water-repellent material. Rayon can be made stronger than cotton and has been widely used in tire cord, although nylon and other synthetic fibers have largely replaced it in this role.

Cellulose Acetate

Cellulose acetate is produced by the reaction of cellulose with acetic anhydride (Jahn 1952).

$$\underset{\text{Cellulose}}{\text{Cell—OH}} + \underset{\text{Acetic anhydride}}{(CH_3CO)_2} \xrightarrow[\text{acid}]{\text{sulfuric}} \underset{\text{Cellulose acetate}}{\text{Cell—O—}\overset{\overset{\large O}{\|}}{C}\text{—}CH_3} + \underset{\text{Acetic acid}}{CH_3CO_2H}$$

Figure 5.5 Triacetate fiber is spun onto beams or large spools, used for shipping the fiber. The major use of triacetate fiber is in easy-care fabrics. (NFB Photothèque. Photo by Chris Lund)

In the final product most, but not all, of the hydroxyl groups are replaced by acetyl groups (CH_3CO—). Retention of some hydroxyl groups improves both strength and solubility in acetone.

Acetate rayon, the principal cellulose acetate product, is prepared by dissolving the cellulose acetate in acetone and forcing the solution through small holes into hot air to evaporate the solvent. Acetate rayon, therefore, is cellulose acetate, in contrast to viscose rayon, which is simply regenerated cellulose. Other cellulose acetate products include plastics, photographic film, safety glass, lacquers, and transparent sheeting.

Cellulose Nitrate

Cellulose nitrate (or nitrocellulose) (Jahn 1952), the oldest cellulose derivative, was discovered in 1845. Celluloid, a cellulose nitrate product plasticized with camphor, was patented in 1864 and is the pioneer plastic material.

Cellulose nitrate is prepared by nitrating cellulose with a mixture of nitric and sulfuric acids. In this process, the hydroxyl groups are largely replaced by nitrate ester (—ONO_2) groups:

$$\underset{\text{Cellulose}}{\text{Cell—OH}} + \underset{\text{Nitric acid}}{HONO_2} \longrightarrow \underset{\text{Cellulose nitrate}}{\text{Cell—ONO}_2} + H_2O$$

Depending on the extent of the nitration, the resulting cellulose nitrate is used in plastics, lacquers, glues, or smokeless powders.

Cellulose Ethers

Methyl, ethyl, hydroxyethyl, and sodium carboxymethyl ethers of cellulose (Jahn 1952) are commercially prepared by reacting alkali cellulose with the corresponding chlorides.

	CH_3Cl Methyl chloride	$\longrightarrow$	Cell—OCH_3 Methyl cellulose
Cell—ONa + Alkali cellulose	C_2H_5Cl Ethyl chloride	$\longrightarrow$	Cell—OC_2H_5 Ethyl cellulose
	$ClCH_2CH_2OH$ Ethylene chlorohydrin	$\longrightarrow$	Cell—OCH_2CH_2OH Hydroxyethyl cellulose
	$ClCH_2CO_2Na$ Sodium chloroacetate	$\longrightarrow$	Cell—OCH_2CO_2Na Sodium carboxymethyl cellulose

Hydroxyethyl cellulose can also be prepared by reacting alkali cellulose directly with ethylene oxide.

Generally, between 0.4 and 2.6 of the 3 hydroxyl groups per glucose unit are substituted. The products most likely to be substituted are those that are soluble in water, giving viscous solutions. Cellulose ethers are widely used as thickening agents, as protective colloids to hold insoluble material in suspension, and as film-forming agents. Examples of products containing cellulose ethers include lacquers, cosmetics, and adhesives.

Charcoal

Charcoal may be considered both a chemical and an energy source. Although it can be obtained under suitable conditions from any wood, yields are in direct relation to the density of wood burned and thus the denser species are preferred. Charcoal can also be manufactured from conifer sawdust, shavings, or chips in especially designed ovens and then formed into briquettes with starch binders. In past years, vapors from the carbonization of charcoal were condensed to recover wood alcohol and acetic acid, but these products are now made entirely from petroleum at a fraction of the cost.

Because of the ready availability of wood and the relatively simple and inexpensive equipment needed to produce charcoal on a small scale, many people have been tempted to become charcoal manufacturers. However, the failure of small charcoal companies has been high, owing to seasonal markets and the proportionately high cost of briquetting, bagging, and marketing. Those companies that can afford larger-scale, continuous carbonization furnaces and expensive briquetting machines (50% of the demand for charcoal is for briquets) and that have large industrial contracts are clearly in a more advantageous position. The Canadian demand for charcoal rose from 30 000 t [33 000 tons] in 1969 to 52 000 t [57 000 tons] in 1971, and the future appears promising (Hallett 1971). Some uses for charcoal, listed by Wenzl (1970), are given in Table 5.3.

Table 5.3 *Commercial application of charcoal*

Uses as domestic and specialized fuel	Metallurgical uses	Chemical uses
Recreational barbecuing	Copper	Carbon disulfide
Curing tobacco	Brass	Calcium carbide
Cooking in restaurants	Pig iron	Silicon carbide
Heating, foundry, plumbing equipment	Steel	Sodium cyanide
Heating in shipyards and citrus groves	Nickel	Potassium cyanide
	Aluminum	Carbon monoxide
	Electromanganese	Activated carbon
	Armor plate	Black powder
	Foundry molds	Fireworks
		Rubber
		Gas adsorbent
		Crayons
		Soil conditioner
		Pharmaceuticals
		Poultry and animal feeds

Wood Resin

To avoid confusion of the meaning of such terms as naval stores, oleoresin, turpentine, rosin, pitch, gum, Canada balsam, and tall oil, all of these products will be discussed under the common heading of wood resin, a comprehensive term for secretions of certain trees that are oxidation or polymerization products of the terpenes.

The composition of wood resin varies both qualitatively and quantitatively from species to species. Wood resin is produced by the live parenchyma cells in the sapwood. It is not directly related to the photosynthetic products found in cell sap, which contribute to the production of wood, bark, and leaves, but is a secondary derivative of them. The parenchyma cells can be stimulated by cutting or wounding the tree, causing the wood resin to exude, or the cuts can be treated with chemicals such as sulfuric acid or paraquat to increase the flow of resin. The mechanism that induces the exudation of the wood resin is still not clearly understood. There seems little doubt, however, that part of the tree's defense mechanism is involved, since the wood resin seals the exposed wound and protects the cambium from further injury.

Interest in wood resins dates back to when tar and pitch from pine trees were used in the construction of wooden sailing ships; hence, the name naval stores is often used for tar and pitch. The United States is the largest producer of tar and pitch, mostly obtained from longleaf and slash pine. These trees contain 10–15% wood resin, which is known as pine oleoresin, or simply oleoresin. It can easily be fractionated into a steam-volatile portion, called turpentine, and a solid fraction, called rosin. Trees of other coniferous species, such as spruce, Douglas-fir, and balsam fir, also produce oleoresin. Oleoresin from spruce is called spruce gum; that from balsam fir is known as Canada balsam. Because of its clarity and refractive index, which is similar to that of optical glass, Canada balsam is used as a cement in assembling lenses and other optical equipment.

The decreased number of mature pine trees in the southern United States has meant a reduction in wood resin obtained from tapping. Part of this loss has been made up by steam distillation and solvent extraction of resin from old, recovered pine stumps.

Byproducts of the kraft pulping industry are now the major source of wood resin. Crude turpentine recovered from digester relief gases is carefully fractionated and chemically treated to remove vile-smelling sulfur compounds and then marketed as sulfate wood turpentine. Many of the nonvolatile resin components are converted into soap during the kraft cook and dissolve in the hot cooking liquor. During subsequent evaporation of the black liquor, the sodium salts of these resin components float to the top and are removed as 'soap skimmings.' These skimmings, together with similar material washed from the pulp, are then acidified with sulfuric acid, which liberates a crude, oily material called tall oil. This name comes from the Swedish word *tall*, which means 'spruce.' After further refining, tall oil is used chiefly in the paint and hot-melt adhesive industries. In 1960–61, production of tall oil in the major pulp-producing countries of the world was reported to be 488 000 t (538 000 tons), of which 357 000 t (394 000 tons) were produced in the United States and 7000 t (8000 tons) in Canada (Wenzl 1970). It can be assumed that production has increased since then. Fractionation of tall oil yields rosin, which is used mainly for sizing of paper, and fatty acids, which are important constituents of protective coatings, printing ink, and many other products.

Maple Sap Products

Canada produces about 75% of the world supply of maple sap products. In recent years, Canadian production has averaged about 14 000 t (15 000 tons) a year. Ninety percent of this output is gathered and processed in Quebec.

Maple Syrup

The sap of all the native species of maple contains sugars, but only the sugar maple produces sap with a sugar content high enough to make it profitable to produce syrup and sugar.

Sap from the sugar maple is composed largely of water and also contains about 3% sucrose, various minor components such as calcium, potassium, iron, and magnesium, a number of organic acids (the principal one being malic acid), and a small percentage of nitrogenous matter.

Sap flows in the early spring, when changes in temperature cause expansion and contraction in the cells and intercellular spaces of the wood, resulting in an alternation of pressure and suction. The exact time of the flow depends on the weather and the latitude, and the run usually lasts several weeks. The flow of sap is directly proportional to the size of the tree's crown and the amount of light the crown receives. The trees must therefore be grown under ideal conditions in order to ensure optimum production.

The usual procedure for making maple syrup is to heat the sap in an evaporator, which consists of shallow evaporating pans, over a wood

Figure 5.6 Collecting sap from sugar maples at Ste Clothilde de Beauce, Quebec. (NFB Photothèque. Photo by P. Gaudard)

fire. The sap is conducted from the storage tank to the evaporator through an automatic regulator so that evaporation may take place as rapidly as possible without burning the sap. When the sap is heated to the boiling point, the nitrogenous matter and other waste material coagulate to form a scum, which comes to the surface and is skimmed off with a perforated metal skimmer. As the liquid becomes syrupy it turns an amber color, and the mineral matter is deposited on the bottom or sides of the pan as 'sugar sand.' This precipitate is composed largely of calcium bimalate.

When the syrup has reached the proper density (determined with a hydrometer or from the boiling point), it is filtered through felt or flannel strainers into cooling vessels. Sometimes milk, egg white, or other substances are used as clarifying agents, but they are not considered essential and are apt to affect the flavor of the syrup.

The more important compounds contributing to the flavor of maple syrup are acetol, isomaltol, cyclotene, α-furanone, hydroxymethylfurfural, vanillin, syringaldehyde, and dihydroconiferyl alcohol (Underwood, Filipic, and Bell 1969). The relative amounts of these compounds in maple syrup are remarkably constant, regardless of geographic location or who makes the syrup. Thus, the profile of these compounds, as shown by gas-chromatographic analysis of a chloroform extract of maple syrup, can help in determining the quality of the flavor and detecting adulteration.

Maple Sugar

Maple sugar is produced by concentrating the syrup to a higher degree. The clear syrup from the filter is poured into a 'sugaring off' pan, which is placed on an arch or stove and boiled until it reaches a proper state of granulation. This point is determined by inserting a thermometer into the syrup or by testing the liquid in cold water, on packed snow, or on ice. When the desired point is reached, the hot syrup is removed from the fire and poured into molds.

Byproducts of Maple Sugar
High-grade vinegar may be made from the sugar contained in the wash water from pans, strainers, and other appliances, together with the 'buddy' sap, which is collected late in the season, when the buds begin to burst, and which is not suitable for making syrup. The sugar solution is collected in a barrel, some yeast is added, and alcoholic fermentation is allowed to take place. The liquid is strained into another barrel, mother of vinegar is added, and acetic acid fermentation takes place, converting the alcohol into vinegar.

The sugar sand that is collected from the evaporating pans, strainers, and other vessels is washed with hot water to remove the sugar. After the washing, the undissolved residue may be used as a source of calcium bimalate, which is a suitable acid constituent for baking powder. The sugar sand may also be used as a source of malic acid.

Saccharification of Wood
At the right temperature and acid concentration, cellulose and other polysaccharides from practically any wood can be converted into simple sugars by treatment with mineral acids. This process is called hydrolysis, or saccharification, and has been used in the commercial production of sugar solutions from wood or wood waste in a number of countries, particularly in Germany, where it is known as the Scholler process.

The Scholler process is essentially a batchlike process in which dilute sulfuric acid is percolated through wood chips in a digester at temperatures of 135–190°C (275–375°F). In the United States during the Second World War, a number of improvements were made to the process when it was changed from batch acid-addition to continuous acid-addition and to continuous drawing off of the sugar solution. With these changes, the time required for hydrolysis was reduced from 12–16 hours to 3 hours. Sugars were obtained in a concentration of 4–5% and in a yield of 50–60% of the dry weight of the wood. Fermenting the sugars obtained by hydrolysis of the wood produced ethyl alcohol in a yield of 210–251 litres per tonne (42–50 gal per ton) of dry wood.

Wood hydrolysis was important during the Second World War for the preparation of ethyl alcohol. Because other low-cost sugars are available today, wood hydrolysis is no longer used commercially to produce sugars or alcohol except in the USSR.

Food and Food Supplements
Wood saccharification, using waste wood from industry, could produce large quantities of sugar (wood molasses), which could become an important source of food. Torula yeast (*Candida utilis*) can be grown on the sugars produced by wood saccharification or on the sugars present in the waste liquor from the sulfite pulping process. Although only hexose sugars can be fermented to alcohol by the yeast, both hexose and pentose sugars can be used in producing the torula yeast. The average yield of yeast is 50% of the weight of the sugars consumed. Since the yeast itself is 50% protein by weight, one-quarter of the weight of sugars consumed can be converted to edible protein. Protein

can also be obtained in even higher yields by growing and harvesting certain bacteria on the wood sugars (Srinivasan 1975).

A more easily harvested protein-rich product can be obtained by growing the fungus *Paecilomyces varioti* in the Finnish-developed 'Pekilo process' (Forss and Passinen 1976). The fibrous mycelium of this microorganism is similar in amino acid composition to torula yeast, but it is easily separated by filtration because of its larger size and shape.

In a process developed by the United States army (Wilke 1975; Gaden et al. 1976), an enzyme produced by a fungus (a mutant *Trichoderma viride*) can be isolated and used to hydrolyze cellulose to sugar. Corrosive acids and high temperatures and pressures are not required. Unfortunately, the cellulose in wood is protected from the enzyme by its close association with lignin. Removal of the lignin or a loosening of its association by a preliminary treatment (e.g., very fine grinding) is required.

Cud-chewing animals, such as cows and sheep, are able to digest cellulose because microorganisms in their rumens produce somewhat similar enzymes that hydrolyze cellulose to sugar. Although lignin makes wood relatively indigestible, animal fodder made from wood that has been suitably treated can be used to supplement normal rations.

Various studies have shown that a few minutes of treatment with high-pressure steam can alter some hardwoods, such as aspen, so that much of the carbohydrate can be digested by sheep and cattle (Bender, Heaney, and Bowden, 1970; Heaney and Bender, 1970). This treated material, used to supplement natural food such as hay, can provide approximately half the energy requirements of overwintering animals. Because the treated wood contains practically no nitrogen and does not provide the necessary food for protein synthesis, a protein or nitrogen source may be necessary, particularly in feed lots where beef cattle are being finished for marketing, but steamed wood is suitable for replacing part of the ration.

Vanillin

Before 1940, vanillin was largely obtained from the tropical vanilla bean. Now practically all vanillin is produced by the alkaline oxidation of lignin sulfonic acid from waste sulfite liquor. The relatively low yields of 6–12%, based on the lignin sulfonic acid, require an elaborate extraction and purification process to produce a marketable product. In 1964, about 680 t (710 tons) of vanillin were produced in North America. Today, Canada alone produces 5400 t (6000 tons) annually, indicating a growing interest in this versatile organic chemical. Vanillin is mainly used as a flavoring agent and as an intermediate in pharmaceuticals. The recent demand for L-dopamine in the treatment of Parkinson's disease is a good example of how vanillin can be used as a starting material in the synthesis of drugs.

Essential Oils

Volatile oils, usually called essential oils, can be obtained from the wood and foliage of most trees by distillation with steam. In Canada, the preferred species for oil production are cedars, pines, spruces, hemlocks, birches, and Douglas-fir. Traditionally, inexpensive equipment and a simple batch method have been used to obtain essential oils. Leaves and twigs cut into lengths of 1.3–2.5 cm (½–1 in.) are heated in a large closed vessel through which steam is passed. The steam sweeps the evaporated oil through a water-cooled condenser, where both the oil and the water condense, and into a receiver where the oil is collected as a floating layer on the surface of the condensed water.

In past years, high labor costs for the collection of foliage have been responsible for the failure of several manufacturers. Now it is hoped that the new concept of whole-tree logging, which will result in the accumulation of foliage at selected sites, will reduce the cost of collection. In addition to essential oils, technical foliage (leaves, needles, and twigs of less than 0.6 cm [¼ in.] diameter) contains 36.8% lignin, 35.0% cellulose, 11.4% extractives, 10% protein, and 6.5% ash. Work in the Soviet Union has shown that the residue from steam distillation of the essential oils can be used as a food supplement for cattle and poultry. This twofold use of foliage should make the economics of recovery considerably more attractive.

Foliage oils are used mainly in perfumes, household cleaners, furniture polishes, veterinary soaps, insecticides, and medical preparations. Although oil from individual species is still in demand, it seems likely that the use of oil from mixed species will increase as a result of the high labor costs of collection and segregation of foliage. Fortunately, the prospects of chemical modification of mixed foliage oil to increase the yield of such desirable components as thujone are encouraging.

Other Chemicals from Foliage

The extractives of technical foliage contain several other chemicals, including chlorophyll, carotene (which is convertible to vitamin A), and vitamin C. Yields of these compounds vary according to species, site, and age, but values of 5%, 0.02%, and 0.15%, respectively, of the weight of leaves and twigs have been reported in Soviet literature (Andersen and Mochylski 1969). These chemicals can be obtained as a mixture in a crude vitamin flour (with or without the removal of the essential oils) or as highly purified individual components. Foliage products, up to 3% of the total food, can be added as protein supplements to cattle and poultry food. Individually, chlorophyll is used in the cosmetics industry in lotions, creams, deodorants, soaps, toothpaste, and other consumer products. Vitamins A and C are used directly as pharmaceuticals. While this approach to foliage use may appear visionary in North America, technical foliage is already being used this way in the Soviet Union.

Bark Tannins

Tannin, a complex, water-soluble phenolic extractive, is most often associated with the leather industry and the tanning of animal hides. Most tannin used in Canada has been imported from South America (quebracho), South Africa (wattle), and the United States (chestnut). Canada has traditionally depended on imported material, even though there are many good sources of tannin available in this country. For example, eastern hemlock bark contains 11%; white spruce bark, 20%; and three species of sumac, 16–24%. In western Canada the pulp and

Figure 5.7 Bark tannins are one of the ingredients of surfactants used to promote wetting, penetration, and color stability of textile dyes. (NFB Photothèque. Photo by Dunkin Bancroft)

lumber mills offer virtually unlimited sources of tannin from the barks of western hemlock and second-growth Douglas-fir. Western hemlock contains about 15% tannin; Douglas-fir, 8–18%. Both species have been reported to have a satisfactory tannin for the leather industry (MacLean and Gardner 1956).

In recent years, bark tannin, from western hemlock especially, has been extracted and marketed as a commercial product (Hergert et al. 1965). Formulations prepared from bark tannin are used to stabilize commercial suspensions of pigments, dyes, and pesticides and to control the viscosity of the 'mud' that is pumped around the bit during oil well drilling. Bark tannin forms complexes with such metals as iron, zinc, copper, and manganese. Some of these metals are required by plants in very small amounts. Applied to the ground, or sprayed directly on the foliage, these bark complexes gradually release the necessary traces of metals to the plant. In addition, chemically modified hemlock tannin is used as a component in phenol, resorcinol, and even urea-formaldehyde resin adhesives.

Fuel

Until well after the Second World War, wood was a major fuel in Canada. Production of fuel wood in 1948, for example, was equivalent to 21.6 million m³ (9.15 billion board feet) of timber, or 23.84% of the total volume cut. Production declined steadily until 1972, the year before the Arab oil embargo, when it hit bottom at 3.38 million m³ (1.43 billion board feet), or only 2.73% of the total cut. This decline was caused by the increasing value of roundwood for lumber and veneer and by the greater convenience of gas and oil as fuel. Since the embargo, use of fuel wood in North America has been rising. In the United States, sales of wood-burning appliances, counting only airtight stoves and furnaces, have doubled each year from 1974 through 1977.

Because it is a bulky solid of relatively low density, wood is awkward and expensive to transport over long distances. Fuel wood has traditionally been used near its growth site, which has usually been a farmer's wood lot. Waste wood, which has already had a 'free ride' to a central location such as a sawmill or a pulp mill, has an economic advantage. Accordingly, in 1973, for example, more than twice as much heat energy was recovered from waste wood in the forest industry as from wood specifically cut as fuel wood. At pulp mills, almost eight times as much heat energy was recovered by burning lignin and other dissolved wood components in spent pulping liquors.

Figure 5.8 Bark tannins and lignosulfates, recovered from spent sulfite liquor, are used to control viscosity of drilling muds pumped around the bit during drilling of oil wells. (NFB Photothèque. Photo by W. Vollman)

The heat energy released by burning one unit of weight of oven-dry wood is approximately the same for all species, roughly 19.8 MJ/kg (8500 BTU/lb). Ordinary firewood, however, is not oven-dry; it contains moisture in widely varying amounts, ranging from about 20% in seasoned wood to 70% or more in green wood. Heat obtained per unit of weight is therefore widely variable. Wood is sold by volume, usually by the standard cord, which is 3.624 m³ (stacked) (128 cu ft) of stacked wood (8 ft × 4 ft × 4 ft), or in the fireplace cord, which is one-third as much (8 ft × 4 ft × 16 in.). The amount of woody material in a cord obviously depends on how closely the pile is stacked and on the relative density of the wood. A cord of birch wood of relative density 0.57 is about twice as valuable as a cord of balsam fir of relative density 0.33 (both calculated on an oven-dry basis).

The heating value also varies somewhat with the composition of the wood. The fuel value of lignin is 25.1 MJ/kg (10 800 BTU/lb), whereas that of cellulose and hemicellulose is 17.6 MJ/kg (7500 BTU/lb). Many extractives have significantly higher heating values. The overall effect of composition, however, is not great, except for those species of high resin content. The heating values of the more important Canadian woods are given in Table 5.4, which is based on data compiled by Hale (1933).

As in Hale's original tables, the cord is assumed to be a standard cord of split wood and to contain 2.549 m³ (90 cu ft) of solid material. Every 100 parts by weight is taken to consist of 20 parts of moisture and

Table 5.4 *Heating values of Canadian woods*

Species	Calorific value: GJ/m^3 of solid wood, air-dry	Calorific value: Millions of BTU/cord, air-dry	Amount of wood equivalent to heating oil: Number of m^3 of solid wood (air-dry) equivalent to 1 000 L of oil	Amount of wood equivalent to heating oil: Number of cords of wood (air-dry) equivalent to 100 gal of oil
HARDWOODS				
Alder, red	7.20	17.4	8.67	1.55
Ash, black	9.35	22.6	6.67	1.19
Ash, green	9.15	22.1	6.83	1.22
Ash, white	10.3	25.0	6.03	1.08
Aspen, largetooth	7.53	18.2	8.29	1.48
Aspen, trembling	7.33	17.7	8.52	1.52
Basswood	7.04	17.0	8.87	1.58
Beech	11.5	27.8	5.43	0.97
Birch, white	9.68	23.4	6.45	1.15
Birch, yellow	10.8	26.2	5.76	1.03
Butternut	7.20	17.4	8.67	1.55
Cottonwood	6.95	16.8	8.98	1.60
Cottonwood, black	6.42	15.5	9.73	1.74
Elm, rock	13.2	32.0	4.71	0.84
Elm, slippery	10.5	25.4	5.94	1.06
Elm, white	10.1	24.5	6.16	1.10
Hickory, bitternut	12.1	29.2	5.17	0.92
Hickory, shagbark	12.7	30.6	4.93	0.88
Maple, Manitoba	7.99	19.3	7.82	1.39
Maple, red	9.93	24.0	6.29	1.12
Maple, silver	8.98	21.7	6.95	1.24
Maple, sugar	12.0	29.0	5.20	0.93
Oak, black and bur	11.7	28.2	5.35	0.95
Oak, red	11.3	27.3	5.53	0.99
Oak, white	12.7	30.6	4.93	0.89
Poplar, balsam	7.12	17.2	8.77	1.56
SOFTWOODS				
Cedar, eastern white	6.75	16.3	10.2	1.82
Cedar, western red	6.95	16.8	9.88	1.76
Douglas-fir (coast)	10.1	24.3	6.83	1.22
Douglas-fir (inland or mountain)	9.19	22.2	7.47	1.33
Fir, amabilis	6.83	16.5	10.1	1.79
Fir, balsam	6.42	15.5	10.7	1.91
Fir, grand	7.20	17.4	9.54	1.70
Hemlock, eastern	7.41	17.9	9.27	1.65
Hemlock, western	7.99	19.3	8.60	1.53
Larch, eastern or tamarack	9.93	24.0	6.91	1.23
Larch, western	11.0	26.5	6.26	1.12
Pine, jack	8.94	21.6	7.68	1.37
Pine, lodgepole	8.32	20.1	8.26	1.47
Pine, ponderosa	9.15	22.1	7.51	1.34
Pine, red	8.15	19.7	8.42	1.50
Pine, western white	7.70	18.6	8.92	1.59
Pine, white	7.08	17.1	9.70	1.73
Spruce, black	7.91	19.1	8.69	1.55
Spruce, Engelmann	7.28	17.6	9.43	1.68
Spruce, red	7.49	18.1	9.17	1.64
Spruce, Sitka	6.58	15.9	10.4	1.86
Spruce, white	6.70	16.2	10.2	1.83

80 parts of oven-dry wood. The calorific values listed represent the available heat, including the latent heat of vaporization of the moisture originally present and of that produced by the combustion.

When wood or other fuel is burned in a furnace or other heating device, only part of the available heat is recovered as useful heat, the latent heat of vaporization being lost up the flue along with the sensible heat of all the hot flue gases. The efficiency of the heating equipment – the percentage of the available heat of the fuel that is actually recovered as useful heat – varies greatly with the equipment and how it is operated. The optimum efficiency of wood-burning equipment is about 45–60% for airtight stoves and 50–60% for wood furnaces (Pratt 1978). Gas or oil-fired furnaces have higher optimum efficiencies of 65–80%.

In Table 5.4 the amounts of wood listed as being equivalent in useful heat to 1000 L, or 100 imp gal, of fuel oil are based on assumed combustion efficiencies of 50%, 55%, and 80% for softwoods, hardwoods, and heating oil, respectively, and on a gross calorific value of 42.9 MJ/L (185 000 BTU/imp gal) for the oil.

Wood could be improved as a fuel. Its bulkiness could be reduced and its energy content increased by pyrolysis. The solid products produced by pyrolysis after removal of the large quantity of water formed during the process, would make an excellent, more cheaply transportable fuel (Shafizadeh and DeGroot 1976). Gaseous fuels could also be produced from wood by several processes, most of which involve high-temperature pyrolysis, with or without added steam, air, or oxygen. Hydrogen, carbon monoxide, and carbon dioxide are major components in most of these products. Methane can also be produced from wood.

Liquid fuels are expensive to make from wood, and efficiencies of energy conversion rarely exceed 33%. Carbon monoxide and hydrogen from pyrolysis can be converted catalytically to methanol, or alternatively, carbohydrates of wood can be fermented to ethanol.

As fossil fuels are depleted and become increasingly expensive to recover, the importance of renewable resources such as wood seems destined to increase. Economics, however, will dictate in large measure the processes to be used and when they will be available.

THE FUTURE OF WOOD CHEMICALS

Wood has been and will continue to be one of our greatest sources of chemicals. These important constituents of wood will be used in steadily increasing quantities in the coming years, although the trend will be to use wood chemicals towards providing more of the replacement feedstock for the petrochemical industry. In general, this task can be accomplished in three ways – by pyrolysis, hydrogenation, and hydrolysis.

Pyrolysis

It has long been known that when wood is heated to temperatures of 1100°C (2000°F), in the absence of oxygen, its cellulose and lignin break down into simple, low-molecular-weight chemicals such as methanol, acetic acid, acetone, furfural, carbon dioxide, carbon monoxide, methane, carbon, and hydrogen. Yields of these chemicals are greatly influenced by such variables as temperature, heating rate, species, moisture content, and types of processing equipment. These variables are being critically examined in Canada and the United States to increase yields of the desired components. Because wood pyrolysis produces combustible, gaseous products such as carbon monoxide, methane, and hydrogen, it can be considered a source of energy as well as of chemicals.

Hydrogenation

Hydrogenation of wood is the treatment of wood with hydrogen and suitable catalysts at elevated temperatures and pressures to produce either a gas, which is enhanced in heating value by large amounts of low-boiling hydrocarbons, or oil. Unlike pyrolysis, which functions at atmospheric pressure, hydrogenation requires high pressures of up to 20.7 MPa (gauge) (3000 psig) and expensive equipment. The advantages, however, are higher yields of hydrocarbons and phenols. At present no commercial processes are available, but the same technology used for the hydrogenation of coal can be applied to wood.

Hydrolysis

The reactions of wood with water, salt solutions, acids, and gases have been described in early reviews of wood chemistry and briefly mentioned in this chapter. Wood's heterogeneous composition of cellulose, hemicelluloses, and lignin offers many reaction possibilities. To date, all commercial processes for wood hydrolysis involve either strong-acid or dilute-acid technology. Although the main products of wood hydrolysis are simple sugars, such as hexoses and pentoses, other valuable byproducts can be obtained by fermentation (ethanol and food yeast) and by dehydration (furfural). Even residual lignin can be converted to substantial amounts of monophenols by hydrogenation in the presence of a catalyst.

Other forest chemicals are now being used to a limited extent. Bark tannins and waste sulfite liquors, for example, are being recovered for use as additives in oil-well drilling and as ingredients in adhesives. Researchers in Canada are examining foliage as a possible replacement for phenol in phenol-formaldehyde adhesives. All options for obtaining useful chemicals from bark, wood, and foliage are being considered in many countries of the world as well as in Canada. Wood's potential as a source of chemicals is as diverse as its many other uses. Its versatility is unparalleled.

REFERENCES

Andersen, R., and Mochylski, I., eds. 1969. *Conference proceedings – Present position and prospects of the utilization of tree foliage*. Latv. Inst. For. Probl., Riga

Barton, G.M. 1973. Chemical color tests for Canadian woods. *Can. For. Ind.* 93 (2): 57–62

Barton, G.M., and Gardner, J.A.F. 1966. *Brown stain formation and the phenolic extractives of western hemlock*. Dep. For. Publ. 1147. Ottawa

Barton, G.M., and MacDonald, B.F. 1971. *The chemistry and utilization of western red cedar*. Dep. For. Publ. 1023. Ottawa

Bender, F. 1963. *Cedar leaf oils*. Dep. For Publ. 1008. Ottawa

Bender, F.; Heaney, D.P.; and Bowden, A. 1970. Potential of steamed wood as a feed for ruminants. *For. Prod. J.* 20 (4): 36–41

Bollen, W.B., and Glennie, D.W. 1961. Sawdust, bark and other wood wastes for soil conditioning and mulching. *For. Prod. J.* 11 (1): 38–46
Bray, M.W., and Martin, J.S. 1947. Sulfate pulping of Douglas-fir, western hemlock, Pacific silver fir, and western red cedar logging and sawmill waste. *Pap. Trade J.* 125 (16), Tappi Section 182–8
Forss, K., and Passinen, K. 1976. Utilization of the spent sulphite liquor components in the Pekilo protein process and the influence of the process upon the environmental problems of a sulphite mill. *Paperi ja Puu.* 58 (9): 608–18
Fraser, H.S., and Swan, E.P. 1972. A chemical test to differentiate *Abies amabilis* from *A. lasiocarpa* wood. *Can. For. Serv. Bi-mon. Res. Notes* 28 (5): 32
Freudenberg, K., and Neish, A.C. 1968. *Constitution and biosynthesis of lignin.* New York: Springer-Verlag
Gaden, E.L.; Mandels, M.H.; Reese, E.T.; and Spano, L.A. 1976. *Enzymatic conversion of cellulosic materials; technology and application.* Biotech. and Bioeng. Symposium No. 6. New York: Interscience
Gardner, J.A.F. 1965. Extractive chemistry of wood and its influence on finishing. *Off. Dig. J. Paint Technol. Eng.* 37 (485): 698–706
Gardner, J.A.F., and Barton, G.M. 1960. The distribution of dihydroquercetin in Douglas-fir and western larch. *For. Prod. J.* 10 (3): 171–3
Hale, J.D. 1933. *Heating value of wood fuels.* Dep. of the Interior, Forest Service, Forest Products Laboratories of Canada
Hall, J.A. 1971. *Utilization of Douglas-fir bark.* U.S. Dep. Agric. Pac. NW For. Range Expt. Stn., Portland, Oreg.
Hallett, R.M. 1971. Manufacturing and marketing charcoal in eastern Canada. *Can. For. Ind.* 91 (8): 56–9
Heaney, D.P., and Bender, F. 1970. The feeding value of steamed aspen for sheep. *For. Prod. J.* 20 (9): 98–102
Hergert, H.L.; Van Blaricom, L.E.; Steinberg, J.C.; and Gray, K.R. 1965. Isolation and properties of dispersants from western hemlock bark. *For. Prod. J.* 15 (11): 485–91
Hillis, W.E., ed. 1962. *Wood extractives and their significance to the pulp and paper industries.* New York: Academic Press
Hoge, W.H. 1954. The resistance of Douglas-fir to sulfite pulping. *Tappi* 37 (9): 369–76
Howsmon, J.A., and Sisson, W.A. 1954. Structures and properties of cellulosic fibers – Submicroscopic structure. In *Cellulose and cellulose derivatives, Part I,* edited by E. Ott and H.M. Spurlin. New York: Interscience
Jahn, E.C. 1952. Cellulose compounds and derivatives. In *Wood chemistry,* 2nd ed., edited by L.E. Wise and E.C. Jahn. New York: Van Nostrand Reinhold
Kline, E. 1954. Xanthates. In *Cellulose and cellulose derivatives, Part II,* edited by E. Ott and H.M. Spurlin. New York: Interscience
Levitin, N. 1962. Extractives of red and white pine and their effect on painted lumber. *Timber of Can.* 23 (6): 66–71
– 1972. The coloring of mineral stained maple. *Wood Sci.* 5 (2): 87–94
MacLean, H. 1970. Influences of basic chemical research on western red cedar utilization. *For. Prod. J.* 20 (2): 48–51
MacLean, H., and Gardner, J.A.F. 1956. *Tannin for the leather industry from sea-water floated western hemlock bark.* Dep. For. Publ. V-1009. Ottawa
Mühlethaler, K. 1965. The fine structure of the cellulose microfibril. In *Cellular ultrastructure of woody plants,* edited by W.A. Coté. Syracuse: Univ. of Syracuse Press
Polcin, J., and Rapson, W.H. 1971. Sapwood and heartwood groundwood of western hemlock and jack pine. Part II. Heat stability of extractives. *Pulp Pap. Mag. Can.* 72 (10): 84–90
Pratt, M., ed. 1978. *Heating with wood.* The Institute of Man and Resources Publ. 1/78. Charlottetown, PEI
Quon, H.H., and Swan, E.P. 1972. Deoxypicropodophyllin and sugiol from *Thuja plicata* bark. *Can. For. Serv. Bi-mon. Res. Notes* 28 (4): 23–4
Rogers, I.H., and Manville, J.F. 1972. Juvenile hormone mimics in conifers, I. Isolation of (-)-*cis*-4[1'(R)-5'-dimethyl-3'-oxohexyl]-cyclohexane-1-carboxylic acid from Douglas-fir wood. *Can. J. Chem.* 50 (14): 2380–2
Rydholm, S.A. 1965. *Pulping processes.* New York: Interscience

Sarkanen, K.V., and Hergert, H.L. 1971. Lignins classification and distribution. In *Lignins – Occurrence, formation, structure, and reaction*, edited by K.V. Sarkanen and C.H. Ludwig. New York: Interscience

Scroggins, T.L., and Currier, R. 1971. Agricultural uses for western red cedar residues from shake and shingle mills. *For. Prod. J.* 21 (11): 17–24

Shafizadeh, F., and DeGroot, W.F. 1976. Combustion characteristics of cellulosic fuels. In *Thermal uses and properties of carbohydrates and lignins*, edited by F. Shafizadeh, K.V. Sarkanen, and D.A. Tillman. New York: Interscience

Srinivasan, V.R. 1975. Production of bio-proteins from cellulose. SITRA *Symp. Enzymatic Hydrolysis Cellulose*: 393–405

Stamm, A.J. 1964. *Wood and cellulose science*. New York: Ronald Press

Timell, T.E. 1957a. Nitration as a means of isolating the alpha-cellulose component of wood. *Tappi* 40 (1): 30–3

– 1957b. Carbohydrate composition of ten North American species of wood. *Tappi* 40 (7): 568–72

– 1967. Recent progress in the chemistry of wood hemicelluloses. *Wood Sci. Tech.* 1: 45–70

Underwood, J.C.; Filipic, V.J.; and Bell, R.A. 1969. Gas-liquid chromatographic flavor profile of maple syrup. *J. Assoc. Offic. Agr. Chem.* 52 (4): 717–19

Wenzl, H.F.J. 1970. *The chemical technology of wood*, translated by F.E. Brauns and Dorothy A. Brauns. New York: Academic Press

Wilke, C.R., ed. 1975. *Cellulose as a chemical and energy resource*. Biotech. and Bioeng. Symposium. New York: Interscience

Wise, L.E., and Jahn, E.C., eds. 1952. *Wood Chemistry*. New York: Van Nostrand Reinhold

6

Lumber Production

ALLAN BUELL

Forestry Consultant, Manotick, Ontario

C.F. MCBRIDE

Retired, formerly with Western Forest Products Laboratory, Vancouver

HISTORY

From Canada's early history, lumber has filled two important roles in the economy. First, it has provided a readily available source of building material for local use, and second, it has been the basis of a large and important export market, valued at $3901 million in 1979. It cannot be emphasized too strongly that this major permanent industry is based on a renewable resource.

Canada's role as an exporter of forest products began in the seventeenth century in eastern Canada when white and red pine were exported in the form of hewn and round timbers. On the west coast it was not until the eighteenth century that limited quantities of Douglas-fir and Sitka spruce were exported for masts and spars for sailing ships.

The first lumber was produced for local use by hand-operated pit saws or whipsaws, with a daily production of 0.25–0.5 m^3 (100–200 board ft) of rough lumber. As machinery and power were developed, small sawmills were built, usually to serve local needs. By the middle of the nineteenth century, exported sawn lumber was bringing in more money than hewn timbers and spars. White pine was still a much-sought-after species.

Slowly machinery was improved and larger sawmills were built. As the number of large mills steadily increased, the number of small local mills decreased. This trend continues today, and more lumber is produced in fewer mills each year. New machinery and technology have been developed to improve the efficiency of the lumber-manufacturing process. The energy to operate mills has evolved from the manual energy of pit sawing to power from water, steam, diesel engines, and, most recently, electricity. Some new mills fill all their energy requirements by burning bark and other residues. The low-energy requirements for the manufacture and processing of lumber,

Figure 6.1 A typical hardwood sawmill in eastern Canada.

compared with those of other building materials, are an important advantage to the industry.

Lumber production in Canada increased rapidly from 1900 onwards, particularly in British Columbia. By 1926 more lumber was produced in that province than in the rest of Canada. In 1978 lumber production in British Columbia was 66% of the total Canadian lumber production of 44.9 million m^3 (19.0 billion board ft).

Early shipments of lumber were made either on a log-run basis with no determination of quality, or by comparison with the quality of earlier shipments. At the beginning of the nineteenth century, both the export and home markets had expanded so that uniform standards for sorting lumber were required to satisfy the needs of buyer, seller, producer, exporter, and importer.

Although not the first published systems, the 'Maine Surveying' in the state of Maine and the 'Quebec Culling' in eastern Canada became the most widely used lumber-grading rules. Current lumber-grading criteria used in North America originated from these two systems. As the lumber industry expanded, regional rules were developed to suit new species and markets, and inspection systems were initiated to provide more uniform interpretation of the rules.

After many attempts to standardize the softwood-grading rules, the lumber industry finally adopted a new rule, the National Lumber Grades Authority 1979 Standard Grading Rules for Canadian Lumber. This rule standardizes the sizes, grade names, and permissible characteristics in each grade for all softwood lumber marketed in Canada and the United States (see also discussion under 'Species' in this Chapter).

Hardwood lumber in North America is marketed under the grading rules of the National Hardwood Lumber Association.

LUMBER MANUFACTURE

Log Quality and Allocation

To obtain maximum value from the forest, trees should be bucked to produce the largest volume of high-quality logs, and logs should be sorted by size and quality for the most appropriate end use. Broad sorts, or classification, may include veneer logs, poles and piling, and lumber and pulp. Further sorting may take place at the manufacturing plant. Large, integrated operations use elaborate sorting operations that facilitate bucking for quality and length and sorting for end product.

At the sawmill the logs may be sorted by species or species groups, diameter, length, and quality. The size and type of mill and the nature of its products dictate how the logs are sorted. Small mills may not sort at all or may sort by species only.

Log quality is measured by the volume and quality of the potential lumber yield. External characteristics of the log such as diameter, surface indicators, and position in the tree (butt logs or uppers), are used to estimate potential lumber quality. To obtain the maximum yield from top-grade logs, it is best to saw them in a mill with a standard headrig and carriage, which is the first machine in a sawmill to start the breakdown of logs into lumber products. The sawyer observes the quality of the log as it is sawn and uses his judgment to obtain the best

NOT RECOMMENDED

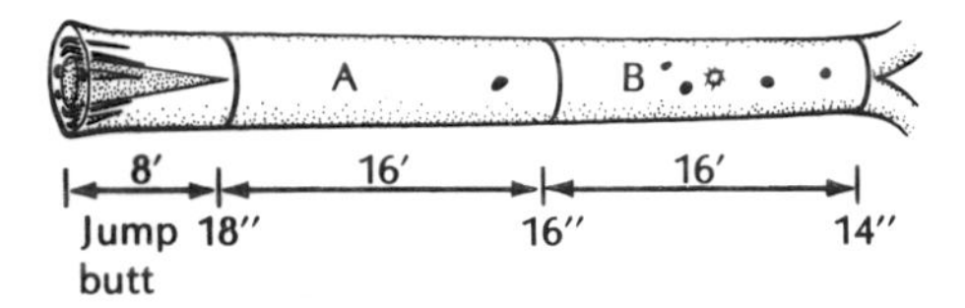

Log	Log grade	Diameter (in.)	Length (ft)	Gross scale (OLR)	Net scale (OLR)	Log value	Lumber value: Per log	Lumber value: Per Mbf net scale	Lumber value: Per gross cu ft of log volume
A	F-2	16	16	162	162	10.53	22.10	136.42	0.87
B	F-3	14	16	121	121	6.65	12.19	100.74	0.62
Total (and average) for tree					283	17.18	34.29	(121.17)	(0.76)

RECOMMENDED

Log	Log grade	Diameter (in.)	Length (ft)	Gross scale (OLR)	Net scale (OLR)	Log value	Lumber value: Per log	Lumber value: Per Mbf net scale	Lumber value: Per gross cu ft of log volume
C	F-1	17	16	185	165	12.37	30.18	182.91	1.06
D	F-2	15	12	106	106	6.89	13.97	131.79	0.83
E	F-3	14	8	61	61	3.35	6.15	100.82	0.67
Total (and average) for tree					332	22.61	50.30	(151.51)	(0.92)

Figure 6.2 Bucking hardwood logs to obtain best yield. In the method that is not recommended, the sawyer assumed rot extended a considerable distance up the stem, so the yield was lower-value F-2 and F-3 logs. In the recommended method, the depth of rot was determined before bucking, and maximum yield and value were obtained from the log.

yields of the top grades. Bucking hardwood logs for quality is discussed fully by Petro (1971). For instance, Petro shows the value per thousand board feet (2.36 m^3) of lumber from a grade F-1 log to be almost twice that of an F-3 log (Figure 6.2), and this differential warrants extra care in processing the high-value log. On the Pacific coast, a No. 1 Douglas-fir log must be over 76 cm (30 in.) in diameter and yield over 50% clear lumber, whereas a No. 3 Douglas-fir log will yield mainly lumber for construction, with very few clears.

Linear-throughput mills where logs are fed continuously through the mill, are designed to process logs of relatively uniform quality, and production per day is the governing criterion. If logs are sorted by diameter, the linear-throughput is increased because logs can be fed end to end without the need for gaps between them to permit changing the settings of the headrig so that it can accommodate each log (Dobie 1970).

In many small-log mills, sorting is done by broad diameter classes dictated by the capacity of the various headrigs in the mill. Mills also sort logs by quality to coordinate the output of lumber grade with marketing schedules (Figure 6.3).

Basic Steps in Lumber Manufacture

Lumber manufacturing is the process by which round, tapered logs are converted into rectangular lumber and other products, such as pulp chips, sawdust, and planer shavings. Each sawmill uses the machinery and techniques that are best suited to the supply, size, and quality of raw material. Given the wide range of log quality and the large number of specialized sawmill machines, it is inevitable that sawmill layouts vary greatly. There is no standard sawmill; therefore, in this chapter, only the basic process and the more important sawmilling machines are

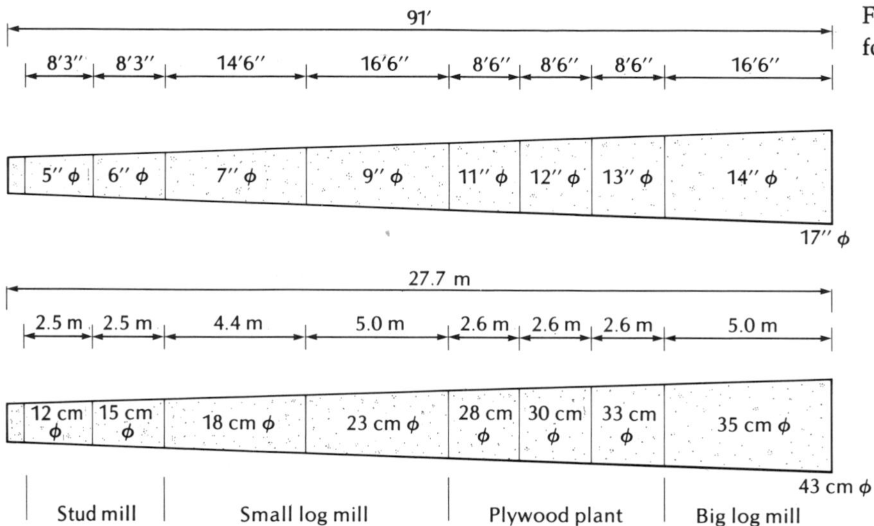

Figure 6.3 Log bucking and assignment, for specific end uses.

discussed. Emphasis is on softwood mills, which produce 93% of the lumber manufactured in Canada.

The first step in the manufacturing process is to remove the bark from the log. The reasons for debarking are to obtain bark-free pulp chips, to increase saw life, and to increase productivity. Debarking is accomplished by powerful water jets, mechanical arms scraping the bark, or abrasion from revolving heads. The mechanical ring barker is the most common type of barker employed in Canada.

The debarked log is sawn at the headsaw into slabs (with one flat face) and flitches (with two parallel flat faces). The slabs go to the chipper, where slabs, edging, and trim ends, which formerly were burned, are converted into pulp chips. The flitches proceed either through a resaw, or directly to an edger, which produces rectangular lumber by sawing off the rounded, tapered edges. Edgers use a fixed saw and a number of movable saws mounted on an arbor or shaft. The distance between the saws is adjusted for each flitch to obtain the width that provides the highest value. The rectangular boards are then fed past trim saws, which cut the pieces to length (including a small overlength allowance). The trim saws may also cut out major defects, producing two short boards.

After trimming, the lumber is sorted by width, length, thickness, and grade. Large modern mills usually have automated sorting devices, but in the smaller mills manual sorting, known as on the 'green chain', is common.

At this stage, if the market is for unseasoned material, the lumber may be planed, after which it is given final grading, trimmed, grade marked, antistain treated, and packaged for shipment. If the mill markets seasoned lumber, the wood is dried, usually in a dry kiln, and then processed through a planer, graded, retrimmed, grade marked, and packaged for shipment.

Sawing Methods

Three basic sawing patterns can be used, depending on the type of machinery in the mill, the quality of the logs, the product mix desired, and management policy.

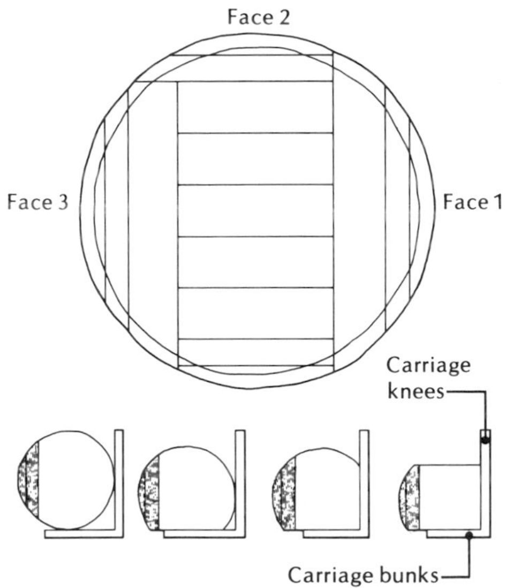

Figure 6.4 Sawing around. The lines show sawing pattern with sawing sequence (below).

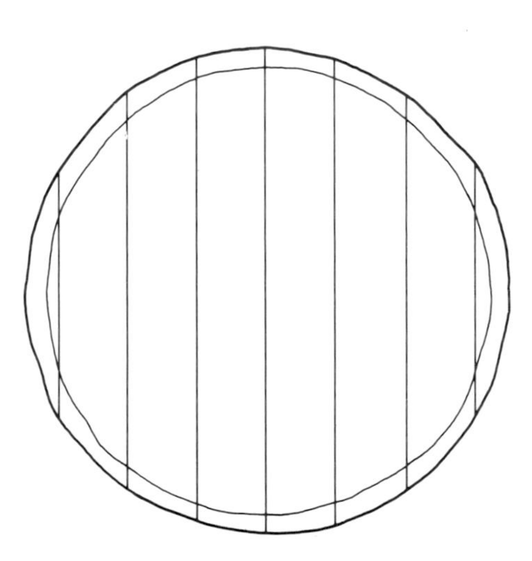

Figure 6.5 Through-and-through (live) sawing. The lines show sawing pattern.

For high-value logs, the common procedure is to saw around the log using a headrig with a carriage (Figure 6.4). The log is slabbed and a few boards cut; then the log is turned through 270° and the process repeated until all four sides have been cut. The remaining center cant is then cut into sizes appropriate to its quality. 'Sawing around' produces the maximum yield of flat-grain lumber. Mills cutting high-quality softwood logs to produce clear, vertical-grain lumber cut thick, flat-grain flitches, which are resawn into vertical-grain material. For hardwood logs of medium and high quality, the poorest face is slabbed first (to provide a steady base), and then the other faces are sawn. In general, the sawyer always saws the face with the greatest potential value, so he must turn the log from time to time.

Through-and-through sawing, or live sawing, involves sawing the log completely with one set of parallel cuts (Figure 6.5). This pattern is seldom used in North America, although it is advantageous in some circumstances because it can help to produce maximum lumber volume and value from some high-quality hardwood logs (Flann 1978). In European mills, a small log is moved through a battery of reciprocating sawblades, or a sash gang, set at a fixed spacing, and is reduced entirely to flitches in one pass. Larger logs are sawn with a circular or band headsaw and carriage; the log is turned once through 180° after the first face has been deeply sawn.

Canting, the third method of sawing, is a combination of the two previous methods, and it is the most common pattern for softwoods. A slab and one or more boards are cut from each of two opposite sides of a log, leaving a cant of the desired depth with two parallel sawn faces and two rounded sides. The cant is then processed through a multiple-saw machine, such as a sash gang, circular gang, or a quad-band resaw, from which it emerges cut into lumber. This sawing method can be used for most types of headrigs.

Saws

Saws are of two basic types, circular and band. Sash gang saws are simply several short straight sawblades fitted into a reciprocating sash.

The circular saw is a circular steel plate, usually ground to a uniform thickness, with cutting teeth either ground into the rim of the plate or attached by a locking mechanism. The saw is attached to an arbor by collars in a headsaw and by a spline in an edger. The advantages of these saws are that they are relatively cheap and rugged and can be maintained without highly skilled labor. The disadvantage is that the kerf, or the cut made by the saw, is usually wider than that of bandsaws. Progress is being made in reducing the kerf and improving the surface quality of the cut (McLauchlan 1972).

A bandsaw is a continuous ribbon of steel mounted over an upper and lower wheel. Either one edge (single cut) or both edges (double cut) may have teeth. Bandsaws generally have a thinner kerf than circular saws, make smoother, more accurate cuts, and can be built to cut larger logs than circular saws. Disadvantages include the high cost of the saws and the bandmill machinery and the specialized staff required to maintain the equipment.

In the past decade, the increasing cost of wood has stimulated interest in reducing saw kerf and in increasing sawing accuracy and the smoothness of the sawn surfaces. High-strain bandsaws and guided circular saws are among the many advances the industry has made. Both the Eastern and Western Forest Products Laboratories of Forintek Canada Corp. have active research programs in these areas.

Headrigs

In sawmill terminology, the first machine that starts reducing a log to lumber is called a headrig. The sawmilling industry uses many types of headrigs. Brief descriptions of the more common types follow.

Single- or Double-Cut Bandsaw with Carriage

This headrig is a collection of equipment used for the initial log breakdown. Included are a carriage running on a track, a bandsaw, a mechanism to position and secure the log on the carriage, and devices to remove the sawn lumber and slabs. An important part of the headrig carriage is the setworks, which controls the thickness of lumber to be cut on each pass of the log through the saw. All the functions of the carriage, loading and turning the log, setting the thickness of cut, and moving the log past the saw are controlled by the sawyer. A single-cut band mill can make a saw cut only when the carriage moves forward. A double-cut band mill can saw on both the forward and the return travel of the carriage because the saw has cutting teeth on both edges (Dobie and Sturgeon 1972).

Figure 6.6 Circular saw with carriage feed.

Circular Saw with Carriage

The carriage arrangement, including setworks, is similar to that of the single-cut bandsaw headrig. Sawing occurs only when the carriage is moving forward. Carriage-fed circular headrigs are of two types: those with a single circular saw, and those with two circular saws operating in the same plane, one (i.e., a 'top' saw) above the other to permit the cutting of large logs. The trend is away from circular headrigs and toward chipper-canters for small logs and thin bandsaws for large logs. The circular headsaw is standard equipment in small mills, however.

Figure 6.7 Round-log twin bandsaw.

Scrag Saws

There are two common headrigs of this type; the first consists of two adjustable circular saws on a single arbor (typically used in stud mills), and the second consists of two or more arbors, each with a pair of saws and each pair cutting different lines. The individual saws may be fixed or may be movable to different settings. A continuous chain carries the logs through the saws. Because these headrigs are not intended to saw for grade, their most common application is in mills specializing in the manufacture of studs from small timber. The greatest advantages of scrag saws are low initial cost, simplicity, and relative freedom from maintenance problems. The disadvantages are the large percentage of sawdust produced and the inability to saw large logs for grade. A typical flow design for a scrag mill is shown in Figure 6.8.

Twin and Quad Bandsaws

Round logs are fed into twin and quad band saws (Figure 6.7) by a sharp chain log feed, or other log-carrying device, which holds the log firmly and guides it accurately through the saws. The saws are set to the desired spacing by hydraulic controls. A twin band usually

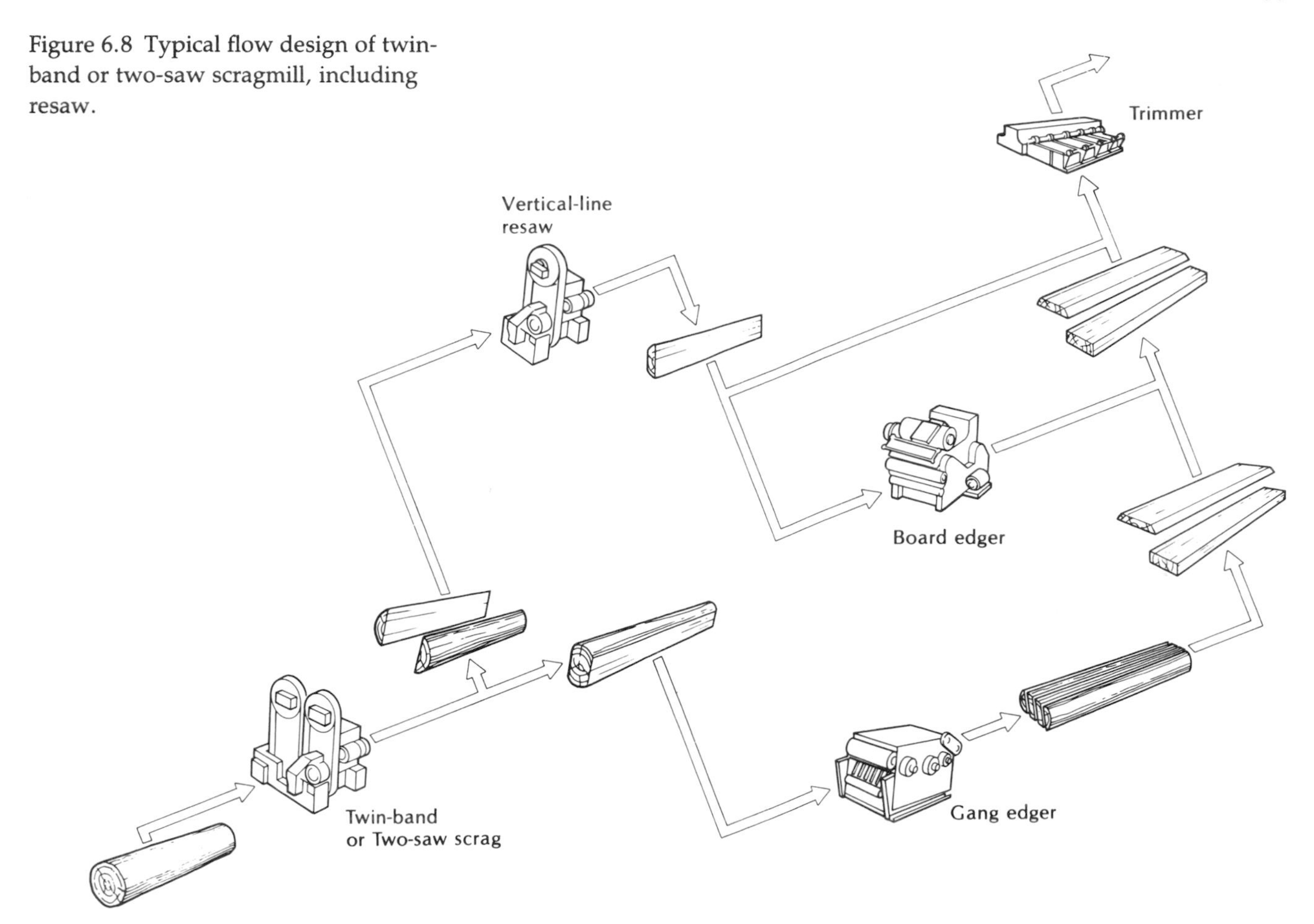

Figure 6.8 Typical flow design of twin-band or two-saw scragmill, including resaw.

produces two slabs and a cant that is further broken down by resaws. A quad bandsaw consists of two pairs of band mills operating in tandem and is capable of making four saw cuts in one pass. Twin and quad bandsaws are also used to process cants.

Gang Saws

In Scandinavia the optimum sawing pattern is determined according to softwood log size, quality, and product price schedule. The logs are sorted into diameter classes and are then processed through a round-log gang saw, followed by a cant gang saw, often called the sash gang or frame saw. This type of saw is seldom used in Canada. In Canada much less sorting is done, and the logs are broken down into waney planks, or flitches, of uniform thickness as in cant sawing or through-and-through sawing. The advantages of a sash gang are narrow kerf and accurate cutting, although the newer, thin-kerf, high-strain bandsaws have comparable advantages. The most common gang saw in North America is the multiple circular resaw, often called a bull edger.

Chipping Headrigs

The chipping headrig, also called the chipping canter, is the most important innovation in sawmilling in recent years (Dobie 1967). First introduced in Canada in 1964 (Dobie and McBride 1964), it consists of a number of rotating heads in which chipper knives are inserted. As the logs are fed through the machine, the top and bottom surfaces of the logs are profiled, or chipped, in 51 mm (2 in.) steps by the chippers, and the excess wood on the sides, the slabs, is removed by two more chipper heads. The profiled cant is then sawn into lumber by saws matched to the steps in the log. Figure 6.9 shows examples of the profiled cants produced by a chipper headrig.

Since the slabs are chipped off the log, the recovery of pulp chips is increased and the volume of sawdust is decreased. The system has low labor costs and is particularly suited to small logs. The large capital cost and the fact that only straight logs can be processed are disadvantages. The chipper headrig also has a tendency to increase chip recovery at the expense of lumber production.

Other systems use chipper heads to make cants from logs by chipping slabs from either two or four faces.

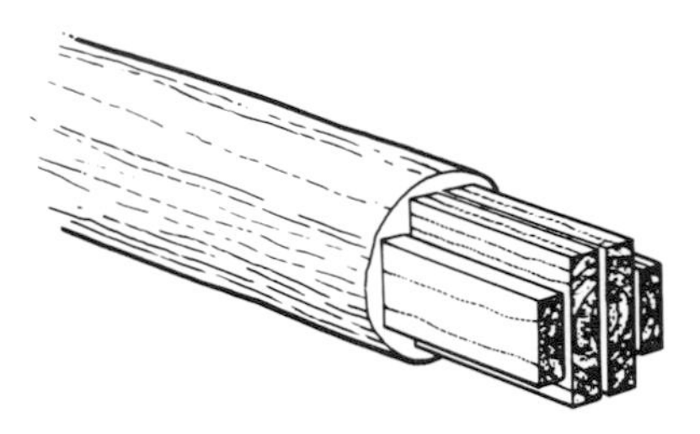

2 - 38 x 140 mm (2 in. x 6 in.)
2 - 38 x 89 mm (2 in. x 4 in.)

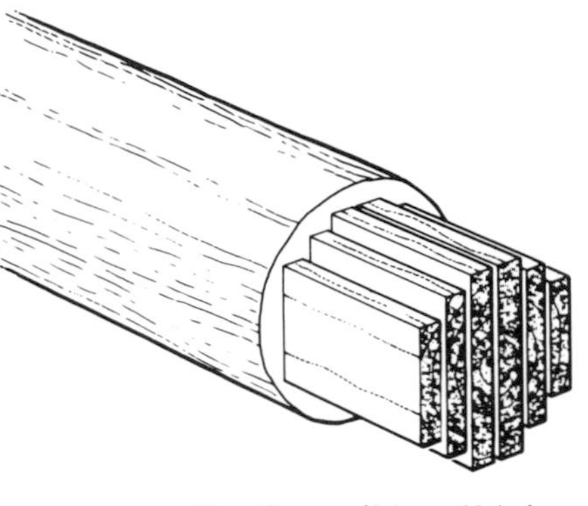

2 - 38 x 286 mm (2 in. x 12 in.)
2 - 38 x 235 mm (2 in. x 10 in.)
2 - 38 x 184 mm (2 in. x 8 in.)

Figure 6.9 Two examples of profiled cants produced by a chipper-headrig.

Secondary Sawmilling Machines

Because the headrig is considered the key to the throughput of a sawmill, reducing the number of cuts per log on the headrig and using secondary machines to do further processing increases the production of a sawmill. Under these circumstances the headrig is used as a breakdown machine, cutting the log into pieces that are multiples of the final sizes. Many auxiliary machines are available, and the combinations are numerous. Gang edgers, cant gangs, and various types of resaws are most commonly used. A horizontal band resaw is shown in Figure 6.10.

Planing Lumber

In Canada about 80% of softwood lumber is planed, and the other 20% is sold in the rough-sawn state. Planing ensures uniformity of widths and thicknesses, and provides a high-quality finish; both are important in the marketing of construction lumber. Because hardwood is normally remanufactured into furniture, pallets, and so on, it is usually sold rough. Most planing machines produce flat, square-edged lumber, although a large amount of dimension lumber is manufactured with eased or slightly rounded edges. The planer-matcher produces a variety of self-matching planed products, such as tongue-and-groove, decking, shiplap, interlocking siding, and flooring. After passing through the planer, lumber is graded, trimmed, grade stamped, side or end marked, sorted and stacked, and packaged for shipment.

A few modern mills with very accurate sawing now use sanders instead of planers. Sanders produce a smoother finish and less chipped grain around knots.

Excellent sources of information on sawmilling are Brown and Bethel (1958), Quelch (1970), Koch (1972), and Williston (1976).

Figure 6.10 Horizontal-band resaw.

HARDWOOD LUMBER

Grading Rules

Hardwood lumber in Canada and the United States is graded according to the rules published by the National Hardwood Lumber

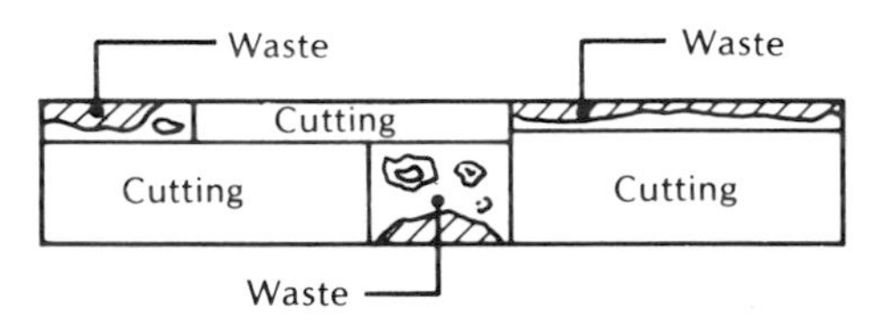

Figure 6.11 Diagram showing cuttings from a piece of hardwood lumber. The amount of usable material, or cuttings, determines the grade of the piece.

Association. Inspectors employed by the association are located throughout each country to serve both producers and consumers. The grade of a board is usually based on its poorer face and is determined by the percentage of the piece that can be made into cuttings of a certain quality and dimensions (width and length). The higher the yield and the larger the size of cuttings that can be cut from the board, the higher the grade. Most cuttings used in the furniture industry, are clear on one face and sound on the reverse face. Figure 6.11 shows a typical board. The two highest grades – First and Seconds – are usually sold together as FAS. The next grade is Selects, followed by No. 1 Common, No. 2 Common, Sound Wormy, No. 3A Common, and No. 3B Common. In general, grades No. 2 Common and Better are used by the furniture industry, and the lower grades are used by the pallet industry.

Hardwood lumber is usually manufactured in random widths in order to ensure the highest lumber value as determined by both surface yield and lumber grade. The grades specify minimum widths of 140 mm (6 in.) for FAS, 89 mm (4 in.) for Selects, and 64 mm (3 in.) for the Common grades.

Dimension Parts

A specialized market has been developed for dimension parts or components. These are sold as rough, semifinished, or finished dimension stock. The rough material is kiln dried and cut to size (including allowance for final machining). The semifinished stock is further processed by surfacing, edge gluing, shaping, and so on. Finished stock is completely machined and sanded.

SOFTWOOD LUMBER

Grading Rules

All softwood lumber in Canada is graded under the National Lumber Grading Authority (NLGA) 1979 Standard Grading Rules for Canadian Lumber. The Canadian Lumber Standards Division (CLS) of the Canadian Standards Association (CSA) is responsible for the control and regulation of the grade stamping of Canadian softwood lumber. Lumber associations and grading agencies must meet certain standards to be certified by CLS so they can direct the grade stamping of lumber in the mills under contract to them.

A lumber grade is based on probable use of the lumber and includes pieces that differ slightly from each other within defined quality limits. The characteristics (or defects) allowable in each grade are enumerated in the grading rules.

The term characteristics is used to describe deviations from the perfect wood structure. These are divided into three categories: (1) *natural*, those that occur in the development of the tree, such as knots, pitch pockets, bark seams, shake, and stain; (2) *manufacturing*, those developed in the manufacture of lumber, such as improper trimming, skips in dressing, and torn grain; (3) *seasoning*, those developed in the drying process, such as checks, splits, warp, and collapse.

Clears such as Finish, Casing, and Panelling are usually appearance grades in which the attractive features of the wood are the important factors. In vertical-grain flooring, wearing qualities are critical; in siding exposed to the weather, both tight construction and appearance are the basic requisites.

Boards and shiplap are a knotty type of construction lumber. They require bracing strength when used as sheathing and tight construction when used as form lumber for concrete. The grade assigned to a piece is governed by the types of defects, their location in the piece, and their effect on strength.

In Factory Lumber, which is to be cut up into clear or almost clear components, the volume and quality of usable cuttings are the governing criteria. Factory Lumber may contain such defects as large knots, bark seams, burls, and other imperfections, provided they can be removed by ripping and/or crosscutting to produce shorter or narrower pieces free of large imperfections.

A.F.P.A.® 00
S—P—F
S-DRY STAND

Alberta Forest Products Association
10428 - 123rd Street
Edmonton, Alberta T5N 1N7

Figure 6.12 Facsimile of a grade stamp.

Species

The NLGA 1979 grading rules apply to all softwood lumber species manufactured in Canada, and they may also be applied to hardwood species manufactured for applications where softwood species are ordinarily used. The northern aspen species group is an example of such a hardwood species. The strength values assigned to each species group are based on the weakest species in that group.

Exception is made for western red cedar, which has a separate rule for all categories except dimension lumber. Certain species such as western white pine, eastern white pine, and red pine have traditional specific grade names that have been retained from the previous grading rules for the 'clear' and 'board' grades.

Species Groups

Lumber is marketed in species combinations, comprised of species that have similar appearance, somewhat similar strength properties, and, usually, similar end uses. It is often difficult to visually segregate lumber by species. The lumber of spruce and balsam fir is difficult to distinguish visually, for example. The various species combinations that may be legally included in the lumber grade stamp for each species group are shown in Table 6.1.

The grade stamps give the following information: grading agency, mill number, species group, grade, surfacing, dryness at the time of surfacing. Figure 6.12 illustrates the typical grade stamp. Associations and agencies involved in grade stamping of softwood lumber are listed in Table 6.5, and the various stress grades for construction are shown in Table 6.2.

Lumber Sizes

Lumber sizes have been given in imperial terms for many years, but on January 1, 1978, the Canadian construction industry began calling for tenders in metric terms. Because a large percentage of Canadian lumber is shipped to the United States, which will not convert to metric sizes for several years, the lumber industry has chosen the soft conversion*

* Soft conversion is the direct conversion from imperial units to SI metric units, retaining accuracy in the converted value but not exceeding the former measurement tolerances.

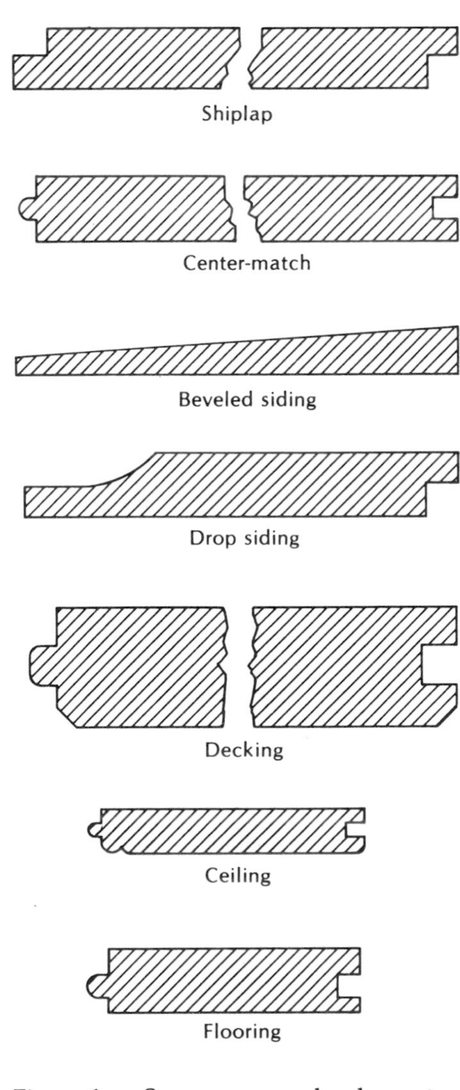

Figure 6.13 Some common lumber patterns.

to metric. This enables the industry to continue to produce lumber in the present sizes and lengths, marketing it in metric terms in Canada, while continuing to market lumber in imperial terms in the United States.

Under the imperial terminology, sizes have two designations, nominal and actual. The nominal size is the size used to name the piece and to calculate the board foot volume. The actual size is the actual measured dimensions, which are somewhat smaller than the nominal dimensions. Thus, a nominal 2 in. × 4 in. piece actually measures $1\frac{9}{16}$ in. × $3\frac{9}{16}$ in. when surfaced green and $1\frac{1}{2}$ in. × $3\frac{1}{2}$ in. when surfaced dry. The difference is the result of the allowance for shrinkage down to 19% moisture content.

The metric nomenclature eliminates nominal size and is based on the actual dry size for thickness and width converted to the nearest millimetre and for length converted to metres to two decimal places. Volumes are in cubic metres to the third decimal place and surface coverage, where applicable, in square metres. A manual entitled *Metric Manual for Wood Products* has been prepared to assist in the conversion process. In metric nomenclature, the nominal 2 in. × 4 in. piece is 38 mm × 89 mm.

Lumber is divided into three categories on the basis of thickness: *boards*, or lumber less than 2 in. in nominal thickness, or 32 mm or less; *dimension lumber*, or lumber from 2 in. to, but not including, 5 in. in nominal thickness, or between 38 mm and 102 mm; and *timbers*, or lumber 5 in. or more in nominal thickness, or 114 mm or more in the least dimension.

Lengths in the present grading rules are in 1 ft or 0.30 metre multiples, although generally lumber is manufactured in even 2 ft or 0.61 metre multiples.

Table 6.4 shows imperial sizes and their metric equivalents.

Surfacing

Over 80% of the lumber produced in Canada is planed, although the west coast mills still sell over half their cut in the rough state. Lumber is planed to produce uniform sizes and a smooth finish for both ease of handling and appearance.

Depending on the end use, surfacing may take various forms, such as surfaced one side (S1S), surfaced two edges (S2E), and many combinations up to the most common, surfaced four sides (S4S). The surfacing process may also include running to a pattern. Typical patterns of shiplap, center-match, beveled siding, drop siding, decking, ceiling, and flooring are illustrated in Figure 6.13.

MARKETING

In 1978 Canada produced 44.9 million m^3 (19.0 billion board ft) of lumber, of which 29.6 million m^3 (12.5 billion board ft) were produced in British Columbia and 7.6 million m^3 (3.2 billion board ft) in Quebec. The value of the lumber exports from British Columbia was 78% of the total exported from Canada in 1978.

The relative importance of the various markets for British Columbia lumber in 1978 is shown in Table 6.3.

Table 6.1 *Species combination*

Strength group*	Commercial species or species group designation	Species† in combination	Grade mark abbreviation	Wood characteristics
A	Douglas-fir–larch (North)‡	Douglas fir Western larch	D Fir-L N	Woods similar in strength and weight. High degree of hardness and good resistance to decay. Good nail-holding, gluing, and painting qualities. Color ranges from reddish-brown to yellowish-white.
B	Hem-fir (North)‡	Western hemlock Amabilis fir	Hem-Fir N	Light woods of moderate strength. They work easily, take paint well, and hold nails well. Good gluing characteristics. Color ranges from pale yellow-brown to white.
C	Eastern hemlock Tamarack (North)‡	Eastern hemlock Tamarack	Hem-Tam N	Moderately strong woods mostly used for general construction. Fairly hard and durable. Color ranges from yellowish-brown to whitish.
D	Spruce-pine-fir	White spruce Red spruce Black spruce Engelmann spruce Lodgepole pine Jack pine Alpine fir Balsam fir	S-P-F	Woods of similar characteristics. They have moderate strength, work easily, take paint easily, and hold nails well. Generally white to pale yellow in color.
D	Coast Sitka spruce	Coast Sitka spruce	C Sitka	A light, resilient wood of moderate strength that works and takes paint easily and holds nails well. Creamy white to light pink in color, with large proportion of clear wood.
D	Ponderosa pine	Ponderosa pine	P Pine	Moderately strong and easily worked to smooth, uniform texture. Takes paints, stains, and varnishes well. Seasons readily. Good nail-holding qualities. Color variable from pale yellow sapwood to deep yellow to reddish-brown heartwood. Considerable figure in heartwood.
E	Red pine	Red pine	R Pine N	A fairly strong and easy-to-work wood that takes a good finish and holds screws and nails well. Moderately durable; it seasons with little checking or cupping. Sapwood is thick, pale, yellow color; heartwood pale brown to reddish tinge.
E	Western white pine	Western white pine	WW Pine	Softest of the Canadian pines, they work and finish exceptionally well. Not as strong as most pines but do not tend to split or splinter. Good nail-holding properties. Low shrinkage; better than all other Canadian species except the cedars. Take stains, paints, and varnishes well. Color of sapwood almost white; heartwood creamy white to light straw brown.
E	Eastern white pine (North)‡	Eastern white pine	East White Pine N (WE Pine N)§	(see Western white pine)
E	Western cedars (North)‡	Western red cedar Pacific coast Yellow cedar	W Cedar N Y Cedar N	Woods with exceptional resistance to decay. High in appearance qualities, they work easily and take fine finishes. Each species has distinct and easily recognizable color. Red cedar varies from reddish-brown heartwood to light sapwood; yellow cedar has a uniform warm yellow color.

Table 6.1 *(concluded)*

Strength group*	Commercial species or species group designation	Species† in combination	Grade mark abbreviation	Wood characteristics
F	Northern aspen‖	Trembling aspen Largetooth aspen Balsam poplar	N Aspen	Light woods of moderate strength; they work easily, finish well, and hold nails well. Generally light in color, varying from almost white to grayish-white.
No allowable unit stress provided for	Black cottonwood‖	Black cottonwood	B Cot	Characteristics similar to those of northern aspen group, but it is lower in strength and stiffness.
E	Northern species*	Any species included in this table except those in the northern aspen and black cottonwood species groups	North Species	See characteristics of previous groups.
D	Coast species**	Douglas-fir Western larch Western hemlock Amabilis fir Coast Sitka spruce	Coast Species	See characteristics within previous groups that these species are listed in.

*Strength group determines allowable stresses.
†Botanical names and further information on species listed are given in Chapter 2, 'Commercial Woods.'
‡Designation 'North,' or 'N,' in the grade mark provides regional identification for lumber exported to the United States.
§Alternative abbreviation for eastern white pine.
‖Northern aspen species group and black cottonwood are technically hardwoods, but they are graded and marketed under softwood standards.
**Allowable stresses for northern species and coast species are based on weakest species in combination for each property.

The coastal mills in British Columbia kiln-dry about 13% of their production, and the rest is shipped unseasoned, mostly by boat, where weight is not a limiting factor. The interior production is 72% kiln-dried, and an unknown volume is air-dried. The shipments from the interior are mostly by rail, where weight is important.

Of the total Quebec production in 1978, approximately 40% went to Ontario, 31% to the United States, and 2% to overseas markets.

Lumber handling has been simplified by packaging, and now most lumber is shipped in standard-size packages; each package usually contains one size and length and consists of one grade or a grade combination that is marketed together.

When unseasoned lumber is packaged, it is antistain treated to prevent development of mold and sap stain. Packaged dry lumber is usually covered with some type of protective wrapping.

Table 6.2 *Stress grades and uses in Canada*

Lumber product	Grade category	Metric size	Nominal size (in.)	Grade	Principal uses
Dimension lumber	Light framing	38–89 mm thick 38–114 mm wide	2–4 in. thick 2–5 in. wide	Construction Standard	Widely used for general framing purposes. Pieces are of good appearance but graded primarily for strength and serviceability.
				Utility	Widely used where a combination of good strength and economical construction is desired for such purposes as studding, blocking, plates, and bracing.
				Economy*	Temporary or low-cost construction where strength and appearance are not important.
	Structural light framing	38–89 mm thick 38–114 mm wide	2–4 in. thick 2–5 in. wide	Select Structural No. 1	Intended primarily for use where high strength, stiffness, and good appearance are desired.
				No. 2	For most general construction uses.
				No. 3	Appropriate for use in general construction where appearance is not a factor.
	Stud	38–89 mm thick 38–114 mm wide	2–4 in. thick 2–5 in. wide	Stud	Special-purpose grade intended for all stud uses.
				Economy Stud*	Temporary or low-cost construction where strength and appearance are not important.
	Structural joists and planks	38–89 mm thick 140 mm and wider	2–4 in. thick, 6 in. and wider	Select Structural No. 1	Intended primarily for use where high strength, stiffness, and good appearance are desired.
				No. 2	For most general construction uses.
				No. 3	Appropriate for use in general construction where appearance is not a factor.
				Economy*	Temporary or low-cost construction where strength and appearance are not important.
	Appearance	38–89 mm thick 38 mm and wider	2–4 in. thick 2 in. and wider	Appearance	Intended for use in general housing and light construction where lumber permitting knots, but of high strength and fine appearance, is desired.

Table 6.2 *(concluded)*

Lumber product	Grade category	Metric size	Nominal size (in.)	Grade	Principal uses
Decking	Decking	38–89 mm thick 140 mm and wider	2–4 in. thick 6 in. and wider	Select	For roof and floor decking where strength and fine appearance are required.
				Commercial	For roof and floor decking where strength is required but appearance is not so important.
Timber	Beams and stringers	114 mm and thicker width more than 38 mm greater than thickness	5 in. and thicker width more than 2 in. greater than thickness	Select Structural No. 1	For use as heavy beams in buildings, bridges, docks, warehouses, and heavy construction where superior strength is required.
				Standard* Utility*	For use in rough, general construction.
	Posts and timbers	114 × 114 mm and larger, width not more than 38 mm greater than thickness	5 in. × 5 in. and larger, width not more than 2 in. greater than thickness	Select Structural No. 1	For use as columns, posts, and struts in heavy construction such as warehouses, docks, and other large structures where superior strength is required.
				Standard* Utility*	For use in rough, general construction.

Note: All grades are 'stress graded,' meaning that working stresses have been assigned (and span tables calculated for dimension lumber) except those marked with an asterisk.

Table 6.3 *Markets for British Columbia lumber in 1978*

	Cubic metres (million)	Board feet (million)	Percentage
Canada	5.37	2 273	18
United States	19.95	8 452	68
W. Europe (excluding United Kingdom)	0.73	310	3
United Kingdom	0.96	408	3
Japan	1.85	785	6
Other countries	0.72	304	2
Total	29.58	12 532	100

Table 6.4 *Metric sizes for dimension lumber and boards*

Nominal size (in.)	Actual size (in.) Dry	Actual size (in.) Green	Metric equivalent (mm) Dry	Metric equivalent (mm) Green	Metric nomenclature (mm)
DIMENSION LUMBER					
2 × 2	1 1/2 × 1 1/2	1 9/16 × 1 9/16	38.10 × 38.10	36.69 × 39.69	38 × 38
3	2 1/2	2 9/16	63.50	65.09	64
4	3 1/2	3 9/16	88.90	90.49	89
5	4 1/2	4 5/8	114.30	117.47	114
6	5 1/2	5 5/8	139.70	142.87	140
8	7 1/4	7 1/2	184.15	190.50	184
10	9 1/4	9 1/2	234.95	241.30	235
12	11 1/4	11 1/2	285.75	292.10	286
14	13 1/4	13 1/2	336.55	342.90	337
16	15 1/4	15 1/2	387.35	393.70	387
3 × 4, etc.	2 1/2 × 3 1/2	2 9/16 × 3 9/16	63.50 × 88.90	65.09 × 90.49	64 × 89
4 × 4, etc.	3 1/2 × 3 1/2	3 9/16 × 3 9/16	88.90 × 88.90	90.49 × 90.49	89 × 89
BOARDS					
1 × 2	3/4 × 1 1/2	25/32 × 1 9/16	19.05 × 38.10	19.84 × 39.69	19 × 38
3	2 1/2	2 9/16	63.50	65.09	64
4	3 1/2	3 9/16	88.90	90.49	89
5	4 1/2	4 5/8	114.30	117.47	114
6	5 1/2	5 5/8	139.70	142.87	140
8	7 1/4	7 1/2	184.15	190.50	184
10	9 1/4	9 1/2	234.95	241.30	235
12	11 1/4	11 1/2	285.75	292.10	286
14	13 1/4	13 1/2	336.55	342.90	337
16	15 1/4	15 1/2	387.35	393.70	387
1 1/4 × 2, etc.	1 × 1 1/2	11/32 × 1 9/16	25.40 × 38.10	26.19 × 39.69	25 × 38
1 1/2 × 2, etc.	1 1/4 × 1 1/2	19/32 × 1 9/16	31.75 × 38.10	32.54 × 39.69	25 × 38

Source: Canadian Wood Council

Table 6.5 *Organizations involved in softwood lumber production and lumber grade stamping in Canada*

Alberta Forest Products Association
204 – 11710 Kingsway Avenue
Edmonton, Alberta

Canadian Lumbermen's Association
27 Goulburn Avenue
Ottawa, Ontario K1N 8C7

Cariboo Lumber Manufacturers' Association
301 – 197 Second Avenue North
Williams Lake, British Columbia V2G 1Z5

Central Forest Products Association, Inc.
14-G 1975 Corydon Avenue
Winnipeg, Manitoba R3P 0R1

Council of Forest Industries of British Columbia
1500 – 1055 West Hastings Street
Vancouver, British Columbia V6E 2H1

Council of Forest Industries of British Columbia
Northern Interior Lumber Sector
514 – 550 Victoria Street
Prince George, British Columbia V2L 2K1

Interior Lumber Manufacturers' Association
295 – 333 Martin Street
Penticton, British Columbia V2A 5K7

Maritime Lumber Bureau
PO Box 459
Amherst, Nova Scotia B4H 4A1

Ontario Lumber Manufacturers' Association
159 Bay Street, Suite 414
Toronto, Ontario M5J 1J7

Saskatchewan Forest Products Corporation
101 First Avenue East
Prince Albert, Saskatchewan S6V 2A6

Quebec Lumber Manufacturers' Association
580 Est, Grande-Allee, Suite 540
Quebec, Quebec G1R 2K2

National Lumber Grades Authority
1055 West Hastings Street
Vancouver, British Columbia V6E 2H1

MacDonald Inspection
125 East 4th Avenue
Vancouver, British Columbia V5T 1G4

Pacific Lumber Inspection Bureau
BC Division
1460 – 1055 West Hastings Street
Vancouver, British Columbia V6E 2G8

NWT Grade Stamping Agency
PO Box 2157
Yellowknife, Northwest Territories X0E 1H0

Canadian Lumber Standards Administrative Board
Office of the Manager
1475 – 1055 West Hastings Street
Vancouver, British Columbia V6E 2E9

REFERENCES

Brown, N.C., and Bethel, J.S. 1958. *Lumber*. 2nd ed. New York: Wiley.
Dobie, J. 1967. How chipper headrigs reduce small log processing costs. *Can. For. Ind.* 87 (8): 60–5
– 1970., Advantages of log sorting for chipper headrigs. *For. Prod. J.* 20 (1): 19–24
Dobie, J., and McBride, C.F. 1964. Lumber and pulp chips from small logs – A new method. *B.C. Lumberman*, Sept. 1964
Dobie, J., and Sturgeon, W.J. 1972. An assessment of the economic benefits of double cut headrigs. *For. Prod. J.* 22 (2): 22–4
Flann, I.B. 1978. *Live sawing hardwoods! Can it mean dollars for you?* East. For. Prod. Lab. Inf. Rep. OP-X-198E. Ottawa.
Koch, P. 1972. *Utilization of southern pine*, Vol. 2. Washington, DC: U.S. Dep. Agric.
McLauchlan, T.A. 1972. Recent developments in circular rip sawing. *For. Prod. J.* 22 (6): 42–8
Petro, F.J. 1971. *Felling and bucking hardwoods: How to improve your profit*. Can. For. Serv. Publ. 1291. Ottawa
Quelch, P.S. 1970. *Armstrong saw filer's handbook*. 2nd ed. (rev.) Portland, Oreg.: Armstrong Mfg.
Williston, E.M. 1976. *Lumber manufacturing: Design and operational sawmills and planer mills*. San Francisco: Miller Freeman

BIBLIOGRAPHY

Aune, J.E., and Lefebvre, E.L. 1975. *Small-log sawmill systems in western Canada*. West. For. Prod. Lab. Inf. Rep. VP-X-141. Vancouver
Bousquet, D.W., and Flann, I.B. 1975. Hardwood sawmill productivity for live and around savings. *For. Prod. J.* 25 (7): 32–7
Calvert, W.W., and Garlicki, A.M. 1975. Improving the quality of winter barking. *Can. For. Ind.* 95 (3): 52–4
Canadian Wood Council. 1979. *Metric manual for wood products*. Ottawa
Dobie, J. 1968. *A comparison of productivity with small logs in various types of sawmills*. Proc. of High Speed Headrig Conf. Nov. 12–15. College of Forestry. Syracuse: State Univ. of NY
– 1973. Economics of scale and trends in sawmill capacity in British Columbia. *For. Chron.* 49 (2): 79–82
Flann, I.B. 1978. *Short-log processing – 1977*. East For. Prod. Lab. Inf. Rep. OP-X-207E. Ottawa
Flann, I.B., and Bousquet, D.W. 1974. *Sawing live vs. around for hard maple logs*. East. For. Prod. Lab. Inf. Rep. OP-X-91E. Ottawa
Huey, B.M. 1969. Advance in precision sawmilling (Thrasher thin saws). *Crows For. Prod. Dig.* 47 (11): 16
Johnston, J.S., and St. Laurent, A. 1975. *Performance of removable cutter circular saw*. East. For. Prod. Lab. Inf. Rep. OP-X-159E. Ottawa
Jones, P.D., and Simons, E.N. 1961. *Story of the saw: Spear and Jackson Ltd. 1769–1960*. Manchester: Newman Neame
Kerbes, E.L. 1969. Narrow kerf, accurate sawing; key to higher lumber recovery. *Can. For. Ind.* 89 (3): 76–9
National Lumber Grade Authority (NLGA). 1977. *Canadian lumber grading manual*. Vancouver
– 1979. *Standard lumber grading rules for Canadian lumber*. Vancouver
Petro, F.J., and Calvert W.W. 1976. *How to grade hardwood logs for factory lumber*. Dep. Fish. Environ. For Tech. Rep. 6. Ottawa
Pnevmaticos, S.M.; Flann, I.B.; and Petro, F.J. 1971. How log characteristics relate to sawing profit. *Can. For. Indus.* Jan. 1971
White, V.S. 1973. *Modern sawmill techniques*. Proc. First Sawmill Clinic, Portland, Oreg. San Francisco: Miller Freeman

Figure 7.1 A large commercial kiln in western Canada.

7

The Drying of Wood

GEORGE BRAMHALL
Retired, formerly with Western Forest Products Laboratory, Vancouver

INTRODUCTION

The Purpose of Drying Wood

Green lumber, as it comes from the tree, contains a high proportion of water, ranging from 30% moisture content in the heartwood of some species to over 300% in the sapwood of some low-density species. The greater part of this moisture must be removed before the lumber is used or less than satisfactory service will result.

Wood is a hygroscopic material, which means that it absorbs or gives off moisture in response to humidity changes in the surrounding atmosphere. It has a natural propensity to swell and shrink as its moisture content increases and decreases in the range from oven-dry to 30% (Figure 7.2). The most important reason for drying lumber is to minimize these dimensional changes after the wood is placed in service. Where dimensional stability is especially important, as in furniture and other high-quality products, wood must be dried to, and be kept at, the moisture content it will attain in service, both before being cut to final size and during the remainder of the manufacturing process.

Lumber for general construction is dried to a moisture content of 20% or less, which provides protection against wood-staining molds and decay-producing fungi. This topic is discussed further in 'Decay and Stain in Wood,' Chapter 8.

Wood products such as railway ties, fence posts, and utility poles, which are used under conditions favoring decay, must be dried to the lowest moisture content required in service before being treated with preservatives. Proper drying induces maximum shrinkage and checking, which permits treatment of all the wood that may later be exposed to decay-producing fungi.

Essentially all the free (liquid) water must be removed from the cell lumens of lumber or wood products intended for fire-retardant treatment. Removal of free water permits unimpeded entry of the chemicals used in these treatments.

Veneer and lumber to be glued into plywood or glued-laminated beams must be dried to a uniform moisture content in the range of 3–12%, depending on the adhesive and the curing temperature chosen. Wood should also be at a suitably low moisture content before paints and other finishes are applied.

Because drying reduces the weight of wood, lumber to be shipped by railroad or by truck is dried at the mill to reduce shipping costs. Transportation costs of lumber shipped by water, are based on volume rather than weight, and therefore lumber shipped by this means is not usually dried.

The moisture contents required for the efficient manufacture and service of various wood commodities are shown in Figure 7.3.

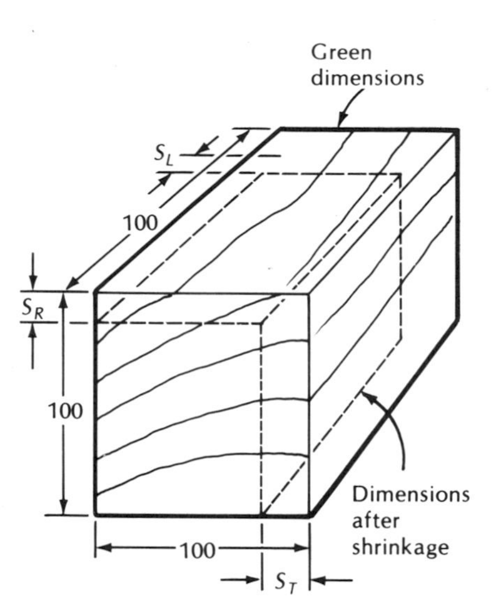

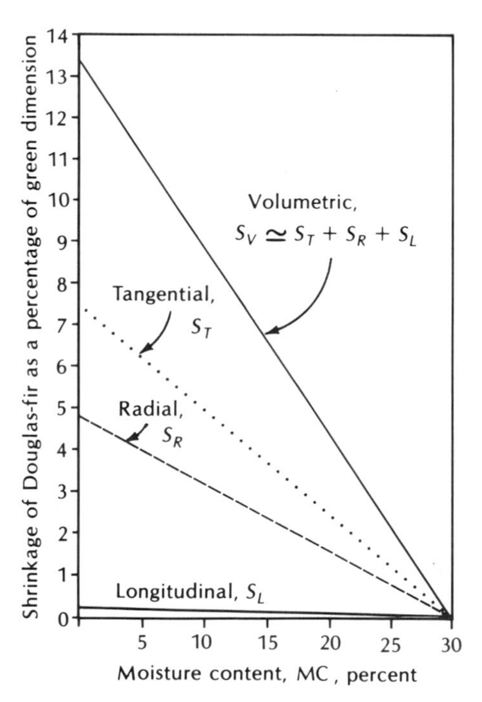

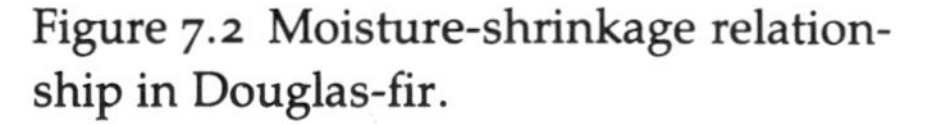
Figure 7.2 Moisture-shrinkage relationship in Douglas-fir.

National Standards

The National Lumber Grading Association 1979 Standard Grading Rules for Canadian Lumber define softwood lumber as dry when its moisture content is 19% or less. To allow for the normal range of moisture contents that may be found in any shipment, up to 5% may fail to conform to this definition; this allowance assumes that the lumber has no other nonconformities (such as being below grade or off size), because the total allowable degrade is 5%. Moisture contents may be determined by oven drying or by electrical moisture meters measuring resistance, power loss, or capacitance. These methods are discussed later in this chapter.

THEORY

Anatomical Considerations

As outlined in Chapter 3, the transportation route for water in the living tree is vertically through the sapwood cell lumens and from cell to cell through the pits connecting the cells. In softwoods, these vertically oriented cells are called tracheids. In addition, a radially oriented system of cells called rays permits communication between the circumference of the tree and points within the stem. In the conversion of sapwood to heartwood in some softwood species, many of the pits and cell cavities become obstructed, reducing permeability to very low values.

The anatomy of hardwoods is considerably more complex than that of softwoods. As in softwoods, however, each hardwood cell consists of a cell wall and lumen, and a system of rays permits radial communication.

Moisture in Wood

Moisture in wood exists in three principal forms: free water, which is liquid water within the cell lumen; water vapor, which, together with air, occupies that part of the cell lumen not occupied by free water; and bound water, which is adsorbed in the cell wall.

Free water is held primarily by capillary forces and is relatively easy to move. Bound water is attached to the cellulose molecules with much stronger molecular forces, termed hydrogen bonds, and requires more

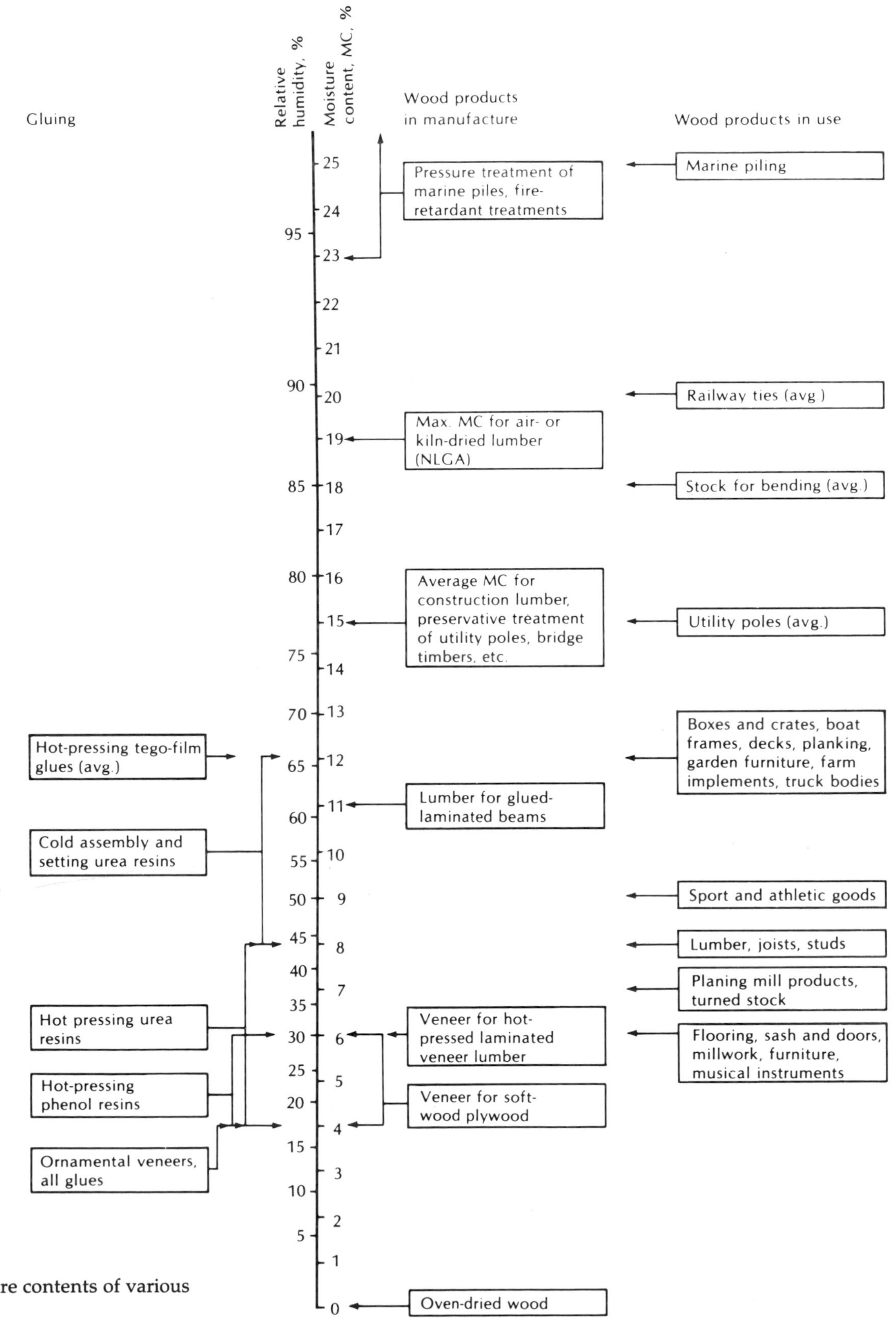

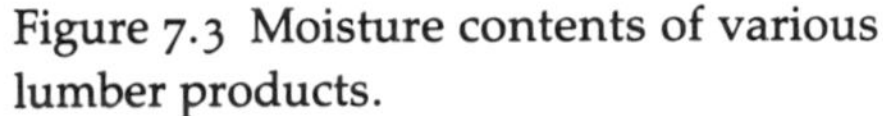
Figure 7.3 Moisture contents of various lumber products.

energy to remove. As a result, when wood is dried, the first water to evaporate is the free water. When all the free water from the cell lumen has evaporated but the cell walls themselves are still fully saturated, the wood has reached the fiber saturation point. At this point the wood usually has a moisture content of 30%, although it can vary from 24% to over 30%, depending on the species.

Equilibrium Moisture Content

Wood left exposed to air at a constant relative humidity and temperature will eventually attain a constant moisture content. This moisture content, which is at equilibrium with its surrounding atmosphere, is known as the equilibrium moisture content. Although the value of the equilibrium moisture content varies with both humidity and temperature, it is affected most by humidity.

Figure 7.4 shows the equilibrium moisture content of 25 mm (1 in.) thick material of various species exposed to outdoor conditions in a covered open shed in Montreal for a period of one year. Since climatic conditions vary across the country, the average equilibrium moisture content of lumber differs with locality. Some typical calculated values of the average equilibrium moisture content of interior woodwork in different cities during the months of June and January are given in Table 7.1; these values exclude other sources of moisture vapor in the building.

Table 7.1 *Estimated average equilibrium moisture content of interior woodwork*

City	Equilibrium moisture content* (%)	
	January	June
Vancouver, BC	6.0	11.2
Prince George, BC	3.7	9.5
Calgary, Alta.	3.0	9.6
Edmonton, Alta.	2.0	7.8
Regina, Sask.	2.1	10.5
Winnipeg, Man.	2.0	11.0
Toronto, Ont.	3.0	9.5
Montreal, Que.	2.5	9.8
Quebec, Que.	2.7	12.7
Saint John, NB	3.7	14.0
St John's, Nfld.	3.5	7.6
Aklavik, NWT		8.9

*Calculations are based on relative humidities and temperatures from meteorological data, adjusted to 20°C (70°F). Equilibrium moisture contents were then taken from standard tables.

Hysteresis

The equilibrium moisture content also depends on whether the wood is being dried (desorbed) or wetted (adsorbed) to reach equilibrium. For most conditions, the moisture content reached during adsorption is about 84% of the value reached in desorption for the same humidity and temperature. This phenomenon is referred to as hysteresis (Stamm 1964) and is illustrated in Figure 7.5.

How Wood Dries

Wood dries as a result of movement of moisture to the surface, where it evaporates and escapes to the surrounding atmosphere. Several processes are involved.

Heat Requirement

Approximately 2.25 megajoules (MJ) of heat must be supplied to evaporate one kilogram of water (about 970 BTU per pound). This is called the latent heat of evaporation. The free water in wood requires essentially the same amount of heat per kilogram (or per pound) to convert it to vapor. Below the fiber saturation point additional heat energy is required to free the bound water, but this heat of wetting can be ignored for most purposes.

Vapor Pressure

An understanding of the term vapor pressure is necessary in order to comprehend the drying of lumber. Vapor pressure determines the rate of moisture movement and therefore the rate of wood drying.

If a closed container without air is partly filled with water, the liquid will evaporate and vapor will occupy the space above the liquid. When

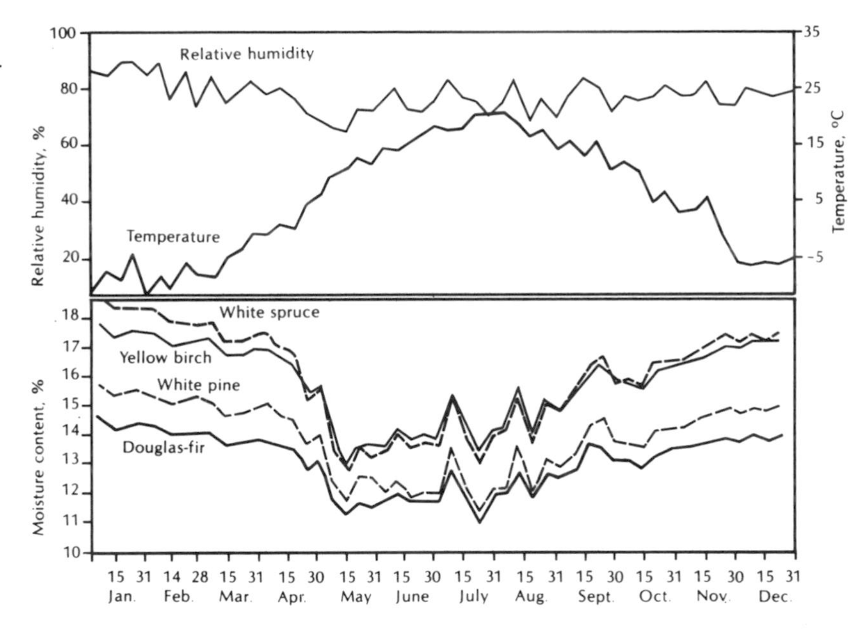

Figure 7.4 Equilibrium moisture content of lumber exposed to outdoor conditions at Montreal.

the space becomes saturated with water vapor, the pressure in the container will be the saturated vapor pressure of water at that temperature. A decrease in temperature will cause some vapor to condense, and a lower vapor pressure will result. An increase in temperature will evaporate more water, and the vapor pressure will increase. For each temperature there is a specific value of saturated vapor pressure; the values are listed in Tables 7.2 and 7.3.

Evaporation

The atmosphere in which wood dries has two functions: the transport of heat to the wood to evaporate moisture and the transport of moisture away from the wood. As the air picks up moisture by evaporation, it becomes cooler and more saturated, until finally evaporation ceases. Unless drying air is continually replenished, it cannot continue to dry the wood.

The rate of evaporation from a wood surface is proportional to the difference in vapor pressure between the wood surface and the kiln atmosphere. Increasing the wood temperature increases the surface vapor pressure and thus increases the drying rate.

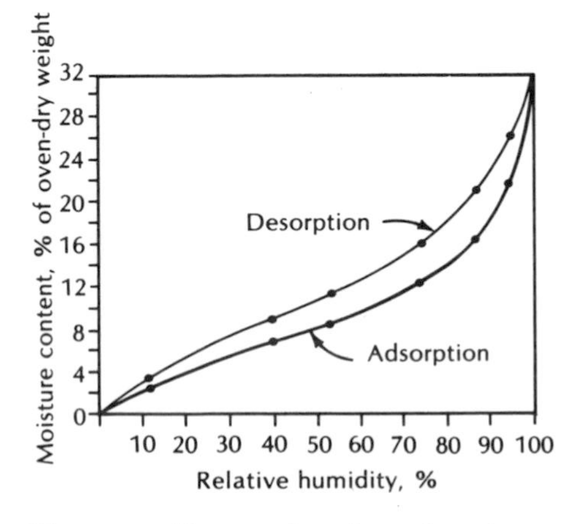

Figure 7.5 Hysteresis: adsorption-desorption isotherms for spruces at 25°C (77°F). (Adapted from A.J. Stamm [1964])

Free-Water Movement

As moisture evaporates from green wood, the liquid surface recedes into the pits and forms spherical depressions known as menisci. Further evaporation causes the menisci to exert a tensile force on the water beneath the surface and pull the water from the interior of the wood. This capillary flow continues to the surface as long as the rate of flow can equal the rate of loss of water by evaporation. For further detail, see Kemp (1959), Cech and Huffman (1968), and Skaar (1972).

Because the contribution of capillary flow to the drying process depends on the permeability of the wood, it varies considerably between species and between heartwood and sapwood of the same

Table 7.2 *Relative humidity and equilibrium moisture content table* and saturated vapor pressure for use with dry-bulb temperature and wet-bulb depression* (°C)

Temp., dry-bulb (°C)	VP (kPa)	Difference between wet- and dry-bulb temperatures in degrees Celsius														
		0.5	1.0	1.5	2.0	2.5	3.0	3.5	4.0	4.5	5.0	6.0	7.0	8.0	9.0	10.0
0	0.6	91	82	73	64	56	47	39	31	22	14					
		20.5	*17.0*	*14.3*	*12.2*	*10.5*	*9.0*	*7.6*	*6.3*	*4.9*	*3.5*					
5	0.9	93	86	79	72	65	58	51	45	38	32	19	6			
		21.6	*18.4*	*15.9*	*14.0*	*12.3*	*10.9*	*9.7*	*8.5*	*7.5*	*6.5*	*4.4*	*1.9*			
10	1.2	94	88	82	76	71	65	60	54	49	44	33	23	14	4	
		22.4	*19.5*	*17.2*	*15.3*	*13.7*	*12.4*	*11.2*	*10.2*	*9.3*	*8.4*	*6.8*	*5.2*	*3.5*	*1.3*	
15	1.7	95	90	85	80	75	70	66	61	57	52	44	35	27	19	12
		23.0	*20.3*	*18.1*	*16.3*	*14.8*	*13.6*	*12.4*	*11.5*	*10.6*	*9.8*	*8.3*	*7.0*	*5.8*	*4.5*	*3.0*
20	2.3	96	91	87	83	78	74	70	66	62	59	51	44	37	30	24
		23.4	*20.9*	*18.8*	*17.1*	*15.7*	*14.4*	*13.4*	*12.4*	*11.5*	*10.8*	*9.4*	*8.2*	*7.2*	*6.2*	*5.2*
25	3.2	96	92	88	84	81	77	74	70	67	63	57	50	44	38	33
		23.6	*21.2*	*19.3*	*17.7*	*16.3*	*15.1*	*14.0*	*13.1*	*12.3*	*11.5*	*10.2*	*9.1*	*8.1*	*7.2*	*6.4*
30	4.2	96	93	89	86	83	79	76	83	70	67	61	55	50	44	39
		23.7	*21.5*	*19.6*	*18.0*	*16.7*	*15.5*	*14.5*	*13.6*	*12.8*	*12.1*	*10.8*	*9.7*	*8.8*	*8.0*	*7.2*
35	5.6	97	93	90	87	84	81	78	75	72	69	64	59	54	49	44
		23.7	*21.5*	*19.8*	*18.2*	*16.9*	*15.8*	*14.8*	*13.9*	*13.1*	*12.5*	*11.2*	*10.2*	*9.3*	*8.5*	*7.8*
40	7.4	97	94	91	88	85	82	80	77	74	72	67	62	57	53	48
		23.6	*21.5*	*19.8*	*18.3*	*17.1*	*16.0*	*15.0*	*14.1*	*13.4*	*12.7*	*11.5*	*10.5*	*9.6*	*8.8*	*8.1*
45	9.6	97	94	91	89	86	83	81	78	76	73	69	64	60	56	52
		23.4	*21.4*	*19.7*	*18.3*	*19.1*	*16.0*	*15.1*	*14.2*	*13.5*	*12.8*	*11.6*	*10.6*	*9.8*	*9.0*	*8.4*
50	12.3	97	95	92	89	87	84	82	80	77	75	71	66	62	58	54
		23.1	*21.2*	*19.6*	*18.2*	*17.0*	*16.0*	*15.0*	*14.2*	*13.5*	*12.8*	*11.7*	*10.7*	*9.9*	*9.2*	*8.5*
55	15.8	97	95	92	90	88	85	83	81	78	76	72	68	64	60	57
		22.8	*20.9*	*19.4*	*18.0*	*16.9*	*15.8*	*15.0*	*14.2*	*13.4*	*12.8*	*11.7*	*10.7*	*9.9*	*9.2*	*8.6*
60	19.9	98	95	93	90	88	86	84	82	79	77	73	69	66	62	59
		22.4	*20.6*	*19.1*	*17.8*	*16.7*	*15.7*	*14.8*	*14.0*	*13.3*	*12.7*	*11.6*	*10.7*	*9.9*	*9.2*	*8.6*
65	25.0	98	95	93	91	89	87	84	82	80	78	74	71	67	64	60
		22.0	*20.3*	*18.8*	*17.5*	*16.4*	*15.5*	*14.6*	*13.9*	*13.2*	*12.6*	*11.5*	*10.6*	*9.8*	*9.1*	*8.5*
70	31.2	98	96	93	91	89	87	85	83	81	79	75	72	68	65	62
		21.6	*19.9*	*18.5*	*17.2*	*16.2*	*15.2*	*14.4*	*13.6*	*13.0*	*12.4*	*11.3*	*10.4*	*9.7*	*9.0*	*8.5*
75	38.6	98	96	94	92	90	88	86	84	82	80	76	73	70	66	63
		21.1	*19.5*	*18.1*	*16.9*	*15.9*	*15.0*	*14.1*	*13.4*	*12.9*	*12.2*	*11.1*	*10.3*	*9.5*	*8.9*	*8.4*
80	47.4	98	96	94	92	90	88	86	84	82	82	77	74	71	67	64
		20.6	*19.1*	*17.7*	*16.6*	*15.5*	*14.7*	*13.9*	*13.2*	*12.5*	*11.9*	*10.9*	*10.1*	*9.4*	*8.8*	*8.2*
85	57.8	98	96	94	92	90	88	87	85	83	81	78	75	72	68	66
		20.1	*18.6*	*17.3*	*16.2*	*15.2*	*14.3*	*13.6*	*12.9*	*12.3*	*11.7*	*10.7*	*9.9*	*9.2*	*8.6*	*8.1*
90	70.1	98	96	94	92	91	89	87	85	84	82	79	76	72	69	67
		19.6	*18.1*	*16.9*	*15.8*	*14.8*	*14.0*	*13.3*	*12.6*	*12.0*	*11.4*	*10.5*	*9.7*	*9.0*	*8.4*	*7.9*
95	84.5	98	96	94	93	91	89	99	86	84	83	79	76	73	70	68
		19.1	*17.7*	*16.4*	*15.4*	*14.5*	*13.7*	*12.9*	*12.3*	*11.7*	*11.2*	*10.3*	*9.5*	*8.8*	*8.2*	*7.7*
100	101.3	98	96	95	93	91	90	88	86	85	83	80	77	74	71	69
		18.5	*17.2*	*16.0*	*15.0*	*14.1*	*13.3*	*12.6*	*12.0*	*11.4*	*10.9*	*10.0*	*9.2*	*8.6*	*8.0*	*7.5*
105	120.8	98	97	95	93	92	90	88	87	84	84	81	78	75	72	69
		18.0	*16.7*	*15.5*	*14.6*	*13.7*	*12.9*	*12.3*	*11.7*	*11.1*	*10.6*	*9.7*	*9.0*	*8.4*	*7.8*	*7.3*
110	143.2	98	97	95	93	92	90	89	87	86	84	81	78	76	73	70
		17.4	*16.2*	*15.1*	*14.1*	*13.3*	*12.6*	*11.9*	*11.3*	*10.8*	*10.3*	*9.5*	*8.7*	*8.1*	*7.6*	*7.1*
115	169.0	98	97	95	94	92	90	89	87	86	85	82	79	76	74	71
		16.8	*15.6*	*14.6*	*13.7*	*12.9*	*12.2*	*11.6*	*11.0*	*10.5*	*10.0*	*9.2*	*8.0*	*7.9*	*7.4*	*6.9*

Table 7.2 *(concluded)*

Temp., dry-bulb (°C)	VP (kPa)	Difference between wet- and dry-bulb temperatures in degrees Celsius														
		11.0	12.0	13.0	14.0	15.0	16.0	17.0	18.0	19.0	20.0	22.0	24.0	26.0	28.0	30.0
0	0.6															
5	0.9															
10	1.2															
15	1.7	4 *1.3*														
20	2.3	17 *4.1*	11 *2.9*	5 *1.6*												
25	3.2	27 *5.6*	22 *4.8*	16 *3.9*	11 *3.0*	7 *1.9*										
30	4.2	34 *6.5*	29 *5.8*	25 *5.2*	16 *4.5*	12 *3.8*	12 *3.0*	8 *2.2*	4 *1.3*							
35	5.6	40 *7.1*	36 *6.5*	31 *5.9*	27 *5.4*	23 *4.8*	20 *4.3*	16 *3.7*	13 *3.1*	9 *2.4*	6 *1.7*					
40	7.4	44 *7.5*	40 *7.0*	36 *6.4*	33 *5.9*	29 *5.5*	26 *5.0*	22 *4.5*	19 *4.1*	16 *3.6*	13 *3.1*	8 *2.0*	2 *0.7*			
45	9.6	48 *7.8*	44 *7.3*	40 *6.8*	37 *6.3*	34 *5.9*	30 *5.5*	27 *5.1*	24 *4.7*	22 *4.3*	19 *3.9*	14 *3.1*	9 *2.2*	4 *1.2*		
50	12.3	51 *8.0*	47 *7.5*	44 *7.0*	40 *6.5*	37 *6.1*	34 *5.8*	31 *5.4*	29 *5.0*	26 *4.7*	23 *4.4*	18 *3.7*	14 *3.0*	10 *2.3*	6 *1.5*	2 *0.6*
55	15.8	53 *8.0*	50 *7.5*	47 *7.1*	43 *6.7*	40 *6.3*	38 *5.9*	35 *5.6*	32 *5.3*	30 *5.0*	27 *4.7*	22 *4.1*	18 *3.5*	14 *2.9*	19 *2.3*	7 *1.7*
60	19.9	55 *8.1*	52 *7.6*	49 *7.0*	46 *6.7*	43 *6.4*	40 *6.0*	38 *5.7*	35 *5.4*	33 *5.1*	30 *4.8*	26 *4.3*	22 *3.8*	18 *3.3*	14 *2.8*	11 *2.3*
65	25.0	57 *8.0*	54 *7.6*	51 *7.1*	48 *6.8*	45 *6.4*	43 *6.1*	40 *5.8*	38 *5.5*	35 *5.2*	33 *4.9*	29 *4.4*	25 *4.0*	21 *3.5*	17 *3.1*	14 *2.7*
70	31.2	59 *8.0*	56 *7.5*	53 *7.1*	50 *6.7*	47 *6.4*	45 *6.1*	42 *5.8*	40 *5.5*	37 *5.2*	35 *5.0*	31 *4.5*	27 *4.1*	23 *3.7*	20 *3.3*	17 *2.9*
75	38.6	60 *7.9*	57 *7.4*	54 *7.0*	52 *6.6*	49 *6.3*	47 *6.0*	44 *5.7*	42 *5.5*	39 *5.2*	37 *5.0*	33 *4.5*	29 *4.1*	26 *3.8*	23 *3.4*	19 *3.1*
80	47.4	61 *7.7*	59 *7.3*	56 *6.9*	53 *6.6*	51 *6.2*	48 *5.9*	46 *5.7*	44 *5.4*	41 *5.2*	39 *4.9*	35 *4.5*	31 *4.1*	28 *3.8*	25 *3.5*	22 *3.1*
85	57.8	63 *7.6*	60 *7.2*	57 *6.8*	55 *6.4*	52 *6.1*	50 *5.8*	47 *5.6*	45 *5.3*	43 *5.1*	41 *4.9*	37 *4.5*	33 *4.1*	30 *3.8*	27 *3.5*	24 *3.2*
90	70.1	64 *7.4*	61 *7.0*	59 *6.7*	56 *6.3*	54 *6.0*	51 *5.7*	49 *5.5*	47 *5.2*	45 *5.0*	43 *4.8*	39 *4.4*	35 *4.1*	32 *3.7*	28 *3.4*	25 *3.2*
95	84.5	65 *7.3*	62 *6.9*	60 *6.5*	57 *6.2*	55 *5.9*	53 *5.6*	50 *5.4*	48 *5.1*	46 *4.9*	44 *4.7*	40 *4.3*	37 *4.0*	33 *3.7*	30 *3.4*	27 *3.1*
100	101.3	66 *7.1*	63 *6.7*	61 *6.3*	58 *6.0*	56 *5.7*	54 *5.5*	52 *5.2*	49 *5.0*	47 *4.8*	45 *4.6*	42 *4.3*	38 *3.9*	35 *3.6*	32 *3.3*	29 *3.1*
105	120.8	67 *6.9*	64 *6.5*	62 *6.2*	60 *5.9*	57 *5.6*	55 *5.3*	53 *5.1*	51 *4.9*	49 *4.7*	47 *4.5*	43 *4.1*	39 *3.8*	36 *3.5*	33 *3.3*	30 *3.0*
110	143.2	68 *6.7*	65 *6.3*	63 *6.0*	61 *5.7*	58 *5.4*	56 *5.2*	54 *4.9*	52 *4.7*	50 *4.5*	48 *4.3*	44 *4.0*	41 *3.7*	38 *3.4*	34 *3.2*	32 *3.0*
115	169.0	69 *6.5*	66 *6.2*	64 *5.8*	62 *5.5*	59 *5.3*	57 *5.0*	55 *4.8*	53 *4.6*	51 *4.4*	49 *4.2*	46 *3.9*	42 *3.6*	39 *3.3*	36 *3.1*	33 *2.9*

Source: Calculated from Hailwood-Horrobin single hydrate equation using parameters determined by Simpson (1973).
*Relative humidity values in roman type; equilibrium moisture content values in italics.

Table 7.3 *Relative humidity and equilibrium moisture content table* for use with dry-bulb temperatures and wet-bulb depressions* (°F)

Temp., dry-bulb (°F)	Difference between wet- and dry-bulb temperatures in degrees Fahrenheit														
	1	2	3	4	5	6	7	8	9	10	11	12	13	14	15
30	89	78	67	57	46	36	27	17	6						
		15.9	*12.9*	*10.8*	*9.0*	*7.4*	*5.7*	*3.9*	*1.6*						
35	90	81	72	63	54	45	37	28	19	11	3				
		16.8	*13.9*	*11.9*	*10.3*	*8.8*	*7.4*	*6.0*	*4.5*	*2.9*	*0.8*				
40	92	83	75	68	60	52	45	37	29	22	15	8			
		17.6	*14.8*	*12.9*	*11.2*	*9.9*	*8.6*	*7.4*	*6.2*	*5.0*	*3.5*	*1.9*			
45	93	85	78	72	64	58	51	44	37	31	25	19	12	6	
		18.3	*15.6*	*13.7*	*12.0*	*10.7*	*9.5*	*8.5*	*7.5*	*6.5*	*5.3*	*4.2*	*2.9*	*1.5*	
50	93	86	80	74	68	62	56	50	44	38	32	27	21	16	10
		19.0	*16.3*	*14.4*	*12.7*	*11.5*	*10.3*	*9.4*	*8.5*	*7.6*	*6.7*	*5.7*	*4.8*	*3.9*	*2.8*
55	94	88	82	76	70	65	60	54	49	44	39	31	28	24	19
		19.5	*16.9*	*15.1*	*13.4*	*12.2*	*11.0*	*10.1*	*9.3*	*8.4*	*7.6*	*6.8*	*6.0*	*5.3*	*4.5*
60	94	89	83	78	73	68	62	58	53	48	43	39	34	30	26
		19.9	*17.4*	*15.6*	*13.9*	*12.7*	*11.6*	*10.7*	*9.9*	*9.1*	*8.3*	*7.6*	*6.9*	*6.3*	*5.6*
65	95	90	84	80	75	70	66	61	56	52	48	44	39	36	32
		20.3	*17.8*	*16.1*	*14.4*	*13.3*	*12.1*	*11.2*	*10.4*	*9.7*	*8.9*	*8.3*	*7.7*	*7.1*	*6.5*
70	95	90	86	81	77	72	68	64	59	55	51	48	44	40	36
		20.6	*18.2*	*16.5*	*14.9*	*13.7*	*12.5*	*11.6*	*10.9*	*10.1*	*9.4*	*8.8*	*8.3*	*7.7*	*7.2*
75	95	91	86	82	78	74	70	66	62	58	54	51	47	44	41
		20.9	*18.5*	*16.8*	*15.2*	*14.0*	*12.9*	*12.0*	*11.2*	*10.5*	*9.8*	*9.3*	*8.7*	*8.2*	*7.7*
80	96	91	87	83	79	75	72	68	64	61	57	54	50	47	44
		21.0	*18.7*	*17.0*	*15.5*	*14.3*	*13.2*	*12.3*	*11.5*	*10.9*	*10.1*	*9.7*	*9.1*	*8.6*	*8.1*
90	96	92	89	85	81	78	74	71	68	65	64	58	55	52	49
		21.3	*18.9*	*17.3*	*15.9*	*14.7*	*13.7*	*12.8*	*12.0*	*11.4*	*10.7*	*10.2*	*9.7*	*9.3*	*8.8*
100	96	93	89	88	83	80	77	73	70	68	65	62	59	56	54
		21.3	*19.0*	*17.5*	*16.1*	*15.0*	*13.9*	*13.1*	*12.4*	*11.8*	*11.2*	*10.6*	*10.1*	*9.6*	*9.2*
110	97	93	90	87	84	81	78	75	73	70	67	65	62	60	57
		21.4	*19.0*	*17.5*	*16.2*	*15.1*	*14.1*	*13.3*	*12.6*	*12.0*	*11.4*	*10.8*	*10.4*	*9.9*	*9.5*
120	97	94	91	88	85	82	80	77	74	72	69	67	65	62	60
		21.3	*19.0*	*17.4*	*16.2*	*15.1*	*14.1*	*13.4*	*12.7*	*12.1*	*11.5*	*11.0*	*10.5*	*10.0*	*9.7*
130	97	94	91	89	86	83	81	78	76	73	71	69	67	64	62
		21.0	*18.8*	*17.2*	*16.0*	*14.9*	*14.0*	*12.4*	*12.7*	*12.1*	*11.5*	*11.0*	*10.5*	*10.0*	*9.7*
140	97	95	92	89	87	84	82	79	77	75	73	70	68	66	64
		20.7	*18.6*	*16.9*	*15.8*	*14.8*	*13.8*	*13.2*	*12.5*	*11.0*	*11.4*	*10.9*	*10.4*	*10.0*	*9.6*
150	98	95	92	90	87	85	82	80	78	76	74	72	70	68	66
		20.2	*18.4*	*16.6*	*15.4*	*14.5*	*13.7*	*13.0*	*12.4*	*11.8*	*11.2*	*10.8*	*10.3*	*9.9*	*9.5*
160	98	95	92	90	88	86	83	81	79	77	75	73	71	69	67
		19.8	*18.1*	*16.2*	*15.2*	*14.2*	*13.4*	*12.7*	*12.1*	*11.5*	*11.0*	*10.6*	*10.1*	*9.7*	*9.4*
170	98	95	93	91	89	86	84	82	80	78	76	74	72	70	69
		19.4	*17.7*	*15.8*	*14.8*	*13.9*	*13.2*	*12.4*	*11.8*	*11.3*	*10.8*	*10.4*	*9.9*	*9.6*	*9.2*
180	98	96	94	91	89	87	85	83	81	79	77	75	73	72	70
		18.9	*17.3*	*15.5*	*14.5*	*13.7*	*12.9*	*12.2*	*11.6*	*11.1*	*10.6*	*10.1*	*9.7*	*9.4*	*9.0*
190	98	96	94	92	90	88	85	84	82	80	78	76	75	73	71
		18.5	*16.9*	*15.2*	*14.2*	*13.5*	*12.7*	*12.0*	*11.4*	*10.9*	*10.5*	*10.0*	*9.6*	*9.2*	*8.9*
200	98	96	94	92	90	88	86	84	82	80	79	77	75	74	72
		18.1	*16.4*	*14.9*	*14.0*	*13.2*	*12.4*	*11.8*	*11.2*	*10.8*	*10.3*	*9.8*	*9.4*	*9.1*	*8.8*
210	98	96	94	92	90	88	86	85	83	81	79	78	76	75	73
		17.7	*16.0*	*14.6*	*13.8*	*13.0*	*12.2*	*11.7*	*11.1*	*10.6*	*10.0*	*9.7*	*9.2*	*9.0*	*8.7*

Table 7.3 *(concluded)*

Temp., dry-bulb (°F)	Difference between wet- and dry-bulb temperatures in degrees Fahrenheit													
	16	17	18	19	20	22	24	26	28	30	34	40	45	50
30														
35														
40														
45														
50	5													
	1.5													
55	14	9	5											
	3.6	*2.5*	*1.3*											
60	21	17	13	9	5									
	4.9	*4.1*	*3.2*	*2.3*	*1.3*									
65	27	24	20	16	13	6								
	5.8	*5.2*	*4.5*	*3.8*	*3.0*	*1.4*								
70	33	29	25	22	19	12	6							
	6.6	*6.0*	*5.5*	*4.9*	*4.3*	*2.9*	*1.5*							
75	37	34	31	28	24	18	12	7	1					
	7.2	*6.7*	*6.2*	*5.6*	*5.1*	*4.1*	*2.9*	*1.7*	*0.2*					
80	41	38	35	32	29	23	18	12	7	3				
	7.7	*7.2*	*6.8*	*6.3*	*5.8*	*5.0*	*4.0*	*3.0*	*1.8*	*0.3*				
90	47	44	41	39	36	31	26	22	17	13	5			
	8.4	*8.0*	*7.6*	*7.2*	*6.8*	*6.1*	*5.3*	*4.6*	*3.8*	*2.8*	*1.3*			
100	51	49	46	44	41	37	33	28	24	21	13	4		
	8.9	*8.5*	*8.1*	*7.8*	*7.4*	*6.7*	*6.1*	*5.4*	*4.9*	*4.2*	*3.1*	*0.7*		
110	55	52	50	48	46	42	38	34	30	26	20	11	4	
	9.2	*8.8*	*8.4*	*8.1*	*7.7*	*7.2*	*6.6*	*6.0*	*5.4*	*4.8*	*4.0*	*2.5*	*1.1*	
120	58	55	53	51	49	45	41	38	34	31	25	17	10	5
	9.4	*9.0*	*8.7*	*8.3*	*7.9*	*7.4*	*6.8*	*6.3*	*5.8*	*5.4*	*4.6*	*3.3*	*2.3*	*1.1*
130	60	58	56	54	52	48	45	41	38	35	29	21	15	10
	9.4	*9.0*	*8.7*	*8.3*	*8.0*	*7.6*	*7.0*	*6.6*	*6.1*	*5.6*	*4.9*	*3.8*	*3.0*	*2.0*
140	62	60	58	56	54	51	47	44	41	38	32	25	19	14
	9.4	*9.0*	*8.7*	*8.4*	*8.0*	*7.6*	*7.1*	*6.6*	*6.2*	*5.8*	*5.1*	*4.1*	*3.4*	*2.6*
150	74	72	60	58	57	53	49	46	43	41	36	28	23	18
	9.2	*8.9*	*8.6*	*8.3*	*8.0*	*7.5*	*7.1*	*6.7*	*6.2*	*5.8*	*5.2*	*4.2*	*3.6*	*2.9*
160	65	64	62	60	58	55	52	49	46	43	38	31	25	21
	9.1	*8.8*	*8.5*	*8.2*	*7.9*	*7.4*	*7.0*	*6.7*	*6.2*	*5.8*	*5.2*	*4.3*	*3.7*	*3.2*
170	67	65	63	62	60	57	53	51	48	45	40	33	28	24
	9.0	*8.6*	*8.4*	*8.0*	*7.8*	*7.3*	*6.9*	*6.6*	*6.2*	*5.7*	*5.2*	*4.4*	*3.7*	*3.2*
180	68	67	65	63	62	58	55	52	50	47	42	35	30	26
	8.8	*8.4*	*8.1*	*7.8*	*7.6*	*7.2*	*6.8*	*6.4*	*6.0*	*5.7*	*5.2*	*4.4*	*3.8*	*3.3*
190	69	68	66	65	63	60	57	54	51	49	44	37	32	28
	8.6	*8.2*	*7.9*	*7.7*	*7.4*	*7.0*	*6.6*	*6.2*	*5.9*	*5.5*	*5.0*	*4.4*	*3.8*	*3.3*
200	70	69	67	66	64	61	58	55	53	51	46	39	34	30
	8.4	*8.1*	*7.7*	*7.5*	*7.2*	*6.9*	*6.4*	*6.0*	*5.7*	*5.4*	*4.9*	*4.3*	*3.8*	*3.3*
210	71	70	68	67	65	63	60	57	54	52	47	41	36	32
	8.3	*8.0*	*7.6*	*7.4*	*7.1*	*6.8*	*6.3*	*5.9*	*5.5*	*5.3*	*4.8*	*4.2*	*3.7*	*3.2*

Source: From U.S. Forest Service Forest Products Laboratory
*Relative humidity values in roman type; equilibrium moisture content values in italics.

species. In permeable wood capillary flow is the fastest of the various mechanisms for movement of moisture to the wood surface.

Vapor Diffusion

Water vapor moves by diffusion. In this process, molecules move in a random manner in all directions. The rate of vapor diffusion is proportional to the difference in concentration of diffusing molecules or, more precisely, to the difference in vapor pressure. The vapor pressure within moist wood increases with increasing moisture content up to the fiber saturation point; thus, while drying is taking place, a vapor pressure gradient exists from the zone within the wood that is at or above the fiber saturation point to the wood surface.

Because a continuous flow path is necessary to permit vapor to diffuse from one zone to another, pure vapor diffusion is effective only in permeable wood. In impermeable wood, most or all passages are blocked so that movement of liquids and gases is obstructed. Vapor diffusion must then operate in conjunction with bound-water diffusion.

Bound-Water Diffusion

The movement of bound water through the cell walls is also a diffusion process. In this process, the molecules are believed to jump from one hygroscopic site to the next. As in vapor diffusion, fewer molecules leave the drier sites than the wetter sites. Consequently, there is a general migration from the wet to the dry parts of the wood. Because the wood surface is usually driest as a result of evaporation, migration occurs from the core to the surface and a moisture gradient exists. For further detail, see Babbitt (1950) and Bramhall (1976, 1979).

The vapor-pressure gradient is responsible for bound-water diffusion as well as for vapor diffusion. Nevertheless, bound-water diffusion is a much slower process than vapor diffusion because more time and energy are required to move bound water through the cell walls.

During drying, neither bound-water diffusion nor vapor diffusion acts alone. As it moves from the core to the surface of the wood, most moisture first passes through the cell walls by bound-water diffusion, then evaporates into the cell lumen, next passes across the lumen by vapor diffusion, and finally is absorbed by the next cell wall. The process is then repeated until the moisture reaches the wood surface.

When drying takes place from the ends of lumber, the migrating moisture must pass through fewer cell walls, and the greater part of the migration is through the cell cavities by the much faster process of vapor diffusion. Consequently drying through the ends is faster than drying through the sides of a piece.

Mass Flow

Another drying process is mass flow during high-temperature drying. When lumber with a moisture content greater than the fiber saturation point is heated above the boiling point of water, some of the water is converted to steam above atmospheric pressure. If the lumber is of a permeable wood, the steam escapes as a result of the pressure differential, and the lumber dries. If the lumber is of an impermeable

Table 7.4 *Shrinkage values for some Canadian species*

Species	% shrinkage from green to oven dry	
	Radial	Tangential
SOFTWOODS		
Cedar, eastern white	1.7	3.6
Cedar, western red	2.1	4.5
Cypress, yellow	3.7	6.0
Douglas-fir	4.8	7.4
Fir, amabilis	4.2	8.9
Fir, balsam	2.7	7.5
Hemlock, western	5.4	8.5
Larch, western	5.1	8.9
Pine, jack	4.0	5.9
Pine, lodgepole	4.7	6.8
Pine, red	3.7	6.3
Pine, western white	3.7	6.8
Pine, ponderosa	4.6	5.9
Pine, eastern white	2.5	6.3
Spruce, black	3.8	7.5
Spruce, Engelmann	4.2	8.2
Spruce, Sitka	4.6	7.8
Spruce, white	3.2	6.9
Tamarack	2.8	6.2
HARDWOODS		
Alder, red	4.2	7.0
Ash, black	4.3	8.2
Ash, green	3.8	5.4
Basswood	6.7	9.3
Birch, white	5.2	7.2
Birch, western white	6.8	9.3
Elm, white	4.4	7.8
Ironwood	4.8	8.0
Maple, broadleaf	4.1	7.6
Maple, Manitoba	3.9	7.4
Oak, bur	4.2	5.4
Poplars:		
trembling aspen	3.6	6.6
largetooth aspen	3.2	6.8
balsam poplar	3.9	6.4
eastern cottonwood	3.1	7.8
black cottonwood	3.6	8.8

Source: Kennedy (1965)

wood, drying can take place only by diffusion, although at this temperature it is a fairly rapid process. Because steam generated in wood heated above the boiling point may have enough pressure to split the wood, the use of high temperatures to dry an impermeable wood usually results in severe degrade and should therefore be avoided.

Since processes that transport moisture are accelerated by increasing temperature, it is tempting to assume that drying can be accelerated simply by raising the wood temperature. While this simple expedient has enjoyed considerable success for several permeable species (Ladell 1957; Koch 1972), it cannot be used indiscriminately. Because the rate of moisture movement by the various processes does not increase at the same rate with increasing temperature, the character of the drying process changes with rising temperature: the surface dries rapidly while the interior remains wet, resulting in high moisture and shrinkage stress gradients which lead to the development of drying defects.

DRYING DEFECTS AND THEIR CAUSES

Shrinkage

When wood is dried below the fiber saturation point, moisture is removed from the cell walls and the wood shrinks. Shrinkage is the cause of most of the problems and loss of lumber value associated with drying.

Because of the nonhomogeneous nature of the cell walls of wood as well as other aspects of wood anatomy, shrinkage depends not only on the moisture content of wood but also on its density. The denser the wood, the greater the shrinkage that can be expected for a given moisture change. Thus, hardwoods as a group shrink more than softwoods, Douglas-fir shrinks more than western red cedar, and maple shrinks more than poplar. A variation in the shrinkage of individual pieces of the same species may also occur, owing to variations in relative density; lighter pieces shrink less than heavier ones. Shrinkage may even vary between earlywood, which is less dense, and latewood, which is more dense, within an annual ring. At the same time, there are woods of equivalent density which differ in shrinkage because of other factors.

On the average, shrinkage in the tangential direction tends to be about twice that in the radial direction, although the ratio varies somewhat between species. For Canadian timbers, radial shrinkage values from green to oven-dry range from about 1.7% to 6.7%, and tangential values range from about 3.6% to over 9.3% (Table 7.4). Normal longitudinal shrinkage is generally disregarded in practice because it amounts to only 0.1–0.2%. The difference of shrinkage ratios in the radial and tangential directions results in differential shrinkage, which causes mixed-grain squares cut in green material to become diamond-shaped when dry, rectangles to cup, and round material to form ellipses (Figure 7.6).

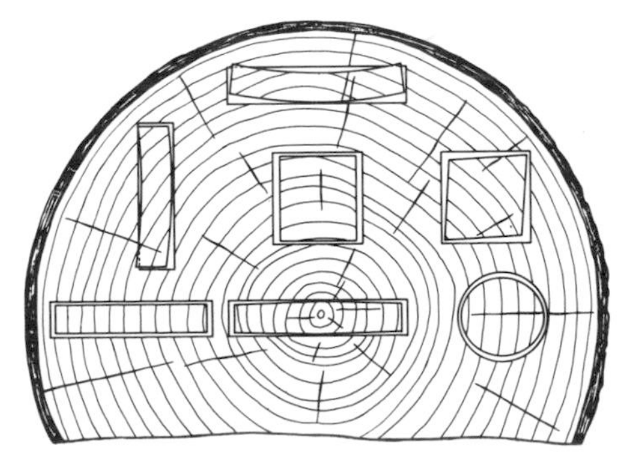

Figure 7.6 Distortion of flat, square, and round wood as affected by the direction of the annual rings. The clear space surrounding each wood section is the shrinkage in drying from green to oven-dry condition.

Other Properties

Like shrinkage, strength properties increase as the moisture content is lowered (see Chapter 4). Wood also becomes more plastic and can be more easily deformed at higher temperatures and moisture contents, a property that is used to control defect development in drying.

Case Hardening

In the initial stages of drying, the surface layers of wood lose water by evaporation and tend to approach the equilibrium moisture content associated with prevailing kiln conditions. Because this condition is necessarily below the fiber saturation point, the wood surface tends to shrink. It is prevented from doing so, however, by the core of the board, which remains for some time at or above the fiber saturation point. The surface then is stretched, while the core is compressed, and the surface is compelled to dry in the swollen or stretched condition. Because the surface has lost most of its plasticity, it remains fixed in the stretched condition when dry. This condition is called tension set.

Subsequently the core loses moisture. As it dries below the fiber saturation point, it also tends to shrink. The tension-set shell (the surface) now prevents the core from shrinking, and so the core is compelled to dry in the stretched condition. The tensile force of the core generates a compressive force in the shell. This condition is called case hardening.

If a case-hardened piece of lumber is rip-sawn, the two halves bow toward each other, binding the saw, and result in bowed stock (Figure 7.8). The test for case hardening is to cut a small piece of lumber into six prongs, remove the second and fifth prongs, and note any bending of the remaining four (Figure 7.7). The value of construction lumber, which is not intended for resawing, is not affected by case hardening. However, case hardening is a serious defect in lumber that will be resawn. It can be removed by steaming or conditioning the lumber in a kiln for several hours at the end of drying. For example, a rule of thumb is to steam softwoods at 10°C (20°F) above the final dry-bulb temperature for 30 minutes for each 25 mm (1 in.) of lumber thickness. Excessive steaming may result in reverse case hardening, in which the tensile and compressive forces are reversed. This condition is as undesirable as case hardening, but it can usually be relieved by additional kiln drying to remove the excess moisture from the surfaces.

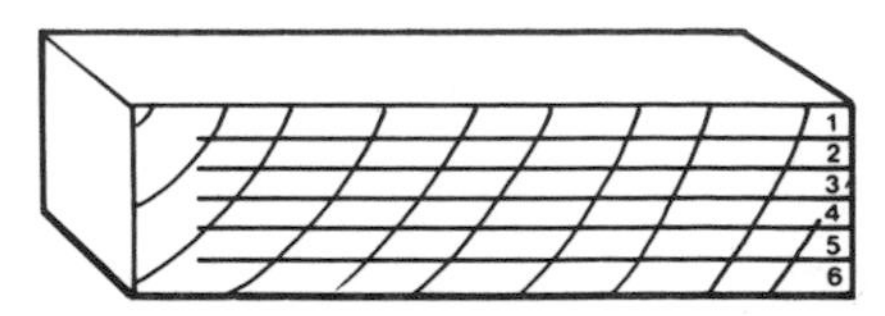

Stock 38 mm (2 in.) and thicker is sawed as shown so as to produce six prongs of equal thickness for case-hardening test. Prongs 2 and 5 are broken out.

Figure 7.7 Prong test for case hardening.

Collapse

According to the Tiemann theory (Kemp 1959), collapse, or the severe distortion or flattening of single cells or rows of cells during drying, usually occurs in very impermeable woods in which large groups of cells are completely saturated with water. Under drying conditions, the moisture leaves the cells by evaporation from a meniscus in an extremely small pit opening. Liquid tensions as high as 3500 kPa (500 psi) can be anticipated in some species. When a cell contains an air bubble, the tension is relieved by expansion of the bubble. When the cell is completely filled with water, however, the tension is transmitted to the cell walls. If the cell walls are strong enough to withstand the tension, the meniscus is pulled into the cell lumen and a bubble is formed to relieve the tension. If the cell walls cannot withstand the tension, they are pulled inward into the lumen. When a large number of cells are affected, visible collapse takes place (Figure 7.9).

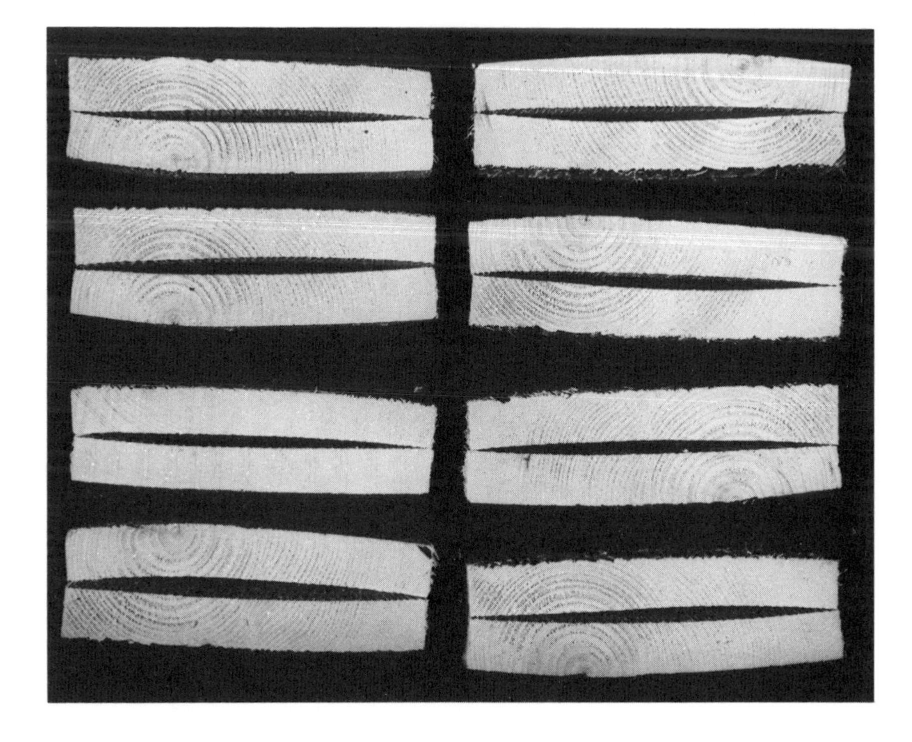

Figure 7.8 Case-hardened lumber.

Another theory of collapse takes into account the forces already described in case hardening. In the initial stages of drying, the surface tends to shrink, whereas the core tends to remain in its swollen state. If a piece of lumber were ripped parallel to its wide faces at this stage, the two faces would spring apart, indicating that the surface is in tension and the core in compression. Under some conditions, it is believed that this internal compression can become so severe that it exceeds the crushing strength of the cells, and they collapse.

Collapse may be visible on the surface and appear as corrugations or as an entire sunken surface, or collapse may be internal and appear in cross-section as irregular-shaped holes. Since wet wood is more plastic at high than at low temperatures, collapse is usually associated with too high a temperature and fast drying in the early stages of drying. Collapse can usually be avoided by slow air drying at normal temperatures until the wood loses enough moisture to permit bubbles to form in the cells or to permit subsequent drying without the development of excessive stresses.

Figure 7.9 Surface corrugations caused by severe collapse in western red cedar.

Checking and Splitting

If the humidity in a kiln is so low that the wood surface, or shell, cannot withstand the tensile forces caused by the tendency to shrink, it will rupture, or check. Excessively dry ends frequently cause checking to extend through the piece, causing a split (Figure 7.10). Checking and splitting are major defects in some hardwoods and in thick lumber where the difference in moisture content between the shell and the core becomes increasingly greater during drying.

Internal Checking

Frequently the tension within the core at the end of drying is such that

Figure 7.10 Checking and splitting, caused by more rapid drying of the ends of a board than the remainder.

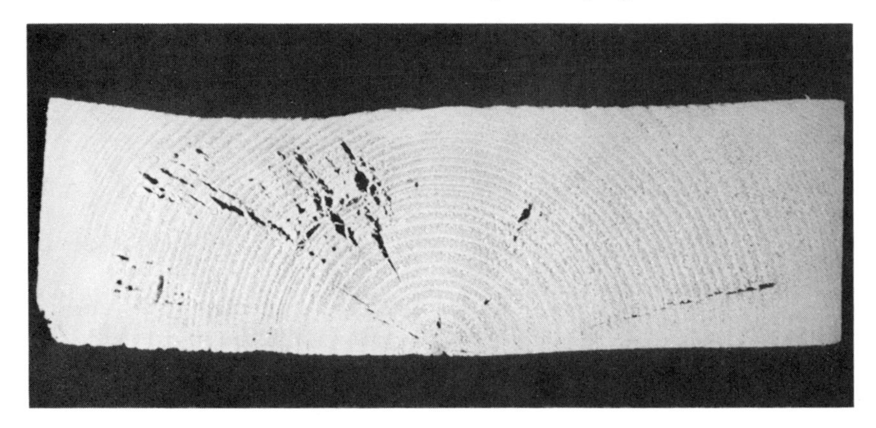

Figure 7.11 Internal checking, commonly known as honeycombing, frequently develops in the core of lumber during the later stages of drying.

it developes internal checking, or honeycombing. Such checks are long and narrow and may develop from surface checks that have penetrated the wood during drying. These surface checks may have closed as a result of the reversal of stresses on the surface layers when the core was dried below the fiber saturation point (Figure 7.11). A more frequent cause of internal checking is the removal of lumber from the kiln before the core is dry, followed by end trimming, which exposes the wet core, and then air drying in a dry atmosphere. Internal checking can be reduced by drying under higher humidity.

Cupping and Roller Check

Cupping, or the distortion of a board so that the forces become concave or convex across the grain or width of the board, may occur (Figure 7.12) as a result of differential shrinkage, particularly in the top several tiers of a drying load where the weight is not sufficient to keep the lumber flat. Because the absolute distortion in cupping varies as the square of the width, it is hardly measurable in narrow lumber but rapidly increases with lumber width. When cupped lumber is fed through a planer, the pressure of the rollers is often enough to split the lumber on the concave side, resulting in the defect known as roller check (Figure 7.13). Cupping can be very costly in some high-density hardwoods when it is unusually severe and when the lumber is very valuable. Cupping can be reduced by spacing the stickers, or spacers, between the tiers of a drying load of lumber, closer together and by placing a heavy load, such as a large, reinforced concrete block, on the lumber during drying.

Anatomy-Based Defects

Lumber defects that appear during kiln drying originate in specialized

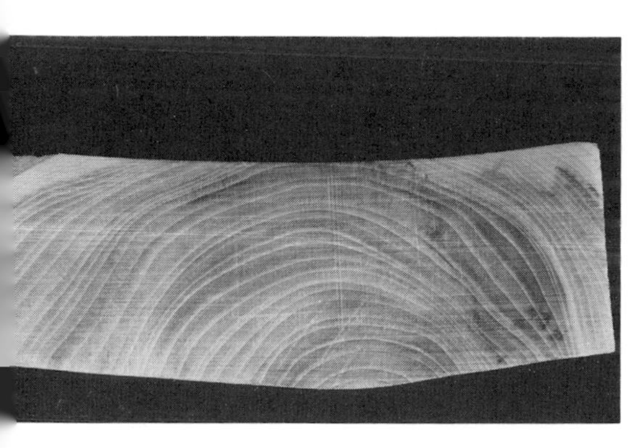

Figure 7.12 Cupping of a wide, flat-sawn board caused by differential shrinkage between the two faces.

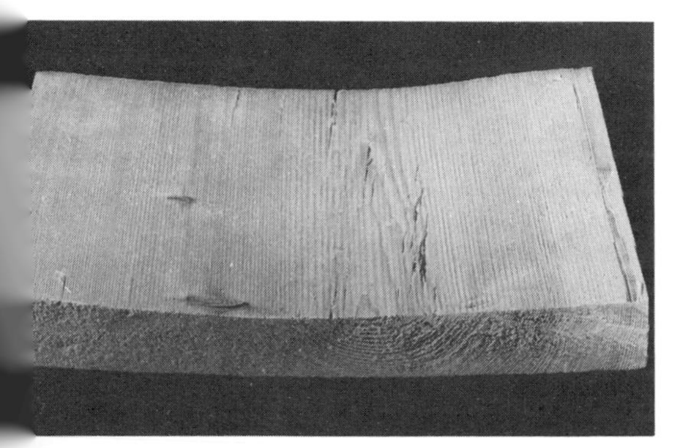

Figure 7.13 When fed through a planer, cupped lumber can split on the concave side. This defect is known as roller check.

wood types in the tree, including reaction wood (compression wood in softwoods and tension wood in hardwoods), juvenile wood, and spiral grain. (For a discussion of these wood types, see Chapter 3). Various types of warp develop in lumber during drying because of the different shrinkage rates of these specialized wood types.

Bow and crook result from the development of a lengthwise curvature along the face and edge of a piece and are caused by the excessive shrinkage of one side compared with the other (Figure 7.14). Twist, a distortion caused by the turning of the edges of a board so that the four corners of any face are no longer in the same plane, may result from the presence of juvenile wood, which is frequently characterized by some spiral grain, or from the shrinkage of spiral grain within normal wood. These defects can be reduced or avoided by drying the lumber under weights or spring-loaded restraints that keep it flat. If wood is dried at higher temperatures and humidities, its plasticity will reduce the stresses responsible for these various forms of warp.

The shrinkage values of knots are different from those of the surrounding wood because of their grain alignment and density. Knots intergrown with surrounding wood may split, whereas dead knots usually shrink excessively and become loose knots.

Shake is the separation of wood along the annual rings. It is usually a characteristic of the species (e.g., some hemlock trees), but it may be accentuated in drying as a result of differences in shrinkage in the radial and tangential directions (Figure 7.15). (See 'Shakes and Checks' in Chapter 3.)

Chemical Stains

Chemical stains often occur in species with a high extractive content because the extractives migrate to the surface, where they are oxidized to a colored product. This defect can sometimes be avoided by more rapid drying in the early stages, which moves the 'evaporation zone' below the wood surface. Thus, the extractives are deposited in this zone and do not reach the surface.

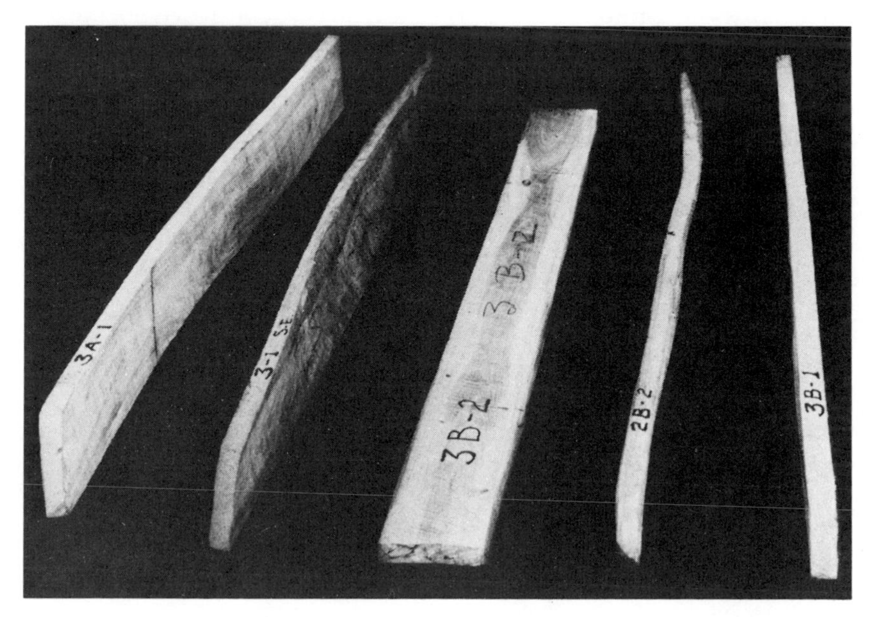

Figure 7.14 Types of warp that develop in lumber during drying. Left to right: bow, combined bow and cup, unwarped, crook, twist.

Fungal Stains
Fungal stains are caused by fungi within the wood. If lumber is dried at temperatures below about 50°C (120°F), any fungi that are present can survive and produce their colored hyphae and fruiting bodies. Fungal growth can be prevented by an initial steaming at about 70°C (160°F) for about two hours to sterilize the wood. Drying can then be carried out at any desired temperature.

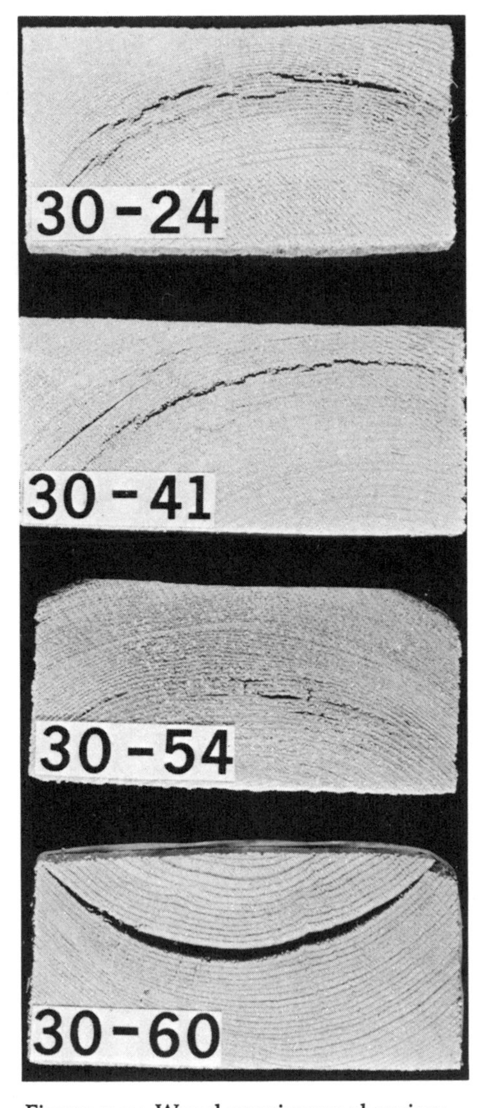

Figure 7.15 Wood specimens showing examples of ring shake developed during drying as a result of differences in shrinkage in the radial and tangential directions.

COMMERCIAL LUMBER DRYING

Lumber should be dried under conditions that will keep the overall cost of drying to a minimum, taking into account the capital costs of equipment, labor, maintenance, and energy; loss of value of lumber due to drying defects; and interest costs of capital when drying is excessively slow. Several methods are well established: kiln drying, air drying, and forced-air drying. Other methods, such as radio-frequency drying, freeze-drying, solvent seasoning, and vapor drying, have been used or considered from time to time. More recently, in order to reduce energy consumption and costs, there has been some consideration of, and experimentation with, solar drying and dehumidification drying. For more detail, see Bramhall and Wellwood (1976).

Drying processes may be divided into two main categories, batch and continuous. The batch process is used in compartment kilns, air drying, and forced-air drying. A complete load of lumber is dried in one operation by this process, the lumber remaining stationary during drying. The continuous process is used in progressive kilns. Loads of lumber, following one after another, are dried as they are moved through the kiln. Most lumber is dried by the batch process; exceptions are certain specialized commodities and lumber dried with new, experimental techniques. Veneer for plywood manufacture is dried by the continuous process.

Stacking
In batch drying, lumber must be properly stacked, and each tier must be separated from adjacent tiers by spacers, or stickers (Figure 7.16). Proper stacking permits air to pass through the pile and make maximum contact with each piece of lumber for fast, uniform drying (Torgeson 1959). The weight of the load and the plasticity of heated lumber reduce the occurrence of warping during drying, ensuring the best possible grade recovery (Mackay 1973). Most efficient drying is achieved when lumber of one species, length, and thickness is dried together. When various lengths must be dried together, 'short circuiting' of the air circulating around the lumber may be avoided by box piling the lumber, placing the longer lengths on the outside of the pile.

Stickers should be of a stiff, strong material. If wood is used, only the stronger species should be chosen. Stickers should be wide enough that neither they nor the lumber in contact with them are crushed, but not so wide that drying underneath them is restricted. A width of 38–76 mm (1½–3 in.) is suggested; the thickness may be 13–25 mm (½–1 in.). The larger thicknesses are recommended for wet or fast-drying softwood lumber, where large quantities of air are required. Thinner stickers may be used for slow-drying material.

Figure 7.16 A properly stacked load of lumber. Each tier is separated by stickers in preparation for drying.

The placement of stickers greatly affects the quality of lumber produced. They should be at intervals of about 0.5 m ($1\frac{1}{2}$ ft) or closer for hardwood and about 1 m (3 ft) or somewhat more for softwood. The stickers must be aligned vertically, directly over the load supports or bunks. In addition, bolsters separating vertically piled forklift loads must be placed directly over the lines of offset stickers in order to prevent warp. Stickers should also be placed flush with the ends of the lumber in order to retard end drying and to exert pressure on the ends, which reduces or prevents checking.

Lumber Segregation

Only lumber of similar drying characteristics should be dried in any one load. Otherwise, the faster-drying material will be overdried, while the slower-drying material may not be dried to the desired moisture content. Permeable species usually dry faster than impermeable species, and species with a high moisture content (or material with a high sapwood content) generally take longer to dry than those with a low moisture content.

The thickness of lumber affects the drying rate. If drying is entirely by diffusion, drying time for any set of conditions will vary as the square of the lumber thickness. If capillary flow is the controlling factor, drying time will vary directly with lumber thickness. Consequently, in practice, when capillary flow controls the early stages of drying and diffusion the later stages of drying, the rate varies as the thickness raised to a power between 1 and 2. Various investigators have determined the exponent as 1.2–1.6, depending on the species dried and the range of moisture content through which it is dried.

Width of lumber is also a factor in drying rate, but for a less obvious reason. When a log is cut, the wider lumber frequently comes from near the center and is heartwood, whereas the narrow widths are usually taken from near the periphery and are sapwood. Depending on the characteristics of the species concerned (the moisture content of the heartwood and sapwood and their drying rates), it may be preferable to dry them in separate loads. More severe drying schedules may be used to achieve faster drying in narrow widths because such defects as cupping and roller checking are less likely to occur in narrow lumber than in wide lumber.

Air Drying

Air seasoning is the oldest method of drying lumber. It requires little capital expense except for space to store the lumber, but the lumber must be held in storage for a long time before it is ready for marketing and one has no control over the drying process. Because drying usually occurs during a few months of summer, lumber cut in the fall must be kept for almost a year before it will dry to a moisture content suitable for use. For many species, air drying cannot reduce moisture content to the required level. The lowest attainable level varies with location but usually is not less than about 10%. In some locations, summer drying is so rapid and uncontrollable that serious lumber degrade occurs. Consequently, for many purposes, air drying is not suitable, and lumber must be dried partially or completely in a kiln.

Although only a very small amount of lumber is now completely air dried, this method can be used to remove moisture from lumber before kiln drying with substantial savings in kiln time and energy. A considerable quantity of hardwood is still partially air dried for this reason. Summer kiln-drying schedules are usually shorter than winter schedules because in summer lumber is air-dried while it is waiting to be placed in the kiln.

Forced-Air Drying

Some of the advantages of both air drying and kiln drying can be obtained by the use of forced-air drying. This method requires a simple, low-cost enclosure with fans installed at one end and adequate venting at the other (Figure 7.17). If the load is properly stacked and baffles are placed to direct air in the enclosure, air can be drawn through the load by the fans. More uniform air movement and drying can be obtained if the air is drawn through the load, because air enters a fan in a streamlined pattern from all angles but is discharged in a more or less direct path in turbulent flow. Electrical energy may be saved if the fans are shut off when the higher air velocity is undesirable or ineffective.

Figure 7.17 Arrangement for forced-air drying of lumber.

The fans have little effect on the drying rate during periods of high humidity when the removal of moisture from the wood is slow. When humidity is excessively low and the lumber is green, the fans may be stopped to prevent rapid drying and consequent surface checking. In a series of experiments (Cech and Huffman 1968), drying times were reduced 22–40% by forced-air drying compared with air drying similar loads of lumber.

Kiln Drying

Kiln drying is the standard method of drying lumber. A kiln consists of an enclosed space with a method of controlling the temperature and fans for circulating the air through the load of lumber. In some kilns, notably in eastern Canada, provision is also made for controlling the humidity of the air. Two types of kilns are found in Canada, the compartment kiln and the progressive kiln.

Practically all softwood kilns in Canada today are classified as compartment kilns and are designed for the batch process; the entire kiln consists of a single compartment that is programmed to one set of conditions at any time and filled with a complete charge of lumber at each operation (Figure 7.18). Commercial kilns are 16–60 m (50–200 ft) in length and accommodate either one or two tracks of lumber 2.5–3 m (8–9 ft) wide and 3–4.5 m (10–14 ft) high (Figure 7.19).

The progressive kiln is designed for the continuous process of drying lumber. The interior of the kiln is divided into several zones. The various zones are operated at the same wet-bulb temperature, but the dry-bulb temperature increases from the infeed to the output end. A green load of lumber, or lumber that is to be dried, is placed in the low-temperature section, or infeed end, and is followed successively by other green loads, all moving through zones of increasing temperature to the output end of the kiln. The last zones may be set at a lower temperature in order to attain equalization of moisture content within each load or to relieve some drying stress (Figure 7.20).

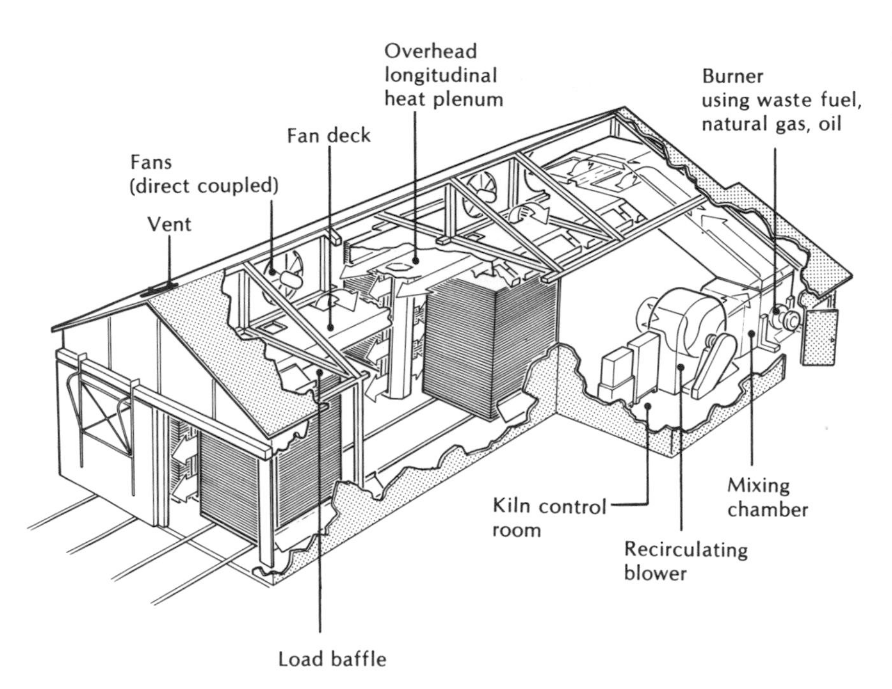

Figure 7.18 A compartment kiln direct fired with external blower and cross-shaft fan arrangement.

Figure 7.19 A commercial kiln of the compartment type.

Automatic and Semiautomatic Control
All modern kilns are equipped with a recorder-controller. Although the form of these instruments varies from manufacturer to manufacturer, their method of operation is practically identical. The wet- and dry-bulb temperature sensors are installed in a location with good air circulation. The dry-bulb sensor consists of a hollow bulb filled with a volatile liquid, such as ethyl ether, and is connected by a capillary to a Bourdon tube or other expandable device in the recorder. As the temperature of the kiln increases, the pressure in the bulb increases, causing the Bourdon tube to expand and to operate a recording pen and controlling device.

The wet-bulb system is in most respects identical with the dry-bulb system, and both are housed in the same instrument case. The wet-bulb sensor is covered with a fabric 'sock' or wick dipped into a container of

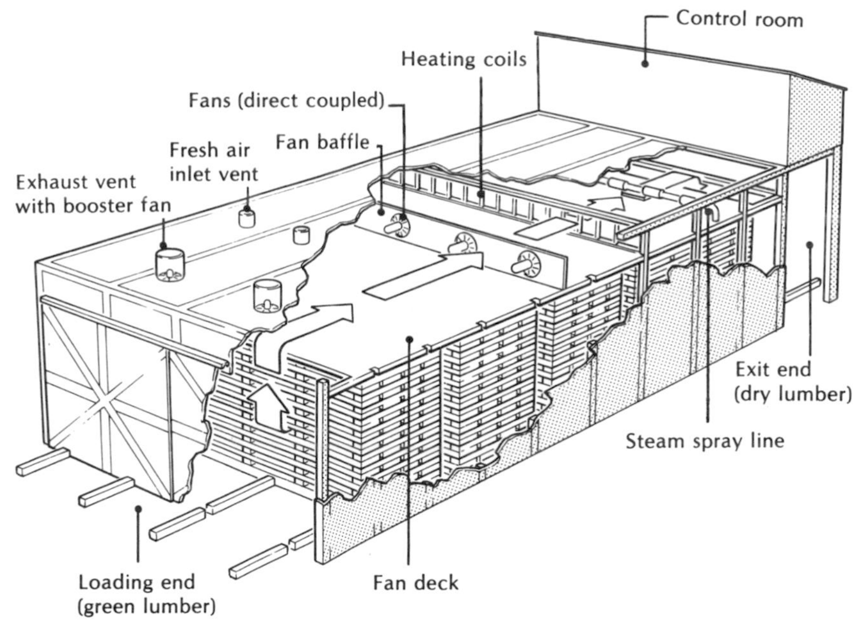

Figure 7.20 Arrangement of a progressive kiln.

water. The water wets the sock by capillary action and evaporates, cooling the sensor. There is an established relationship (Tables 7.2 and 7.3) between the wet-bulb depression (the difference between dry- and wet-bulb readings), the vapor pressure, and relative humidity at any temperature.

The controller aspect of the instrument consists of two pointers or indexes that are set to the desired dry-bulb and wet-bulb temperatures. When the dry-bulb temperature is below its index setting, the controller operates switches or valves to introduce heat into the kiln. When the indexed value is reached, the switches or valves are closed. Similarly, when the wet-bulb temperature is below the index, humidity is introduced. When the index value is reached, the humidity is cut off. If the index value is exceeded, kiln vents are opened to permit the excess humidity to escape (McIntyre 1974).

Kiln schedules, which are predetermined programs of dry-bulb and wet-bulb temperatures that change with time or wood moisture content, may be introduced by manual control of the index points, or, in the case of time-based schedules, by means of cams cut to the desired temperature-time relations. Cam control has the advantage of permitting continuously changing conditions to be introduced, leading in many cases to faster, more efficient schedules.

Methods of Heating Kilns

Traditionally kilns have been heated by radiators supplied by steam. For many years plain iron pipes were the standard material for radiating surfaces, but now finned pipe heating coils are used almost exclusively, placed directly in the air path. The source of steam is a boiler operated on oil, gas, or wood residue. Wood residue is the most common fuel for boilers in western Canada, whereas oil is the preferred fuel in the east.

Although a wood-residue steam boiler might seem to be an

economical source of heat, the cost of steam installations is very high. In addition, the provinces require boiler attendance by a qualified stationary engineer. Operating costs become so substantial, therefore, that they can normally be justified only in larger mills producing great quantities of steam.

Another installation, similar to a steam plant, makes use of hot water. The primary differences are that the system is operated at a maximum temperature of 120°C (250°F), and hydraulic pressure is maintained between 140 and 1100 kPa (20 and 160 psi) to prevent a steam phase from forming. Not only is such a hot-water system less costly to install than a steam plant, but its operating costs are much lower because a stationary engineer is not required. The greatest problem is that a maximum kiln temperature of only about 85°C (185°F) is obtainable, after the various heat losses that are incurred during use. Several installations of this type are used successfully in British Columbia where a high-quality product is desired.

The high cost of a steam boiler for small kiln operations has led to the concept of eliminating steam or hot-water radiators entirely in favor of "direct-fired" or "hot-air" kilns. These kilns use natural gas, propane, butane, or wood residue as fuel, and the products of combustion, almost entirely nitrogen, carbon dioxide, and water vapor, are conducted directly into the kiln. In some kilns the heat energy from the burners is transmitted through a heat exchanger to the circulating air to prevent combustion gases from entering the kiln. In the operation of these kilns, the combustion products, which may reach 1650°C (3000°F), are mixed with a portion of recycled kiln air to reduce the temperature to about 200°C (390°F) before they enter the kiln. When the kiln is heated, there is a net inflow of combustion gases, which is compensated for by a net discharge of the same volume of air from the kiln. The disadvantages of this type of operation are the lack of control over kiln humidity, particularly at the end of the drying cycle, and the cost of gaseous fuel or of an efficient wood-residue burner. These disadvantages, however, are offset by high fuel efficiency and low maintenance and labor costs.

Dehumidification of Kilns

As noted previously, modern kilns are equipped with vents, usually located in the roof on both the suction and the discharge sides of the circulating fans. When the wet-bulb temperature exceeds its indexed value, the vents open and the circulating fans draw air in through the vents on the suction side, expelling the humid kiln air through the vents on the discharge side.

Some alternatives to this method are in use. One is power venting, which consists of a fan controlled by the wet-bulb temperature controller. When the humidity exceeds the indexed value, the fan is turned on to extract air from the kiln. Normally a vent in another part of the kiln is opened at the same time to replace the air that has been removed.

Humidification of Kilns

The steam or water vapor in kilns comes from various sources, and attempts to increase humidity, particularly in hot-air kilns, vary widely in efficiency, depending on the methods used.

In hot-air kilns heated by the combustion products of natural gas, propane, butane, or wood residue, water vapor from the fuel makes up about 40% of the water vapor in the kiln atmosphere; the remainder is provided by evaporation from the lumber. Perfect combustion with no excess air normally results in combustion products having a wet-bulb temperature of about 55°C (130°F); in normal practice, however, the value is somewhat lower.

Because of the additional moisture evaporated from the wood, it is not unusual for gas-fired kilns to attain wet-bulb temperatures of 70–75°C (160–170°F) during periods of rapid drying. Toward the end of drying, when evaporation decreases and the wet-bulb temperature falls, it is often desirable to maintain or even increase the wet-bulb temperature in order to attain a more uniform final moisture content and to relieve drying stresses (case hardening). In such cases a supplementary source of moisture is required.

Steam is the most common form of supplementary humidity, but its use is limited to installations having a boiler. This may be the same boiler that provides steam for drying the lumber, or it may be a relatively small boiler used only to maintain kiln humidity. Steam at very low pressures and temperatures, such as exhaust steam, is the best for providing the required supplementary humidity. In gas-fired kilns, where steam is not available, the most favorable results in increasing kiln humidity come from the use of a fine spray in the hot gases coming undiluted from the burner. The most unsatisfactory results are obtained by spraying cold water into the kiln atmosphere.

Fan Speed and Drying Uniformity

The dry- and wet-bulb conditions recorded and controlled in the kiln are the conditions at the sensors, usually near the point where the air enters the load. As the air passes through the load, it gives up heat and becomes cooler, and picks up moisture and becomes damper. Consequently, at the exhaust side of the load, its drying power is considerably reduced, and so is the rate of drying.

For any given rate of evaporation, the reduction of drying power across the load is inversely proportional to the volume rate of air flow (Koch 1972). Consequently greater uniformity and faster drying can be achieved by reversing the fans at appropriate intervals and by increasing the air flow rate, particularly near the start of drying when evaporation is fastest (Salamon 1973). This procedure involves some cost, however, since doubling the air speed requires eight times the motor capacity and electrical energy.

Low-Temperature Drying

Low-temperature drying means drying at 25–50°C (80–120°F) with as much as a 15°C (27°F) wet-bulb depression. It has the advantage of lower capital cost of equipment and is especially applicable to high-value species that are difficult to dry without degrade. The process yields a relatively low range of final moisture contents, thus providing a uniform product. Its principal disadvantage is the slow drying rate that results in drying times several times longer than in conventional schedules for most species.

Superheated-Steam Drying

Wood products can be dried in an atmosphere of water vapor alone if the kiln temperature is maintained above 100°C (212°F), because the vapor pressure of the moisture in the wood at these temperatures exceeds atmospheric pressure, and moisture is lost by mass flow of the water vapor. In practice, the drying process is dependent on the circulation rate, which affects the efficiency of heat transfer between the kiln atmosphere and the wood surface by the circulating gas.

Veneer for the manufacture of softwood plywood has traditionally been dried by superheated steam at temperatures of about 185°C (365°F). At this temperature, veneers of 2.5 mm ($\frac{1}{10}$ in.) thickness are dried in about eight minutes. Since the rate of heat transfer is the major limiting factor in such drying, drying time is almost proportional to total moisture content per unit area and is approximately doubled for a twofold increase in moisture content or thickness.

Although some air is usually found in veneer dryers, its presence does not increase the drying rate. Rather it is a disadvantage in that it is a major factor in the occurrence of dryer fires (Bramhall 1967). On entering the dryer at the infeed end, the air cools adjacent surfaces, causing condensation of volatile resins and subsequently supporting combustion of these resins when other conditions are favorable.

Special Drying Techniques

Drying by Dehumidification

A more sophisticated system of lumber drying using refrigeration units built into the kiln has recently been introduced in North America (Ullevålseter 1971). An independent source of heat is used to bring the kiln temperature to about 30°C (85°F); then a heat pump similar to an electric refrigerator comes into operation. As kiln air passes over the cold coils of the heat pump, the moisture in the air condenses on them and is drained off. The heat given up by the condensing moisture is transferred to the refrigerant and pumped to the hot coils of the pump. The cold, dried kiln air is passed over the hot coils, reheated, and recirculated through the lumber. The electrical energy needed to operate the pump is ultimately converted to heat, which increases the kiln temperature. The system is especially efficient because it can also be used to extract heat from the atmosphere and release it into the kiln. For every kilowatt of electricity used, 3–4 kilowatts of heat can be released into the kiln. Because present units do not function over 50°C (120°F), the drying rate for most species is seriously limited. However, a modification is now being developed to raise the kiln temperature to about 100°C (212°F).

Radio-frequency Drying

Since the 1940s there has been an interest in drying wood by means of radio-frequency energy. In this procedure the wood is placed between the plates of a condenser supplied with an alternating current of very high frequency. The current induces a rapid rotation of the dipole units in the wood, primarily water molecules, resulting in the generation of heat. The peculiarity of this phenomenon is that the heat is not conducted into the wood but is generated within the wood, and this is done most intensely in zones of high moisture content. The wood can be heated to the boiling point of water and even higher. Permeable wood will lose moisture by mass flow when the vapor pressure exceeds atmospheric pressure. In impermeable species, the pressures generated may be sufficient to shatter the wood. Consequently improper use of this form of drying can result in degrade (Miller 1971).

So far, the economics of this process have been unattractive except for special applications, such as golf club heads and pencil slats. Not only is the capital investment for equipment high, but the energy costs are high as well. Radio-frequency generators seldom exceed 50% efficiency and operate on relatively expensive electrical energy.

Radio-frequency drying has been proposed as a final stage in the drying of veneers for plywood. In this case the rather high costs may be justified by the low amount of moisture, which is critical in preventing the formation of blistered plywood in subsequent hot-pressing.

Freeze-drying

Even when wood is frozen, it will still lose moisture when its vapor pressure exceeds the vapor pressure of the moisture in the surrounding atmosphere. Under normal conditions this moisture loss is a very slow process, but it can be accelerated considerably by placing the wood in an evacuated container and drawing a vacuum that is higher (has lower absolute pressure) than the vapor (Choong, Mackay, and Stewart 1973). This method is effective only when heat can be conducted to the wood being dried. The method is unsuitable for commercial use and is of academic interest only.

Solvent Seasoning

In solvent seasoning, lumber is placed in a tank and flushed with the vapors of organic solvents that are miscible with water. The latent heat of condensation of the organic vapors provides the heat necessary to evaporate the water. In addition, some water is removed in solution in the organic liquid.

Although solvent seasoning results in fast drying of the lumber, high lumber quality, and the removal and recovery of resins, the cost of equipment, solvent losses, and other factors have not made it economically attractive (Stamm 1964).

Vapor Drying

A similar process using the vapors of immiscible solvents has also been developed. The solvent used must boil at more than 100°C (212°F) at atmospheric pressure. It functions exclusively as a heat transfer agent, evaporating from the boiler, condensing on the wood, and giving up its latent heat of evaporation. The heat raises the temperature of the wood and evaporates moisture. The mixture of water and solvent vapors is drawn off, condensed, and separated by gravity, and the solvent is returned to the boiler and reevaporated. The wood temperature and rate of drying are regulated by the rate of boiling the solvent.

Vapor drying differs from solvent seasoning in two important respects: it cannot extract water from the wood, since the solvent and water are immiscible, and it provides a surface equilibrium moisture content that is effectively zero. As a result, considerable fine checking occurs in the dried product. For this reason, and because some solvent is left in the wood, the process is used almost exclusively in the drying of railway ties and utility poles before they are treated with preservatives.

Determination of Moisture Content

The moisture content of wood is measured for several reasons. The first, and most common, is to determine that the moisture content of the wood is within the limits specified for any particular use (see 'The Purpose of Drying Wood' and 'Equilibrium Moisture Content' earlier in this chapter). The second reason is to monitor the drying of wood in order to avoid degrade, which can be caused by improper drying practice.

Oven-Drying Method

This method gives an accurate value of the moisture content of a sample, but it requires considerable labor and time (24 hours) to make a determination. The calculated moisture content is representative of the original wood only if the sample has been carefully chosen and prepared. Because of the labor involved, relatively few determinations can be made per charge, and the estimate of the moisture content of a charge can therefore be in error.

The oven-drying method is widely used to monitor the drying of hardwoods. In this procedure, the initial green moisture contents of several test boards are obtained by cutting a representative sample from each. The samples are trimmed to remove potential slivers, weighed, and dried in an oven at 102°C (215°F) to a constant weight, and that weight is recorded. The moisture content can then be calculated using the equation

$$\text{moisture content (\%)} = \frac{\text{green wt} - \text{dry wt}}{\text{dry wt}} \times 100. \qquad (1)$$

By weighing the undried test boards from which the samples were cut, their dry weights can be calculated from the equation

$$\text{dry wt} = \frac{\text{green wt} \times 100}{100 + \text{moisture content (\%)}}. \qquad (2)$$

The test boards, which are dried in the load of lumber, are removed from the kiln from time to time, weighed, and their moisture contents determined from equation (1). In this way, the moisture content of the load can be monitored during drying.

Electric Moisture Meter Methods

Because some electrical properties of wood vary with its moisture content, it has been possible to make meters of various types to provide a very rapid estimate of moisture content. These fall into three main types: resistance, power-loss, and capacitance moisture meters.

In both the softwood and hardwood industries, the electrical resistance moisture meter is used to make many moisture determinations on kiln-dried lumber to determine whether or not it meets the standard. This meter measures the resistance of the wood to the flow of electricity between needles mounted 2.5 cm (1 in.) apart in a probe and

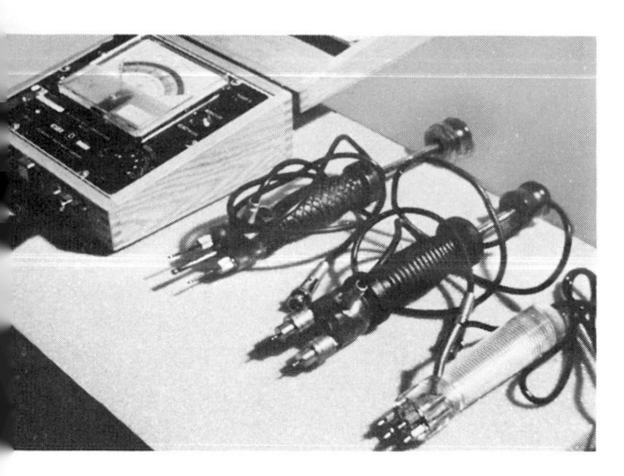

Figure 7.21 A resistance moisture meter with three probes. *Left*: probe with two insulated needles. *Right*: two types of probes with two pairs of uninsulated needles.

Figure 7.22 A power-loss moisture meter, which measures the power-loss factor when radio-frequency current passes through the wood being tested. This measurement is used to estimate moisture content.

driven into the wood. The scale is calibrated directly in moisture contents, but the readings must be corrected for species and for temperature. Corrected tables for various species and temperatures have been prepared (Bramhall and Salamon 1974; Pfaff 1974).

Although many different types of probe are available, only two are widely used. The first consists of two pairs of uninsulated needles 1 cm ($\frac{3}{8}$ in.) long, which provide an estimate of the highest moisture content encountered, whether it is water on the surface of the wood or a wet area within the wood. The second type consists of two insulated needles approximately 2.5 cm (1 in.) long with only the tips uninsulated. These provide an estimate of moisture content only at the tips and are used to measure the surface and the core moisture contents separately (Figure 7.21).

Resistance moisture meters have the advantage of direct reading (after suitable species-temperature correction) and can be used to measure the moisture content at various depths, without specimen cutting or sample preparation. The major drawback of these meters is that the measurement of resistance is affected by other factors besides moisture content, so there is an inherent but unknown error in all readings. If all necessary precautions have been taken, it can usually be assumed that within the range of 7–25% moisture content, the corrected readings will be within 2% of the true moisture content. Beyond this range, the accuracy is substantially less. Other disadvantages are that the probes leave small holes in the lumber and may therefore detract from the value of finished lumber, and where hundreds of readings must be taken considerable physical effort is required in driving and extracting the needles. Resistance moisture meters are used primarily by kiln operators to determine average moisture content after kiln drying and, if insulated probes are used, to determine if the core of the lumber is dry.

A modification of the resistance meter has been used to monitor lumber being dried in the kiln. Needles or electrodes connected by cables to an instrument in the kiln control room are driven into lumber before it is placed in the kiln, enabling the operator to estimate when the lumber is dry. This application is less accurate than the first method of determining moisture content described above.

A second type of electrical meter measures the power loss in the grid circuit of a high-frequency oscillator that uses the wood as a capacitance in a tuned circuit. Because this meter has no wood-penetrating needles and its electrodes are placed in contact with the wood surface, its readings are strongly influenced by the moisture content at the surface and are hardly affected by that of the core (Mackay 1976). Corrections are required for temperature and species (Bramhall and Salamon 1974). The advantages and disadvantages of this meter are similar to those of the resistance meter, except that it has little depth of penetration, the influence of temperature is more complex, and its inherent error is about 5% moisture content. It provides faster determinations with less physical effort, leaves no holes in the wood, and provides readings at moisture contents lower than 7%. For these reasons it is used by many inspection agencies (Figure 7.22).

Another type of capacitance moisture meter uses the change of capacitance with the changing moisture content of wood that is placed between two plates of a condenser. This principle has recently been used to monitor the moisture content of lumber in kilns during the drying process. Thin metal plates are placed in the load, and the capacitance between these plates and the metal truck carrying the lumber is measured. The charge is considered dry when the capacitance reaches a predetermined value.

Hand-operated moisture meters are easy to use, but they cannot be used as a screening instrument to detect all nonconforming lumber. 'In-line' moisture meters, using resistance or capacitance principles or combinations of both, may be built into the dry chain or other conveyor in the mill (McLauchlan 1975). They are used to scrutinize all lumber and to mark nonconforming pieces on the chain. As a result, where suitable automatic handling equipment is available, kilns may be operated to dry softwood construction lumber to a higher than normal final moisture content of 16–17%, and those specimens that are over 19% may be rejected. The kiln schedule can thus be significantly shortened, with savings of time, heat, energy, and electricity. In addition, since degrade increases as lumber is dried to a lower moisture content, degrade is reduced.

The disadvantages of this system lie in the cost of recycling lumber and in the problems of redrying partially dried lumber with a wide range of moisture contents. Obviously the method is expensive to use unless the entire recycling system is automated.

REFERENCES

Babbitt, J.D. 1950. On the differential equations of diffusion. *Can. J. Res.* 28A: 449–74

Bramhall, G. 1967. *Cause and prevention of veneer dryer fires.* West. For. Prod. Lab. Inf. Rep. VP-X-20. Vancouver

– 1976. Fick's laws and bound water diffusion. *Wood Sci.* 8 (3): 153–61

– 1979. Sorption diffusion in wood. *Wood Sci.* 12 (1): 3–13

Bramhall, G., and Salamon, M. 1974. *Combined species-temperature correction tables for moisture meters.* West. For. Prod. Lab. Inf. Rep. VP-X-103 (rev.). Vancouver

Bramhall, G., and Wellwood, R.W. 1976. *Kiln drying of western Canadian softwoods.* West. For. Prod. Lab. Inf. Rep. VP-X-159. Vancouver

Cech, M.Y., and Huffman, D.R. 1968. *Accelerated air-drying of spruce and balsam fir lumber.* Can. For. Branch Dep. Publ. 1247. Ottawa

Choong, E.T.; Mackay, J.F.G.; and Stewart, C.M. 1973. Collapse and moisture flow in kiln-drying and freeze-drying of woods. *Wood Sci.* 6 (2): 127

Kemp, A.E. 1959. Factors associated with the development of collapse in aspen during kiln drying. *For. Prod. J.* 9 (3): 124–30

Kennedy, E.I. 1965. *Strength and related properties of woods grown in Canada.* Can. Dep. For. Publ. 1104. Ottawa

Koch, P. 1972. Drying southern pine at 240°F – Effects of air velocity and humidity, board thickness and density. *For. Prod. J.* 22 (9): 62–7

Ladell, J.L. 1957. *High temperature drying of Canadian softwoods; drying guide and schedules.* For. Prod. Labs. Can. Tech. Note 2. Ottawa

Mackay, J.F.G. 1973. The influence of drying conditions and other factors on twist and torque in *Pinus radiata* studs. *Wood Fiber* 4 (4): 264–71

– 1976. Effect of moisture gradients on the accuracy of power-loss moisture meters. *For. Prod. J.* 26 (3): 49–52

McIntyre, S. 1974. *Operation and maintenance of pneumatic record-controllers for dry kilns.* West. For. Prod. Lab. Inf. Rep. VP-X-92 (rev.). Vancouver

McLauchlan, T.A. 1975. *Continuous moisture meters for kiln dried dimension lumber.* West. For. Prod. Lab. Inf. Rep. VP-X-147. Vancouver

Miller, D.G. 1971. Combining radio-frequency heating with kiln-drying to provide fast drying without degrade. *For. Prod. J.* 21 (12): 17–21

Pfaff, F. 1974. *Electric moisture meter correction factors for eastern Canadian species*. East. For. Prod. Lab. Inf. Rep. OP-X-89F. Ottawa

Salamon, M. 1973. Comparison of kiln schedules for drying spruce. *For. Prod. J.* 23 (3): 45–9

Simpson, W.T. 1973. Predicting equilibrium moisture content of wood by mathematical models. *Wood and Fiber* 5 (1): 41–9

Skaar, C. 1972. *Water in wood*. Syracuse Wood Science Series 4. Syracuse: Univ. of Syracuse Press

Stamm, A.J. 1964. *Wood and cellulose science*. New York: Ronald Press

Torgeson, O.W. 1959. *Circulation of air in a lumber dry kiln*. U.S. Dep. Agric. For. Prod. Lab. Rep. 1678. Madison, Wis.

Ullevålseter, R.O. 1971. *Lumber drying by condensation with the use of refrigerated dew point*. Inst. of Wood. Tech. 1432 ÅS-NLH. Norway

8

Wood Protection

A.J. DOLENKO AND J.K. SHIELDS
Eastern Forest Products Laboratory, Ottawa

F.W. KING
Retired, formerly with Eastern Forest Products Laboratory, Ottawa

J.W. ROFF
Retired, formerly with Western Forest Products Laboratory, Vancouver

D. OSTAFF
Canadian Forestry Service, Fredericton

Wood is a versatile building material that can last indefinitely when used wisely. Dry wood specimens found in ancient Egyptian tombs and dated 2600–1000 BC were still sound when examined in modern times. Japanese temples contain samples of wood that has been in service for about 1300 years in both structural and decorative uses.

Although wood can have a long life if properly used and protected, it is subject to degradation by weathering, fire, and insect or fungal attack if adverse conditions prevail. Some species have only a moderate degree of resistance to degradation, but their durability can be greatly enhanced by suitable protective treatments.

The durability of different species of wood is genetically determined. A number of Canadian woods, such as the cedars and white oak, have high durability, but only the heartwood of these species is durable; the sapwood generally has negligible decay resistance. Further, wood taken from trees of the same species may have wide differences in durability, samples from the same tree may sometimes show differences as large as those observed between species. Additionally, because the age of trees harvested has fallen dramatically over the past 20 years, a higher proportion of the less durable sapwood is being marketed today. These and other factors have prompted development of preservative coatings and treatments that can dramatically increase a wood's durability. Preservatives often impart a greater permanence than that found in naturally durable woods.

For all their advantages, however, protective treatments are not a substitute for good design, which is still necessary to ensure the continued excellent performance of wood and wood products. A large amount of design information is available to architects and designers that can be effectively employed with the appropriate preservatives and coatings to extend the life of wood in structures (Scheffer and Verrall 1973; Tayelor 1974).

Figure 8.1 A country church built of wood – Christ Church near Burritts Rapids, Ontario. (NFB Photothèque. Photo by H. Hallendy)

Good design can overcome problems of decay in the exterior wood components of buildings exposed to moisture from precipitation and condensation (Verrall 1962). It can also meet new problems arising from changes in our lifestyle, such as the increasing use of humidifiers in our houses in winter, which introduces the danger of increased moisture condensation in the walls. Sound building practice combined with the use of water-repellent preservatives will eliminate moisture accumulation in crawl spaces (Hansen 1963), basements, and attics, and will prolong the service life of millwork exposed to periodic wetting (Scheffer and Clarke 1967; Sedziak, Shields, and Krzyzewski 1970). And wood in ground contact, such as preserved wood foundations or wood foundation piles, can be protected by suitable heavy-duty preservative treatments (Sedziak et al. 1969; Sedziak and Unligil 1973; Sedziak, Shields, and Johnston 1973).

As the world demand for wood increases, it is becoming increasingly important to make the best use of our forests. One way of expanding this resource is to prolong the life of wood products by protective treatments. Another is through the use of new wood products, such as preserved wood foundations and fiberboard for house cladding, which are gaining greater acceptance.

Figure 8.2 Many wood species are degraded by staining when exposed outdoors without protective treatment. *Left*: treated wood; *right*: stained, unprotected wood.

PROTECTIVE COATINGS AND SURFACE TREATMENTS FOR WOOD

Requirements of Wood Finishes

Finishes are applied to wood to protect it from the environment and to enhance its natural beauty or to provide color. Because of its chemical and physical properties, wood is susceptible to deterioration as a result of various agents: the action of weathering; dimensional changes due to moisture cycling; staining due to chemical changes in extractives; biological attack by insects, molds, and other fungi; soiling, staining, and spotting by dirt and liquids; destruction by fire. Any or all of these factors can reduce the service life and economic value of wood (Stamm 1964; Goldstein and Loos 1973), but their affect can be diminished by applying suitably formulated surface coatings or treatments. The level of protection required depends on how the wood is to be used (e.g., whether for exterior cladding, joinery components, interior wall panelling, furniture). It also depends on the particular wood used because the susceptibility to degradation varies considerably with wood species.

Besides providing protection, wood finishes must meet certain decorative requirements, such as intensifying the wood figure or providing color. Finishes should be highly durable and not prone to blistering, cracking, or peeling. They should require little maintenance and be relatively inexpensive. Low cost is especially important for exterior finishes because wood products are competing for exterior use with metals and plastics, which are finished with highly durable coatings designed to last for the lifetime of the product. The higher initial price of these products is offset by lower maintenance costs.

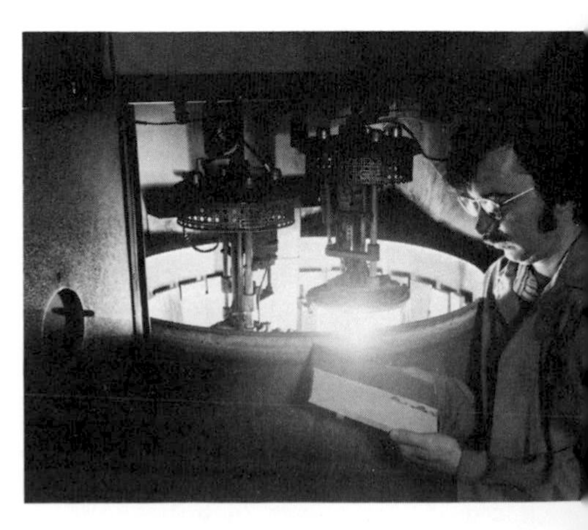

Figure 8.3 Testing the adhesion of wood finishes.

Wood-Finishing Systems

Finishes for wood products may be divided into two broad categories: finishes that form a relatively thick, continuous layer or film on the wood surface (paints, varnishes, lacquers, and overlays), and finishes that penetrate into the surface wood cells, leaving no surface layer (water-repellent preservatives, stains, and sealers). To an increasing extent, the various finishes that fall into these categories may be obtained as water-based systems (except overlays) as well as more conventional systems based on organic solvents. In addition, these finishes may be formulated for interior or exterior applications and may be pigmented or unpigmented, depending on the desired result. The variety of finishes available is rapidly increasing as new ones are developed to meet specific job requirements.

Exterior Finishing

If the surface of unpainted or untreated wood is exposed outdoors, it will rapidly become discolored and roughened by weathering (U.S. Forest Products Laboratory 1966). This deterioration results from the combined chemical and physical action of light, oxygen, air pollutants, and water. Ultraviolet radiation in sunlight induces chemical reactions in the lignin and extractives, converting them into soluble compounds that are leached out by rain. As a result, the surface fibers are liberated and eroded, causing surface roughening. The remaining wood cellulose, along with deposits of dirt, partially accounts for the eventual gray color. Because wood is a hygroscopic and porous material, it readily absorbs and loses liquid water and water vapor. The accompanying swelling and shrinking do not occur in a uniform manner, and surface checking results.

Wood exposed outdoors is also susceptible to staining by extractive leaching and fungal attack. Certain wood species, such as western red cedar, contain large amounts of water-soluble extractives. If these extractives are not adequately sealed within the wood, they will be leached out and stain the natural or painted surface. Staining due to biological attack occurs when wood is exposed to conditions of high humidity and temperature for extended periods of time (U.S. Forest Products Laboratory 1966).

Figure 8.4 Outdoor exposure tests are used to develop improved protective coatings for wood destined for exterior locations.

Natural Finishes

Wood is often used in exterior locations where the surface must be protected without obscuring its natural color and grain. Coatings may be used to achieve this effect (Hill 1972), although the clear varnishlike finishes that are now available have a short service life – approximately one to two years when fully exposed to the weather (Ashton 1974). Once deterioration has begun, it is difficult and expensive to restore the original appearance.

Deterioration occurs because clear finishes transmit ultraviolet radiation and do not prevent degradation of the surface wood cells (Miniutti 1967). Microscopic cracks, which form in the surface film, allow penetration of water, resulting in lifting and peeling of the film and discoloration of the exposed wood. One solution to this problem is to incorporate in the coating chemical additives capable of absorbing ultraviolet radiation. The protection afforded is only temporary because the additive itself is soon rendered inactive by the radiation. Although clear coatings with a service life of five years or more are expected to be available in the near future, the problem of maintenance still remains.

Water-repellent preservative finishes containing fungicides are widely used to achieve a natural appearance. Because they penetrate into the wood without leaving a surface film, they are not susceptible to cracking and peeling. A small amount of pigment is often incorporated, partly for decorative purposes, but mainly to increase exterior durability by absorbing and scattering the ulraviolet radiation. The clarity of the wood figure may be somewhat obscured, depending on the amount of pigment added. Although refinishing is required every two to three years, it is relatively simple, as little surface preparation is required.

The U.S. Forest Products Laboratory has developed a semitransparent oil-based stain, which has been modified by the Eastern Forest Products Laboratory (King and McKnight 1965). This stain, commonly known as the Madison formula, is designed for use as a natural finish for western red cedar.

Pigmented Finishes

Pigmented finishes, such as paint, provide the greatest degree of protection against ultraviolet radiation and offer a wide selection of decorative colors. Traditionally these finishes have been based on resins – for example, alkyds – thinned by organic solvents and dissolved in film-forming drying oils, usually linseed or tung oil. When properly applied and cured, these coatings form a continuous, nonporous film that retards moisture penetration and thereby provides some degree of dimensional stabilization to the wood as long as the film remains intact. Because these coatings reduce the cycling of moisture, the wood is less susceptible to staining by water-soluble extractives than when it is uncoated. However, these coatings do have a number of disadvantages, particularly for on-site finishing. They are slow drying, and the moisture content of the wood must be relatively low (12%) before they are applied to ensure proper adhesion. In addition, water may condense under the film, causing blistering and peeling, which may result in a difficult and costly refinishing job.

Water-based latex paints, based on synthetic polymers (usually acrylic or acrylic copolymers) in the form of an emulsion, are increasingly used for exterior applications. Their increased use has come about partly as a spinoff of the trend toward using more environmentally acceptable coatings systems, and partly because water-reducible coatings possess certain inherent advantages. Because they are water-based, these coatings are not particularly sensitive to the moisture content of the wood at the time of application, and because they dry quickly, exterior painting jobs can be completed more rapidly. Unlike paints based on resins dissolved in organic solvents, the emulsion (or latex) paints form discontinuous or porous films, which are more permeable to moisture. As a result, the latex paints are not very susceptible to blistering, but they do show a greater tendency to retain dirt. Although emulsion paints impart less dimensional stability to the substrate than oil-based paints do, they retain much greater flexibility and have less tendency to chalk and erode.

The rate of deterioration of a paint film depends to a large extent on the quality of the coating used and the care with which it has been applied. Conventional solvent-based coatings like alkyds must be applied to dry wood, preferably with less than 12% moisture content, in accordance with the manufacturer's instructions for surface preparation and application rate. Over the long term, a high-quality finish is more economical than one of lower quality, which will require more frequent maintenance. The cost of the labor is generally several times more than the cost of the paint.

Other factors affecting coating performance include the type and quality of the wood used and the design, engineering, and construction of the building. Provision for adequate drainage and ventilation as well as the proper installation of vapor and barriers will significantly increase the service life of exterior wood finishes.

Interior Finishing

Furniture, wall panelling, kitchen and vanity cabinets, floors, doors, and joinery constitute the majority of finished wood products for interior use, and they can usually be protected quite easily. Very often all that is required is a varnish or lacquer coating that can be brushed or sprayed on all panelling, doors, and floors to provide a hard-wearing, natural finish with reasonable stain and abrasion resistance. These coatings are available in high-gloss to matte finishes. In other instances, painting the wood surface is sufficient, providing the manufacturer's instructions for application and surface preparation are carefully followed.

Composite products such as hardboard, particleboard, and inexpensive plywood veneers form the bulk of the wall-panelling and cabinet components that are factory prefinished on automated, high-speed production lines. By this means, the appearance and texture of a finished, expensive hardwood veneer can be reproduced in a few

Figure 8.5 Finished wood products are widely used in the modern home. (Photo: Canadian Wood Council)

minutes in a series of operations that includes filling, groundcoating, sealing, grain printing, embossing, and topcoating (Rogers 1973). The formulation of the coatings used in these operations is critical, however, because they must impart the required protective and decorative properties and yet offer trouble-free application with high-speed production equipment.

Recent building codes require that the wood used in public buildings, hospitals, schools, and mobile homes be treated to meet stringent fire-protection standards. The application of fire-retardant chemicals improves the resistance of wood to ignition and retards the surface spread of flames, thus allowing extra time for extinguishing a fire. In many instances, the required level of fire protection can be achieved through the application of specially formulated surface coatings.

Furniture Finishing

Finishes that preserve or enhance the natural beauty of expensive wood veneers are desirable for furniture and wall panelling (Leary 1970). Such finishes must exhibit good adhesion and resistance to abrasion, yellowing, water, and heat. Skilled craftsmen, employing specially formulated finishes and sophisticated application equipment, are usually required to achieve these goals (Rogers 1973), and finishing is best carried out at the factory. For example, application of the final clear finish may involve many separate steps to develop the natural beauty in the wood as well as to complement or accent the various design features of a piece of high-quality furniture (Collier 1972; Whaley 1972). Some of the basic steps are outlined below.

Bleaching

Bleaching is done to achieve the desired depth and uniformity of color in finished wood products by subduing the natural color variations that often occur. This goal may be accomplished by a variety of chemical bleaches, which must then be completely removed so as not to affect the performance of subsequent coatings.

Staining

Staining imparts the desired undertone color to the wood. Stains may be divided into two broad groups: soluble colors in which the coloring material is dissolved in a suitable solvent and insoluble pigment colors that are dispersed in a liquid suspension. The pigments are finely ground and, when used in small amounts, produce a nearly transparent effect.

Filling

The large open pores of woods such as oak, walnut, and mahogany may be filled to provide a smooth surface. Fillers permeate and level the pores and impart color to them that usually contrasts with the undertone color. Fillers are prepared in a pastelike consistency to prevent settling during storage and consist of about 75% pigment and 25% vehicle, or liquid. In furniture finishing, fillers are normally thinned before use and applied by spraying. They are allowed to penetrate until a soft, moist residue remains; then they are packed into

Figure 8.6 Applying final protective finish to furniture.

the pores in a 'padding' operation in which the surface is rubbed in a circular motion with tow or burlap pads. Excess filler is removed by first wiping across the grain and then carefully wiping with the grain, using clean cloths.

Washcoating

Furniture finishing involves the application of a very thin coat of sealer, or washcoat, in order to subdue the color of the filler – that is, to control the staining action of the filler. In addition, the washcoat provides a hard surface for wiping and, if properly formulated, improves the adhesion and toughness of the total finishing system. Specially formulated nitrocellulose lacquers are often used for this purpose.

Sealing

Sealers are required to build a smooth and level surface for subsequent topcoats. For medium- and high-sheen finishes, sealers with maximum film build and good sanding properties are needed, although these sealers are no longer widely used as a result of the current popularity of low-sheen, low-build finishes. Glazing sealers provide a smooth, nonabsorbing surface for the application of a wiping glaze, which softens or blends the original color without obscuring it. Glazes are applied before the topcoat.

Topcoating

Topcoating provides the final protective film for the finish. Topcoats fall into two basic types: mainly nitrocellulose lacquers, which dry exclusively by evaporation of the solvent; and conversion coatings, which dry by solvent evaporation and polymerization of the film-forming resin by the action of a catalytic agent or heat. Polymer resins commonly used for this type of coating are the alkyd-aminos, vinyl-aminos, polyesters, epoxies,and urethanes. Drying oil or oleoresinous varnishes may also be included, since they polymerize on drying. Such factors as the end use of the article, the desired effects to be achieved, and the cost influence the choice of topcoat.

Future Developments in Wood Finishes

Tremendous strides have been made in the development of wood finishes over the past 30 years. To a large extent, these advances have paralleled the development of new synthetic resins and improved pigments (Prane 1972). While progress is continuing in these areas, other factors are influencing new developments in the wood-finishing industry. Environmental considerations, the increasing trend toward automation, and the rising cost of petroleum-based products are having a profound effect on the industry.

Increasingly stringent legislation to protect the environment is restricting the use of noxious chemicals and factory emission of organic solvents. These restrictions have spurred massive efforts to develop new finishes and processes that will fulfill the protective and decorative requirements while eliminating the associated pollution problems. Significant among these nonpolluting systems are the water-based finishes, which include both water-soluble and water-dispersed resins (emulsions). Other nonpolluting systems that are finding increased industrial use include the solventless or 100% solids coatings. These coatings include resin precursors that can be converted from a liquid to solid film in a few seconds by ultraviolet or electron-beam radiation (Levinson 1972b), and powder coatings in which the resin is applied as a finely divided powder and subsequently fused onto the surface of the substrate by heat (Levinson 1972a).

In the forest products industry today the trend is to prefinishing on automated production lines. Factory finishing not only provides greater economy and quality control, but also permits a greater choice in the type of finish and application and curing techniques that can be used.

Wood products such as hardboard, particleboard, and flakeboard are being used to a greater extent in the building and furniture industries as composite-board technology improves and as solid-wood and plywood veneers become scarcer. The surfaces of such composite-wood products are very different from those of natural wood because of the many different grain orientations of the particles or fibers. These substrates may therefore require specially formulated coatings and application methods. For example, particleboard may be given a smooth, hard surface by the use of a polyester resin filler that is cured by ultraviolet radiation (Bardin 1973). The filled board can then be finished to simulate a natural-wood surface suitable for use in furniture components. The increased consumption of these substrates through advances in composite-board and coatings technology will result in greater and more efficient use of our country's forests by using more of the wood substance leaving less as waste.

Figure 8.7 A line of telephone poles silhouetted against the sky over Saskatchewan. Preservatives can greatly extend the life of wood in such exposed locations. (NFB Photothèque. Photo by Chris Lund)

WOOD-STABILIZATION TECHNIQUES

Wood is hygroscopic, and when it is dried and converted to products, it retains the tendency to absorb liquid water or water vapor. Deterioration of wood products caused by weathering is largely due to this characteristic. The repeated swelling and shrinking cycles result in raised grain, checking, cupping, and warping, with the attendant loosening of joints.

Surface coatings that form continuous films and water-repellent treatments impart some degree of stabilization. Considerable research effort is being expended in the search for methods to overcome this instability and thereby increase the serviceability of wood products. Several techniques have been developed that capable of providing a fairly high level of dimensional stabilization (Stamm 1964; Palka 1970). At present, these stabilization techniques are not economically feasible except for specialty applications that can bear the increased cost. Increasing costs of labor and materials, however, may soon justify the higher cost of producing wood products that require little or no maintenance and have a longer service life.

The bulking effect of extractives causes some woods, such as eastern white cedar, to shrink and swell much less than others. These non-volatile chemicals remain adsorbed on the cellulose molecules when water is removed during drying, thereby restricting the normal tendency of the cellulose molecules to shrink. Not all extractives act as bulking agents; this action is confined to those nonvolatile molecules that are small enough to occupy the adsorption sites between the cellulose molecules. One stabilization technique used to overcome shrinking in permeable woods is to deposit nonvolatile chemicals in the cell wall, where they act as bulking agents by maintaining the wood in its swollen state (Cech 1968). Either nonpolymerizable chemicals, such as water-soluble salts and sugars, or nonvolatile liquids, such as polyethylene glycol, may be used. The latter have been used in producing wooden gunstocks, bowls, steak plates, and a variety of other products that benefit from stabilization.

Another technique involves the impregnation of the cell lumen and walls with chemicals that are converted into insoluble polymers through the action of heat or radiation. The presence of the polymer in the cell can retard the movement of water vapor and substantially increase the mechanical properties of the wood (Palka 1970). These wood-plastic composites have found some commercial application in hardwood flooring, stabilized hardwood veneers, and specialty items such as billiard cues and golf-club heads.

DECAY AND STAIN IN WOOD

Many species of primitive plants, known as fungi, live in wood. Some of these organisms use only the food that is stored in the wood (molds and sap stains), while others (wood-destroying fungi) attack the cellulose or lignin and ultimately rot the wood. Under suitable conditions these wood-destroying organisms may attack both trees and manufactured wood products. Except in tropical regions where insects are more active, fungi are the chief cause of the biodeterioration of wood. These specialized plants attack the structural elements and cause deterioration of wood (decay or rot), thus degrading its natural properties. Specific forms of deterioration can also be ascribed to bacteria, but their attack is far less extensive (Wilcox 1970; Rossell, Abbot, and Levy 1973).

Because we are concerned with the functional use and appearance of wood, the activities of all wood-inhabiting organisms are considered harmful. Estimates of losses from biodeterioration of logs and timber are higher than those from wildfire in the forest and have exceeded 20% in North American stands. Fungi may also be responsible for extra labor and production costs when business activity is disrupted by replacement of timber that has deteriorated in service. For example, in one special case, it has been estimated that over $6000 was expended when one decay-weakened transmission pole was replaced and power had to be curtailed. Between 85 000 and 100 000 poles are replaced annually in Canada because of decay, most of them in urban areas where costs of machines, poles, and labor can average $500 or more per pole.

For certain uses, chemicals can extend the service life of wood. Untreated wood used for poles, railway ties, bridges, and boats and in other exposed locations has been known to fail in three years. With preservative treatment, however, the service life of wood can be extended to 35 years or more. When preservatives are properly applied to wood before it is used in construction, there is little risk of subsequent damage to the environment. When properly fixed, such preservatives remain within the wood indefinitely and form an effective barrier against attack by wood-destroying organisms.

Figure 8.8 Fruiting body of the fungus *Merulis lacrymans* on a basement ceiling in a cool, wet location.

Growth of Fungi

The body of a fungus is composed of slender, tubelike conductive strands called hyphae, which secrete enzymes, or biochemical catalysts, to enable the fungus to digest the walls and contents of cells. Reproductive structures, such as fruiting bodies, or sporophores, may be produced from the hyphae and are the means of identification for the many thousands of existing fungal species (Figure 8.8). There are numerous forms of decay fungi, many of which are best known as mushrooms or conks. They may appear on piled logs and timber, on timber in service, on wood buried in soil, or on trees in the forest. At maturity, they can produce millions of spores. These microscopic, seedlike bodies may be dispersed for long distances, carried in the air or in water droplets, or sometimes by insects feeding on the mycelium – the mass of filamentous hyphae. Under favorable conditions the spores that reach moist wood germinate and develop new areas of decay. Fungal damage may also be spread by direct contact between decayed material and sound wood when hyphal strands (Figure 8.9) develop at the surface between the adjacent pieces.

The natural durability of wood, or its ability to resist fungal attack, depends on properties associated with the growing tree. Certain species, such as western red cedar and oak, are well known for their high durability, but this applies only to the heartwood because the sapwood is not durable. Decay resistance is provided by fungicidal chemicals formed as the tree grows, and these substances are deposited in the heartwood cells as the tree matures. In most woods these chemicals are more concentrated in the outer heartwood than in wood near the pith; thus, in many cases the inner heartwood tends to be reasonably low in resistance to fungal attack. The relative decay resistance of some species that grow in Canada is shown in Table 8.1. Chapter 2 also contains an appraisal of the durability of Canadian species.

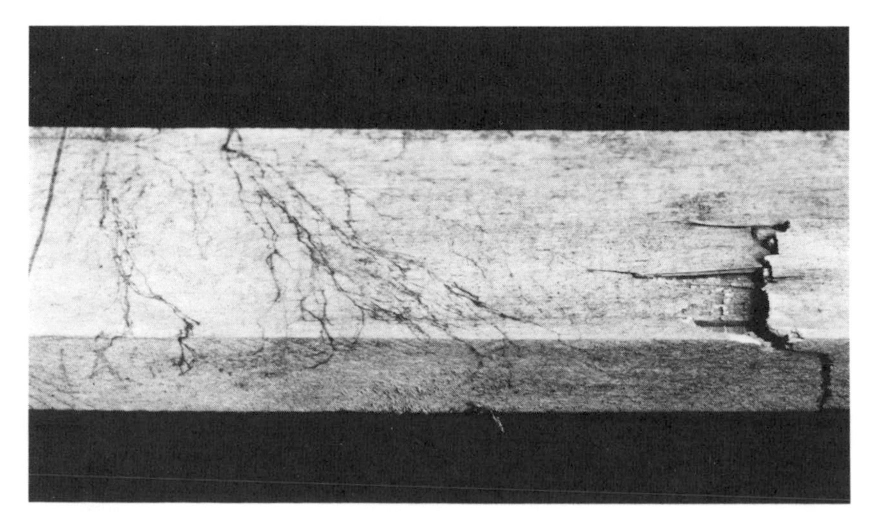

Figure 8.9 Hyphal strands of a wood-destroying fungus (*Coniophora puteana*) on the surface of hardwood lumber.

Table 8.1 *Relative decay resistance of heartwood*

Resistant or very resistant	Moderately resistant	Slightly resistant or nonresistant
Cedar	Douglas-fir	Alder
Cherry, black	Honeylocust*	Ash
Chestnut, sweet	Larch, western	Basswood
Cypress	Pine, eastern white	Beech
Juniper	Tamarack	Birch
Locust, black		Butternut
Mulberry, red*		Elm
Oak, bur		Hemlock
Oak, white		Hickory
Sassafras		Maple
Walnut, black		Oak, red and black*
Yew, Pacific		Pine (other species)
		Poplar
		Spruce
		Willow

Source: Adapted from U.S. Dep. Agric., Forest Serv., Forest Prod. Lab. Tech. Note 229, May 1961

*These woods have higher decay resistance than most others in their group.

Both the development and the spread of wood-inhabiting fungi require food, air, moisture, and warmth. Food is supplied by the wood components, including cellulose, lignin, and the cell contents. While certain woods are more susceptible to attack (i.e., are less durable) than others, all species will eventually be destroyed by fungi under suitable conditions.

The supply of oxygen contained in the woody cells of forest products is usually adequate to support fungal growth. However, if it is removed, as in water-saturated wood (submerged logs), or contaminated by the introduction of fungicidal gases (fumigation), decay will be arrested.

Moisture is most important in the development of fungi and, in combination with air, governs the extent of decay in wood. Infection of wood through germination of spores or by penetration of the mycelium requires both a moist surface and a humid environment. Once the hyphae penetrate the surface, a balance of oxygen and moisture is needed to maintain the respiration that supports the enzymic decomposition of the cellular material.

In living trees the moisture-air balance is significant in that the heartwood can support decay organisms (causing heart rot), whereas the sapwood remains free of decay as long as the bark is intact because the sapwood is saturated with moisture. Once the tree is felled and the log is converted into wood products, air replaces some of the water in the sapwood. Deterioration caused by fungi can occur and will continue until the wood dries to a moisture content below 20%. Wood is considered immune to decay as long as its moisture content remains below 20%, although fungi present in the wood will not necessarily be killed under that condition; certain fungi can remain dormant for very long periods of time and revive when sufficient moisture again becomes available.

Fungi can tolerate a wide range of moisture content in wood and grow rapidly in wood containing approximately 50% moisture; the exact amount of growth depends on the species of fungus and on the density of the wood. The denser woods are more resistant to decay because there is more wood substance in these species and less void space than in the less dense woods. Under high-moisture conditions, with more water occupying the cell lumens, dense wood becomes more saturated than less dense wood. Consequently less oxygen is available for fungal growth, and, as a result, the denser wood has better decay resistance.

Fungi grow more rapidly in warm weather than in cold weather and become dormant under freezing conditions. The optimum temperature for most wood-destroying fungi is around 22°C (70°F). Although a small group of fungi called thermophiles is still active above 35°C (95°F) (Cooney and Emerson 1964; Smith and Afosu-Asiedu 1972), most species are inactive at this temperature. Fungi are killed by prolonged exposure to temperatures above the maximum at which their growth ceases, particularly in a humid atmosphere. Conditions in commercial dry kilns using temperatures above 65°C (150°F) will kill wood-inhabiting fungi.

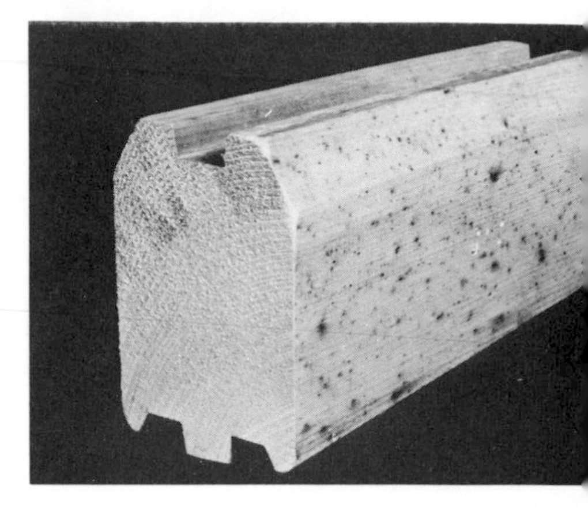

Figure 8.10 Black discolorations caused by mold (*Aureobasidium* sp.) in white pine.

Characteristics of Mold, Sap Stain, and Decay

Wood-inhabiting fungi can be broadly classified into different types, depending on their capacity to break down and digest the lignin-cellulose complex of cell walls. One very large group includes molds and sap stains (Figure 8.10) that cause little or no damage to structural elements because they obtain nourishment mainly from stored food (sugars and starches) contained within the cells. These fungi do not reduce strength significantly and, in some species, appear to inhibit decay to some extent (Shields and Atwell 1963; Hulme and Shields 1972). Mainly they produce unsightly discolorations in shades of blue-black through green and yellow to pink.

Figure 8.11 Brown cubical rot in western red cedar.

The discoloration resulting from the growth of molds (including species of *Aureobasidium*, *Chaetomium*, *Cephaloascus*, *Penicillium*, and *Trichoderma*) is largely superficial and can often be removed by brushing or planing the wood. In contrast, sap-staining fungi (including species of *Ceratocystis*, *Fusarium*, *Stemphylium*, and *Chlorosplenium*) penetrate deeply, and the stains are harder to remove.

Seen under the microscope, the discoloration is often associated with characteristic dark-colored hyphae, which develop extensively in the wood rays, where reserve food is most abundant. The hyphae pass from cell to cell, mainly through the pits. In some cases where they penetrate cell walls, the normally thick hyphae become constricted to fine threads within the tissue (Roff 1964).

Spores of both molds and sap-staining fungi are produced in copious numbers, either directly from the mycelium or sometimes from the ends of erect, microscopic, hairlike structures that give a fuzzy appearance to the mycelium mat. Under warm, humid conditions (particularly if unseasoned wood is freshly cut), mold and sap stain can become visible within one day, the initial rate of development being many times that of decay-causing fungi.

The significance of bacteria in the breakdown of woody tissue is becoming more evident, particularly in wood piles used in the support

Figure 8.12 Decay in a porch deck caused by long exposure to moisture.

of buildings. Many such piles have a high moisture content or are submerged in water, conditions conducive to bacteria attack that is dependent on the presence of free water in the cells. Under these conditions, material in the cell walls and in the pits is selectively degraded, and the wood becomes more porous (Ellwood and Ecklund 1959).

Most wood-decaying fungi show a marked preference for a specific type of wood. Knowledge of this preference, plus a general appreciation of their appearance and the damage they cause, can greatly assist in estimating the severity of the damage that has ensued.

True wood-destroying fungi possess enzyme systems that hydrolyze cellulose and other polysaccharides of woody cells into glucose and other simple nutrients. Many species in this group cannot attack the lignin portion of the cell walls, but after they have removed the accessible carbohydrates the remaining material may be dry, shrunken, often darker in color, and broken into small, crumbling, brick-shaped pieces. This condition is known as brown rot, or brown cubical rot (Figure 8.11). This term also includes the condition incorrectly referred to as dry rot, which results from similar fungal processes.

Under wet conditions, deterioration of the surface of wood resembling brown rot may occur. This condition, called soft rot, is caused by certain microorganisms, including bacteria and fungi, in the superficial layers of the wood. It does not proceed as rapidly as other types of fungal rot.

Other wood destroyers not only attack cellulose, but also can break down lignin by means of oxidizing enzymes. These rots are more varied and are usually lighter in color than the surrounding wood. Known as white rot, they can be fibrous or stringy, or the affected portions may occur in discrete pockets or concentric bands surrounded by firm wood. The pockets may appear pitted, honeycombed, mottled, or as a soft, spongy mass.

From the scores of fungi usually present in rotting wood, certain species have been selected for the following discussion because their

Figure 8.13 Soft rot on timber from a freshwater jetty showing erosion caused by water and sand.

pattern of attack is characteristic of the rot type to which they belong and because they are economically important in the forest or in wood products.

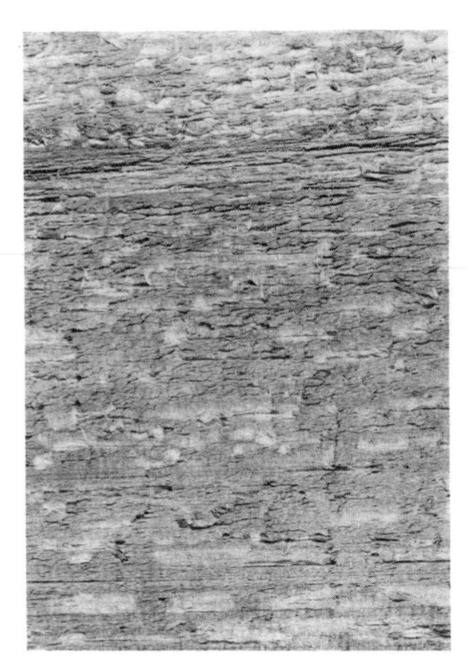

Figure 8.14 White pocket rot in white pine caused by the fungus *Fomes pini*.

Fomes pini Karst. (white pitted rot, white pocket rot, conk rot)
This fungus causes one of the most widespread diseases in coniferous forests. It causes a white rot in heartwood (Figure 8.14), which in the advanced stage appears as small, spindle-shaped cavities, often filled with fibrous material. It may involve the entire heartwood portion or be confined to a few annual rings. In the early stage of attack (incipient decay), the wood has a pinkish to reddish-purple stain, which is called red heart stain in many woods. Wood with advanced white pitted rot is relatively weak, but in the incipient stage it is as strong as unstained material (Atwell 1948; Kennedy and Wakefield 1948). The decay does not advance in service, and except where strength requirements are critical, red-heart-stained lumber is widely accepted in permanent construction. Wood with white pitted rot has also been used as decorative panelling.

The bracketlike fruiting body develops on trees. It is perennial and has a gray-brown upper surface with yellow-brown on the underside.

Fomes igniarius (L. ex Fr.) Kickz (white trunk rot, white heart rot)
Fomes igniarius is a major cause of losses in hardwood forests. It produces a white rot of heartwood that is yellowish to white in color, has a soft and punky texture, and contains irregular patterns of dark lines. The incipient stage appears as gray-yellow and brown to gray-black zones in the wood. The presence of this stain does not indicate significant weakening of the wood, according to work done at the Forest Products Laboratories of Forintek Canada Corp. The perennial sporophores appear on trees and on down timber. They are woody and generally hoof-shaped with a brown undersurface; the upper surface is black and deeply fissured when old.

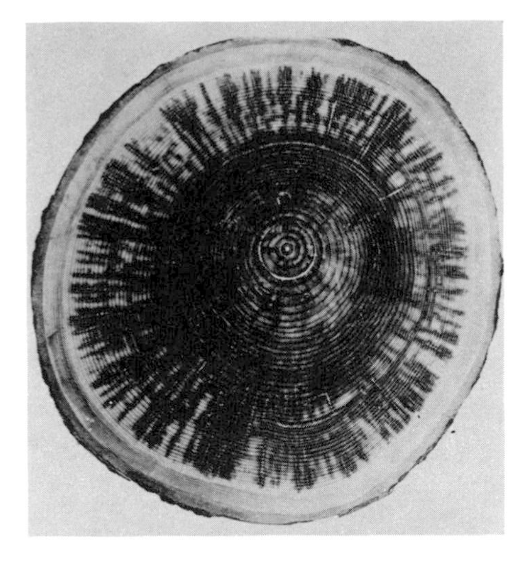

Figure 8.15 Red heart rot in balsam fir associated with the fungus *Stereum sanguinolentum*.

Stereum sanguinolentum (Alb. and Sch. ex Fr.) Fr. (red heart rot, red-brown stain)
Stereum sanguinolentum is widespread in coniferous forests and in down timber. It causes a white rot that is light to red-brown in color, dry, and somewhat stringy in texture; in logs it is usually a circular mass about the pith. Red heart rot is particularly damaging in the true firs (Figure 8.15). Incipient decay develops as a firm red-brown stain, often in streaks; on the end grain it usually appears as rays extending from a stained area. The decay does not advance in wood in service, and the stained wood is not significantly weaker than unstained material (Roff and Whittaker 1963); nor is the pulp strength likely to be affected. The annual sporophores occur profusely, appearing as thin, leathery sheets, often with the upper part curled to form a narrow shelf. When bruised, a fresh fruiting body exudes a blood-red stain.

Lenzites spp. (brown pocket rot)
Two *Lenzites* species, *L. trabea* Pers. and *L. saepiaria* Fr., are similar in growth and appearance, and both occur throughout the world. These brown cubical rots occur in softwoods, and less frequently in hard-

Figure 8.16 Sporophore and mycelium growth of the fungus *Lenzites saepiaria* on a roof sheathing board.

Figure 8.17 Sporophore of the fungus *Lenzites* sp. (*top view*) taken from top surface of a fence post.

Figure 8.18 Bottom surface of sporophore of *Lenzites* sp. showing radiating gills and irregular pores.

woods, in dead timber in the forest and in wood in service. Damage is extensive in exposed wood in buildings, poles, posts, and ties; the decayed wood is broken into small, cubical pieces, sometimes with thin strands of brown mycelium present (Figure 8.16). The fungi survive for very long periods in air-dry wood.

Wood with incipient decay may appear with yellowish to yellow-brown patches or sometimes with blackened areas. The incipient-decayed material, like that caused by other brown cubical rots, is seriously weakened, gives low pulp yields, and should be avoided. The annual sporophores are abundant. They may be semicircular or narrowly shelflike with a velvety, cinammon-brown upper surface (Figure 8.17). The yellow-brown lower surface has radiating gills or irregular pores (Figure 8.18).

Merulius lacrymans (Wulf) Fr. and **Poria incrassata** (B. and C.) Curt. (dry rot, building rot)

Both *Merulius lacrymans* and *Poria incrassata* cause a similar type of brown cubical rot, mainly on coniferous timbers in damp, unventilated situations. These two species, detailed below, cause the most extensive losses from decay in wooden buildings in North America. Once established, the decay can spread rapidly. Early infection is often invisible or may be seen as a superficial, thin, silky gray mycelium with patches of yellow, or as a fan-shaped, lilac-colored mycelial mat. Characteristic thick strands, which are brown to black, may develop on the wood. Because they can also be found in adjacent brickwork, the decay is very difficult to eradicate (Figure 8.19).

M. lacrymans occurs primarily in eastern Canada and the northern United States. The typical sporophore is a thick, pale gray, platelike or bracket structure with a wrinkled surface that may exude water droplets. In time the conk becomes rusty red as a result of the myriads of spores produced. These conks appear in areas adjacent to the decay.

P. incrassata sporophores are similar in shape and are pale olive-gray with a pale yellow margin when young. With age, the sporophore becomes crustlike and brown to black in color, sometimes with masses of orange mycelium. This fungus is reported more commonly in the Pacific coast region of North America and in the southern United States.

Treatments to Control Stain and Decay

Wood that is kept dry – that is, below 20% moisture content – will not deteriorate because of fungi and will be protected for as long as dry conditions last. Similarly, growth of terrestrial, wood-deteriorating fungi will cease on wood stored underwater because no oxygen is available. Under outside service conditions, however, it is not always possible to keep the wood either dry enough or wet enough to control decay (Figure 8.20). To obtain satisfactory service life, the wood must be treated with protective chemicals. Although a preservative may be effective against fungi, it cannot adequately protect wood in service unless the treating solution penetrates deeply into the outer zone of the timber in sufficient concentration to inhibit fungal growth.

A good preservative must be a strong fungicide, possess good penetrating qualities, be leach resistant, and should not adversely affect such properties of wood as swelling, shrinkage, and strength. Treated wood should have a clean appearance, resistance to glowing or visible combustion, low mammalian toxicity, and low cost. It should be water-repellent and fire resistant, and it should not change color as a result of preservative treatment. The treated surface should also take paint well. No single preservative now in use can fulfill all these requirements.

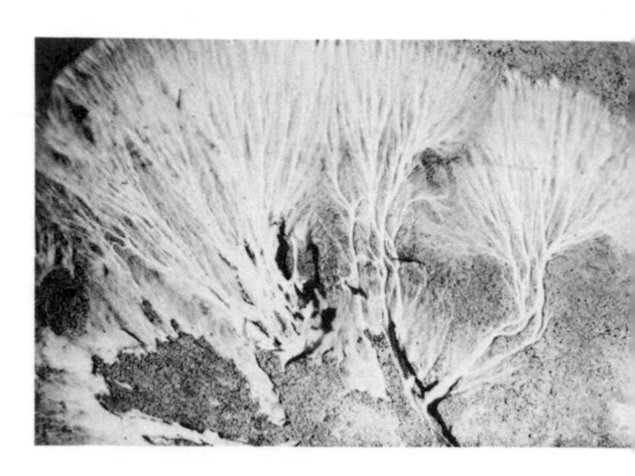

Figure 8.19 Strands of the fungus *Merulius lacrymans* can grow over masonry to attack new wood.

PRESERVATIVE TREATMENT OF WOOD

The modern wood-preserving industry started in the eighteenth and nineteenth centuries when major efforts were made to prevent wood decay in British warships. British railway companies began treating their crossties soon after the patent for pressure-impregnating wood with coal-tar creosote was obtained by Bethell in 1883. The first treating plant of this type in the United States started operations in 1865. In Canada treated crossties were used by about 1906, and the first commercial treating plant for this purpose was built in 1911 in eastern Canada. However, the first operation in Canada using a treating cylinder was established in 1910 at Dominion Mills, Vancouver, for the treatment of wood paving blocks.

The wood-preserving industry has continued to expand and has contributed a great deal to Canada's economic growth. The chief advantages of using treated timber are the substantial savings in labor costs for replacing untreated structures that have become defective and the reduced maintenance costs of treated wood in service.

Specifications covering the quality and types of treatments for poles, crossties, piling, and other wood products are given in standards issued by the Canadian Standards Association (1974) and the American Wood Preservers Association (1974).

Testing New Preservatives

The efficacy of new preservatives is first tested in the laboratory by either the standard soil-block method (American Wood Products Association 1974) or the more rapid respirometry technique developed in Canada (Smith 1975). The former method is based on weight-loss measurements of small, treated wood blocks incubated for 12 weeks on actively growing test fungi. The second method uses automated gas chromatography to measure the carbon dioxide produced under similar test conditions and requires a much shorter time. The minimum retention of preservative (threshold value) required to protect the wood blocks from decay can be determined by either method.

If the laboratory results are satisfactory, wooden stakes and posts, treated with the preservative to various levels of retention, are exposed to natural decay hazards in outdoor field trials. Similarly, tests with small field stakes will indicate in the shortest time whether the preservative treatment will provide satisfactory performance for its intended use.

Wood Preservatives

Creosote is one of the most widely known preservatives and one of the earliest used to prevent decay in wood under service conditions. In addition to creosote, coal tar and petroleum oils are used as solvent

Figure 8.20 Storage of logs under sprinklers protects wood against insects and decay.

systems for other fungicidal chemicals, such as pentachlorophenol. Other types of preservatives use water as the solvent system. These preservatives may be single-salt systems, such as copper sulfate, or multisalt systems containing various chemicals, such as salts of copper, chromium, and arsenic.

Oil-Borne Preservatives

Creosote

Creosote is obtained from the distillation of coal tar, which in turn is a byproduct of the carbonization of coal. Another type of creosote can be obtained from the destructive distillation of wood, but it is not highly regarded as a preservative. Because coal-tar creosote is actually a mixture of several compounds, the composition of creosotes used by the wood-preserving industry varies. Whatever the coal source of creosote, it is considered one of the most effective preservatives and has been widely used. Because of its resistance to leaching by water and its low volatility, it has been used to protect timber submerged in sea water from marine borers as well as to prevent decay in timbers exposed to the weather. Wood treated to high creosote retention is apt to have a 'dirty' surface and cannot be painted with conventional paints. The amount of creosote required for adequate protection varies with the species of wood and the conditions of service. For marine piling, complete full-cell saturation of the sapwood is needed and over 256 kg/m^3 (16 lb/cu ft) of creosote is required. For building timbers, 96–128 kg/m^3 (6–8 lb/cu ft) for above-ground exposure and 128–160 kg/m^3 (8–10 lb/cu ft) for soil contact are appropriate. For larger and smaller timbers, the absorption should be decreased and increased, respectively, in accordance with the change in ratio of surface area to volume.

Figure 8.21 Pressure-treated with creosote, this wood-stave duct was built for Weyerhaeuser Canada Ltd. at Kamloops, British Columbia, to deliver pulp mill gases almost a half mile to a dispersion flue on a nearby mountain. Pressure-treated wood was chosen for this unique application because of its ability to meet the demanding conditions of service. (Photo: Canadian Institute of Timber Construction)

Creosote-Tar and Creosote-Petroleum Mixtures

Dilution of creosote with either coal tar or petroleum (heavy fuel oil) has been used in the past to reduce the cost of the preservative and to prevent excessive splitting or checking of railway crossties. Products treated with these mixtures have a dirty appearance as a result of the bleeding of the preservative. Recommended retentions for creosote-tar mixtures are similar to those for creosote: 112–144 kg/m^3 (7–9 lb/cu ft) for crossties and above-ground timbers, and 192 kg/m^3 (12 lb/cu ft) minimum for ground contact.

Naphthenates

Copper naphthenate is obtained by reacting copper salts with naphthenic acids, which are byproducts of petroleum refining. The dark green preservative is usually applied by brushing or dipping the wood to be used in above-ground locations. For these treatments, the preservative usually has a copper content of 1–3% in a light petroleum solvent. Copper naphthenate is often used to treat wood in greenhouses and poultry houses. Recommended retentions from pressure treatments using copper naphthenate in petroleum oil (0.74% copper equivalent) are 96 kg/m^3 (6 lb/cu ft) for wood above ground to 160 kg/m^3 (10 lb/cu ft) for wood in ground contact. Zinc naphthenate is also used to treat wood. It is almost colorless and is not quite as effective against fungi as copper naphthenate.

Chlorinated Phenols

Pentachlorophenol is the most widely used of the chlorinated phenols in wood preservation. It is a highly effective organic chemical and is used at concentrations of 3–5% dissolved in heavy petroleum oils. Its low water solubility is an important factor in its effectiveness against wood-rotting fungi and insects. The minimum net retentions of pentachlorophenol (equivalent to 5%) should be 96 kg/m^3 (6 lb/cu ft) of solution for wood not in contact with the ground and 160 kg/m^3 (10 lb/cu ft) for wood in ground contact. In recent years liquefied petroleum gases and methylene chloride have been used as solvents for pentachlorophenol in place of oils. Pentachlorophenol is also used in water-repellent solutions at 5% concentration for dipping millwork.

Organometallic Compounds

Organotin compounds in petroleum solvents are a comparatively recent addition to the wood-preservation industry. Laboratory studies

Figure 8.22 Heavy timber piling pressure-treated with creosote can withstand marine hazards for decades – the 2591 m (8500 ft) Annieville dyke on the Lower Fraser River, British Columbia. (Photo: Canadian Institute of Timber Construction.

have shown that these compounds, trialkyltin derivatives (particularly tri-*n*-butyltin oxide) are the most effective against wood-rotting fungi (DaCosta and Osborne 1972). Tri-*n*-butyltin oxide is not recommended, however, for preserving wood in ground contact because decay has appeared in field stake tests conducted under standard P11-70 (American Wood Preservers Association 1974). It has been used extensively in Europe for treating millwork by dipping and vacuum treatments. For this purpose, a 0.3–1.0% solution of tri-*n*-butyltin oxide in organic solvents is used.

Organochlorine Compounds

From 0.5 to 1.0% of the gamma isomer of benzene hexachloride (gamma-BHC) in fuel oil has been used as a spray for logs and other wood products to protect them from insect attack during storage. It also reduces the incidence of blue fungal stain associated with insect borings in pine sapwood (Savory, Pawsey, and Lawrence 1965). The addition of 2.5% pentachlorophenol has been recommended to increase protection against stain fungi. An emulsified form of this preservative, known as lindane, is also available.

Figure 8.23 A familiar use of pressure-treated timber is for railroad ties. When adequately treated with creosote-petroleum mixtures, ties have a service life of about 40 years. (Photo: Canadian Institute of Timber Construction)

Quinolinolates

Copper-8-quinolinolate is a yellowish-green preservative that has some effectiveness against decay fungi. Because of its low toxicity to mammals, it is sometimes used to treat fruit and vegetable harvest baskets and other wooden containers used to store food. It has also been used to protect lumber from staining fungi. Wood products are treated by both dipping and brushing as well as by pressure processes.

Water-Repellent Preservatives

These preservatives are principally used to treat millwork and consist essentially of a fungicide dissolved in mineral spirits or light petroleum oil; antiblooming agents and waxes are added to make the preservatives water-repellent and to prevent crystallization, or blooming, of the fungicide on the surface of the treated wood. The wood is treated by dipping, brushing, or a vacuum process. Although 4–5% pentachlorophenol is the most common fungicide used in these systems, tri-*n*-butyltin oxide at 0.3–1.0% and copper naphthenate at 1–2% concentrations of copper have also been used effectively.

Modifications of these systems, usually containing 2–3% pentachlorophenol, have been used as surface protective agents against wetting and fungal attack of seasoned lumber during transit and storage. Some concentrations of these protective systems may affect gluing and painting because of the high amount of waxes used.

Water-Borne Preservatives

Certain water-borne salt systems have gained wide acceptance as effective heavy-duty preservatives for wood. Although some of the single-salt systems have low resistance to leaching (e.g., borax and copper sulfate), the multisalt systems are usually well fixed in the wood. Water-borne treatments have the advantage of being generally clean and odorless, and the dry, treated wood can be readily painted. Disadvantages include dimensional changes, which may occur during treating, and the necessity of redrying the timber.

Figure 8.24 Water-borne preservatives are used to ensure long-term protection of preserved wood foundations.

Borax

Sodium octoborate tetrahydrate is the most commonly used borate in the diffusion treatment of green lumber (explained under 'Treating Processes' in this chapter) to protect against decay fungi and insects in buildings. Concentration of the borate may be 20–40% in diffusion treatments, depending on lumber thickness. The addition of 0.8% sodium pentachlorophenate to the borate solution has been recommended to control the growth of molds until the wood is seasoned. The

Figure 8.25 Water-borne preservatives, along with oil-borne preservatives, are widely used to treat (*left*) utility poles and (*right*) highway safety structures. (Photos: Canadian Institute of Timber Construction)

recommended net dry-salt retention is 5.3 kg/m^3 (0.33 lb/cu ft) boric-acid equivalent. The borate is leachable because it is not fixed in the wood; therefore, the treated wood is not recommended for use in ground contact or where frequent wetting occurs.

Borax is used together with sodium pentachlorophenate in antistain products intended to protect unseasoned lumber from stain and mold (Verrall and Mook 1951) and has also been valuable in controlling deterioration in stored wood chips (Hulme and Shields 1973).

Copper, Zinc, Chromium, and Arsenic Mixtures

Various formulations have been developed using mixtures of copper salts with or without arsenic and chromium. Some of these systems, such as acid copper chromate (ACC), ammoniacal copper arsenate (ACA), and chromated copper arsenate (CCA), are very effective preservatives against common types of wood-destroying fungi. They are widely used in pressure treatments where good performance of wood products is required under severe decay-hazard conditions. Ammoniacal salt compositions being studied at the Eastern Forest Products Laboratory of Forintek Canada Corp., incorporating copper with arsenic, or copper and zinc with arsenic, have shown very good leach resistance and the capability to penetrate difficult-to-treat wood species, such as spruce. The copper-arsenic preservative has been accepted as a modified ACA in the CSA standard O80 Wood Preservation;

the copper-zinc-arsenic preservative (CZAA) is undergoing further study and evaluation in field trials.

Ingredients used in the various formulations in this group of preservatives include ammonia, ammonium carbonate, arsenic acid, arsenic trioxide, arsenic pentoxide, basic copper carbonate, chromic acid, copper hydroxide, copper oxide, copper sulfate, sodium arsenate, and sodium dichromate. Recommended minimum retentions by weight for timber such as poles and posts in ground contact are 8 kg/m^3 (0.5 lb/cu ft) for ACC (posts only), 6.4 kg/m^3 (0.4 lb/cu ft) for ACA, and 6.4 kg/m^3 (0.4 lb/cu ft) for CCA (Canadian Standards Association 1974; American Wood Preservers Association 1974).

Copper Sulfate

Copper sulfate has been widely used in rural areas to dip-treat local wood species for use as fence posts. The freshly cut posts are placed in a tub or barrel, usually containing a 1% solution. The main disadvantages of copper sulfate are that it is corrosive to iron and steel containers, it leaches readily, and it is not as effective against wood-decaying fungi as some of the multisalt systems are.

Fluor-chrome-arsenate-phenol Mixtures (FCAP)

Formulations using this combination of salts contain sodium fluoride (20–24%), arsenic pentoxide (22–28%), chromate (33–41%, oxide basis), and dinitrophenol (14–18%). Some forms of this preservative are applied as a paste to the surface of freshly cut wood, and the salts are allowed to diffuse into the wood. Recommended minimum retention for FCAP is 3.5 kg/m^3 (0.22 lb/cu ft) for above-ground use. FCAP is not recommended for ground contact because the treated wood is not as resistant to leaching or to attack by soft-rot fungi as is wood treated by some of the other multisalt systems.

Sodium Fluoride

Sodium fluoride is rarely used alone to preserve wood because of its leachability, but it has been used in combinations with other ingredients in the pressure and diffusion processes. It is mainly used in combination with creosote or pentachlorophenol as a supplementary treatment applied as a paste to the ground-line area of standing poles. It is noncorrosive to metals. Sodium fluoride has been used successfully to prevent the development of a chemical brown stain, called kiln burn, during kiln drying of eastern white pine and is comparable in effectiveness to the more hazardous sodium azide used for the same purpose (Cech 1966).

Sodium Salts of Chlorinated Phenols

Either sodium tetrachlorophenate or sodium pentachlorophenate is used with other ingredients in water as a dip treatment for unseasoned timbers, particularly softwood lumber, to prevent the development of sap-stain fungi and molds during air seasoning in mill yards. These superficial treatments have proved to be very effective in controlling fungal discolorations (Cserjesi and Roff 1964; Roff, Cserjesi, and Swann 1974). Concentrations of sodium pentachlorophenate used in the aqueous dip solutions may vary from 0.25 to 0.75%, depending on climatic conditions in the seasoning yards.

Zinc Salts

Zinc salts are preferred if more or less colorless treatments for wood products are desired. Although it is no longer used alone, zinc chloride had long use as a wood preservative and was used in concentrations of up to 5% in pressure treatments. It was generally considered moderately effective as a preservative, but it had low resistance to leaching. Because of its acidic nature, it caused wood deterioration in high concentrations and was corrosive to steel containers. Chromated zinc chloride (CZC), alone or with cupric chloride, is more resistant to leaching and gives greater protection against decay than zinc chloride. The composition of chromated zinc chloride is 18.5% sodium dichromate and 81.5% zinc chloride, while the copperized formulation contains 7% cupric chloride, 20% sodium dichromate, and 75% zinc chloride. A minimum retention of 16 kg/m^3 (1 lb/cu ft) is recommended for both forms.

Zinc meta-arsenite (ZMA) is composed of 60% arsenic trioxide and 40% zinc oxide held in solution by acetic acid. Recommended retention is 5.6 kg/m^3 (0.35 lb/cu ft). This preservative is seldom used now and may be considered obsolete.

Zinc oxide has been found to be effective in paints to reduce mildewing of painted wood on buildings (Sell 1968). It has low toxicity to mammals.

Boliden salt (CZA), composed of 43% zinc sulfate, 16% sodium dichromate, 21% sodium arsenate, and 20% arsenic acid, makes an effective preservative. In the copperized form half of the zinc sulfate is replaced by copper sulfate. The minimum recommended retention for these forms of Boliden salt in pressure treatments is 16 kg/m^3 (1 lb/cu ft). This compound has been substantially replaced by Boliden K33, a nonleachable oxide formulation of copper, chromium, and arsenic (CCA).

Supplementary Treatments

Preservatives used to treat standing poles are highly viscous in order to facilitate their application by trowel or stiff brush to the base of the pole before application of the protective bandage. Such preservatives diffuse or soak into wood. They contain mixtures of creosote and pentachlorophenol with sodium fluoride or sodium borate. Other salts, oils, and solvents have also been used.

For effective protection against decay in service, all machining of preframed and prebored wood must be carried out before treatment. If the wood is bored or cut after treatment, the newly exposed surfaces should be brushed with at least two coats of preservative. Preservative should be poured into bolt holes and other openings. Hot preservatives are preferred in all cases. In addition to creosote, pentachlorophenol in light petroleum oil, tri-*n*-butyltin oxide, and copper and zinc napththenates have been used as supplementary treatments.

Marine piling should not be cut or bored below the high-water line. If it is necessary to bore a pile above the water line, the exposed cut surfaces should be brushed liberally with two coats of hot creosote followed by a coat of coal-tar pitch.

TREATING PROCESSES

Various processes are used to protect wood, depending on the degree of hazard and quantity of preservative needed. Light retentions permit brushing, dipping, spraying, or diffusion treatments. Heavy retentions require preservative impregnation under high pressure in enclosed cylinders.

Conditioning Timber for Treatment

Before treatment, wood must be conditioned to ensure optimum penetration of preservatives by the various processes. For diffusion processes using water-soluble salts, the wood should be in an unseasoned condition.

For effective treatment by other processes, poles, crossties, and other timbers should be debarked, and then the moisture content can be reduced by air seasoning, kiln drying, or boiling under vacuum in the treating cylinder. As a general rule, the drier the wood, the better the penetration of preservatives. The extent of seasoning should be such that no subsequent checking of the wood will take place in service and expose untreated wood to fungal colonization.

The removal of bark is particularly important, since its presence retards seasoning and impedes the penetration of preservatives. For poles, a full year's seasoning is desirable. In central and eastern Canada, railway ties that are cut and piled to air-dry during the winter and early spring can be pressure-treated by late summer or early autumn. The moisture content of the wood should not be greater than 30% for water-borne preservatives or more than 25% for creosote pressure processes.

Vapor drying of unseasoned timber using organic solvents, such as xylene, has been carried out in some operations before treatment with preservatives (Hudson 1969). In this process, vapor from the boiling organic liquid condenses on the cooler timbers, heating the wood and driving off moisture.

In commercial treating plants, green or partially seasoned timbers can be heated in the treating cylinder and then given a vacuum treatment to evaporate moisture in the wood. Another conditioning method, known as Boultonizing, is to remove water by boiling green timbers in the treating oil under vacuum.

The sapwood of almost all wood species can readily be penetrated by preservatives. In contrast, most heartwood and the dried sapwood of spruce, hemlock, and some Douglas-fir are difficult to treat by pressure processes. Such wood is referred to as refractory (some of the reasons for this condition are discussed in 'Cell-Wall Structure' in Chapter 3). To obtain better penetration, squared heartwood crossarms, poles, and bridge timbers are incised before treatment by passing them between a series of toothed rollers.

Penetration of preservative solutions into the sapwood of difficult-to-treat wood species, such as spruce, can be considerably improved by ponding the poles, or storing them underwater, or by continually sprinkling them with water for several weeks. The poles are then removed from water storage, peeled, and seasoned before treating. In a

Figure 8.26 Samples of spruce poles showing deep penetration of preservatives after ponding.

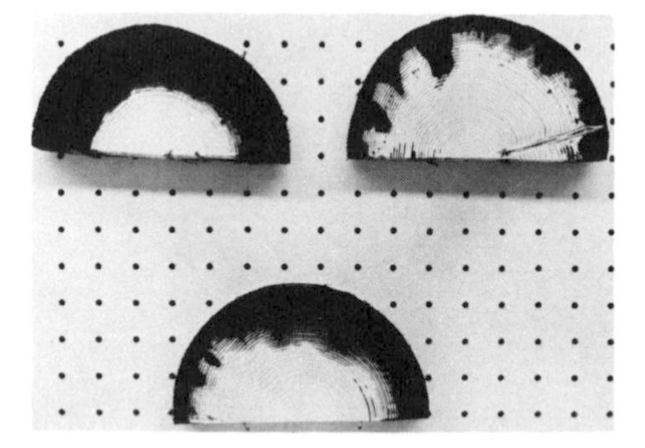

Figure 8.27 Charge of treated piles after withdrawal from a pressure cylinder. Bark has been removed to permit better penetration and distribution of the preservative. (Photo: Canadian Institute of Timber Construction)

study it was shown that ponding spruce poles for five weeks is sufficient to obtain full sapwood peneration by creosote in air-seasoned material (Unligil 1972). The retentions of the creosote in the butt portions of ponded logs were 72–128 kg/m³ (4.5–8 lb/cu ft), while retentions in the butt portions of nonponded logs were 32–40 kg/m³ (2–2.5 lb/cu ft). Spruce sapwood becomes more permeable in ponded logs because of the bacterial degradation of bordered-pit membranes in the cell walls of the wood.

Nonpressure Processes

Brush Treatment

Brush treatment refers to application of the preservative, usually one of the oil-borne types, by brush to the surface of dry wood. For best results the preservative should be heated to 88°C (190°F)* before being applied, and two coats should be brushed thoroughly into all cracks and joints in the wood. Because the treatment is superficial, accidental damage to the surface can expose the underlying untreated wood to decay fungi. Brush treatments add only a few years to the life of timber, which may require retreating every year. Applying the preservative by spraying results in similar protection and service.

Dipping

Dipping is a superficial treatment that consists of submerging the wood for a short period, usually in an oil-preservative mixture. The wood should be thoroughly dry when treated, and for best results the temperature of the preservative should be maintained around 88°C (190°F).* A large container is required to permit total submersion of the wood, and some means of heating the container is necessary. For large quantities of timber this process can be more economical than brushing or spraying because it is faster. Dipping also ensures better coverage and penetration of checks by the preservative.

Steeping (cold soaking)

Soaking seasoned or unseasoned wood in an unheated oil-borne or water-borne preservative solution for one to several days is a superior treatment to brushing, spraying, and dipping. This process is suitable for small amounts of timber in construction work and for the treatment of fence posts on farms where other methods are not practical.

Thermal Process (hot and cold bath)

This process, also known as the open-tank treatment, involves placing seasoned timbers upright in a tank containing pentachlorophenol in oil or creosote that is maintained at a temperature of about 105°C (220°F).* The posts are removed after at least six hours in the hot solution, and their butts are immediately submerged in a tank of cold preservative for at least two hours. Water-borne preservatives may also be used in this process at lower temperatures, providing they are stable at the selected temperature. This process is superior to cold soaking because the preservative penetrates deeper into the wood upon immersion in the cold solution.

Diffusion Processes

There are several methods of treating unseasoned wood with toxic water-borne salts. In these processes the salts are applied to the surface of unseasoned wood and diffuse into the water inside the wood.

In the sap-displacement method, the butts of unbarked green posts or poles are placed in a solution of salts during a period of good drying weather. As the exposed tops of the poles dry out, the sap travels up the pole from the butt end and is replaced by the preservative solution. The bark is not removed until the treatment is completed (see also 'Special Processes' in this chapter.)

* CAUTION: Care should be taken in heating preservatives in solvent systems with a very low flash point because of danger from fire, explosion, or production of toxic gases.

Figure 8.28 Posts impregnated with preservative provide durable boundary fences for farms and other purposes. (Photo: Canadian Institute of Timber Construction)

In other treatments, the salts are combined in a paste and applied to the surface of debarked and unseasoned timber. The timbers are close-piled and covered tightly with a waterproof cover to prevent any drying out. They are left for several weeks, during which time the salts diffuse into the wood.

The double-diffusion method involves immersing unseasoned posts in a solution of copper sulfate until the chemical has diffused into the wood. The posts are then soaked in a second solution containing sodium arsenate or sodium chromate. Reaction occurs between the different salts, and leach-resistant precipitates are formed in the wood (Gjovik, Roth, and Davidson 1972).

In another method, different toxic salts are placed in alternate holes previously bored into the butts of green posts, and the holes are plugged. In a similar method, a powdered salt is applied to the surface of green posts, which are wrapped for a few days, and then a second salt is applied in a like manner. The treated poles are kept under a waterproof cover for two to four months.

The diffusion principle is also employed in the treatment of unseasoned lumber with sodium octoborate. Freshly cut lumber is dipped into a hot, concentrated solution of the borax salt and close-piled under cover for several weeks. Solution concentration, dipping time, and length of storage depend on the lumber thickness. Lumber treated this way is resistant to insects and decay fungi in buildings. Because borates are leachable, the lumber must not be exposed to subsequent wetting.

Ground-Line Treatments

Various treatments have been used to prolong the service life of standing poles by applying the preservative to the ground-line area of the poles (Krzyzewski 1972). The soil is dug to a depth of 0.3–0.6 m (1–2 ft) around the base of the pole, and obvious rot and dirt on the surface of the butt are removed. The gelled preservatives are then applied by brush or trowel. The active ingredients of the treatment emulsion gradually diffuse into the wet wood. Heavy, water-resistant paper is wrapped around the treated area, and the soil is replaced. In other methods, a collar filled with sand, peat moss, or sawdust is placed around the butt of the pole just below the ground line. Creosote is then poured into the collar substrate, and the collar is top-sealed around the pole.

Another type of ground-line treatment is the Cobra process. In this process, a special injecting tool is used to inject a preservative paste into a series of holes made in the ground-line area of poles. In a newly developed variation of this process plastic capsules that slowly release gaseous fungal sterilants are inserted (Cooper, Graham, and Lin 1974).

Pressure Processes

Full-Cell Process

The full-cell process is now generally used with water-based salt systems. Vacuum is applied to the timber inside the treating cylinder and maintained for up to one hour at 74.5–81.3 kPa (22–24 in.) of mercury. The cylinder is then filled with the preservative solution while still under vacuum, and pressure is applied to the solution up to 1379 kPa (200 psi). Temperatures vary from 49 to 66°C (120 to 150°F). A final short vacuum period removes excess preservative solution from the wood.

A special form of this process is the treatment of Douglas-fir for marine piling. The poles are submerged in creosote and boiled under vacuum, or Boultonized, to remove moisture as well as air from the sapwood. Pressure is then applied while the temperature is maintained at 88°C (190°F).

The oscillating-pressure method is a modification of the full-cell process. In this system several short, alternate cycles of vacuum and pressure are applied to the wood.

Empty-Cell Processes

The objective of these processes is to obtain the deepest penetration of preservative in timber with the most economical use of the preservative. Two processes, the Lowry and the Rueping, are commonly used. In both, compression of air inside the wood allows the preservative to be forced deep into the wood. A vacuum cycle then causes the air to expand and to expel excess preservative from the wood. This excess preservative can then be recovered for further use.

In the Lowry process, seasoned timber is placed inside the treating cylinder, which is then closed and filled with preservative heated to 82–99°C (180–210°F). When the cylinder is filled, a pressure of 689–1241 kPa (100–180 psi) is applied, depending on the wood species to be treated. This procedure is followed by draining, and then a vacuum of 81.3–88 kPa (24–26 in.) of mercury is drawn in the cylinder. This treatment penetrates as deeply as the full-cell process does, but it leaves the wood cells nearly empty of preservative solution.

Figure 8.29 Treated timber ready to be removed from a treating cylinder.

Timber to be treated by the Rueping process should be well seasoned, or, if unseasoned, it should be conditioned first by steaming or boiling under vacuum in the cylinder. The timber is then treated with compressed air at 207–414 kPa (30–60 psi) for up to one hour. While still under pressure, the cylinder is filled with preservative solution, and the pressure is raised to 862–1379 kPa (125–200 psi), depending on the wood species, until the desired gross absorption is reached. The temperature of oil-based preservatives is maintained at about 88°C (190°F) during this period. After draining, a final vacuum of at least 67.7 kPa (20 in.) is applied, allowing the air in the timber to expand and to force excess preservative out of the wood.

Organic-Solvent Processes

In one of the organic-solvent procedures, the Cellon process, a liquefied petroleum gas (butane) replaces oil as the carrier for the pentachlorophenol preservative. Because butane is a poor solvent for 'penta,' a cosolvent, such as isopropyl ether, is also used in the system. After the treatment process, the highly volatile solvents are recovered, leaving the pentachlorophenol in the wood. Wood treated by this process can be painted without difficulty. The full-cell treatment is used more frequently than the empty-cell method in this process.

In other solvent-treatment processes, light mineral spirits with a nonvolatile cosolvent, or methylene chloride, are used as carriers for the pentachlorophenol. These light solvent treatments are called LST processes.

Special Processes

End-penetration treatments, which use water-soluble salts, depend on the movement of moisture through the green sapwood of round timbers. The Boucherie process was used for many years in Europe to treat poles with copper sulfate. In this process, a liquid-tight cap is tightly clamped over one end of the unbarked pole, and the preservative solution is brought into this cell from an overhead tank. Hydrostatic pressure of the column of liquid gradually forces the preservative solution into the wood, replacing the natural water. Complete sapwood penetration can be obtained by this method.

In the Prescap process, a modification of the Boucherie method, caps are fitted to the ends of unseasoned logs and pressures on the preserving solution of up to 1379 kPa (200 psi) are applied. With this method a pole can be impregnated in one day instead of two weeks. Treating time can be further reduced to about two hours by removing the caps, cutting a small disc from the ends of the pole, and reapplying the pressure.

Treatments for Primary Wood Products

Primary wood products, such as logs, chips, and lumber, that are unseasoned require temporary protection against deterioration during transport or storage.

Logs

Unseasoned logs should be kept under continual water sprinkling or in ponds, if they cannot be used quickly, to reduce microbiological deterioration (Roff and Dobie 1968). They can also be protected against insects, and subsequent blue stain fungi carried by insects, if the logs are sprayed with the gamma isomer of benzene hexachloride, either in fuel oil or as a water emulsion. Logs intended for use in the construction of cabins and other structures experience less biodeterioration if they are peeled and stored on open decks during winter months. In warm weather, freshly peeled logs can be placed on decks, where air circulation will promote adequate surface drying of each log, and treated with a fungicide, applied by brushing or dipping, to retard the development of superficial molds and staining fungi.

Figure 8.30 Fungal attack in unprotected logs during storage results in destruction of sapwood before heartwood.

Lumber

Freshly cut lumber from unseasoned logs may become stained by fungi if it is not kiln dried or air seasoned within a short time. The lumber must be treated with fungicide if kiln-drying facilities are not available and warm, humid conditions prevail in the seasoning yard. Usually the lumber is immersed for a brief time in aqueous sodium pentachlorophenate, sodium tetrachlorophenate solution, or an emulsion of copper-8-quinolinolate before it is stacked in the yard for seasoning. Alternatively the lumber may be treated with one of several commercial antistain products to provide superficial protection during the drying period. Unseasoned lumber may also be protected against decay by the boron-diffusion process, but some sodium pentachlorophenate must be added to retard development of staining fungi and molds.

Pulpwood Chips

Chips stored outside, particularly those of hardwood species, should be used within four months. Fungal deterioration can be retarded by building a chip pile during the winter months when temperatures are below freezing and by fungicidal treatments such as with borax (Hulme and Shields 1973). Sodium pentachlorophenate has also been used effectively in experimental studies, but is not considered safe because of its high toxicity to mammals.

Trends in Preservation

In order to compete successfully in the market, wood products must have long service life, good appearance, and competitive pricing. The use of pentachlorophenol in organic solvents and of fixed salt treatments is increasing because they leave the surface of wood products in a clean condition ready for painting.

Petroleum-derived chemicals, which are used as preservatives and carrier solvents, are in short supply and are rapidly increasing in cost. Consequently water-borne salts are being used in greater amounts.

Concern over accidental contamination of the environment with chemicals has prompted research to find preservatives of low toxicity to mammals and fish.

In addition, wood-preservation companies are reducing pollution by developing better methods for disposal of wastes (Thompson 1970). These methods include closed steaming in the retort after treated timber has been removed and the use of gravity separator tanks where sludge is removed by skimming the surface of the effluent, leaving the solid wastes at the bottom. Additional wastes are removed in the flocculating tanks and during subsequent filtering of the effluent through sand or charcoal. The waste products collected during these procedures are incinerated, while treating ingredients are recovered for reuse in the preservation processes.

FIRE PROTECTION

Over the centuries man has used wood for two different purposes: for construction of his home and to meet his fuel needs. For the first purpose it is preferable that wood does not burn; for the second, it should. The unique burning characteristics of wood allow both purposes to be satisfied reasonably well.

The first attempts to reduce the combustibility of wood appear to have been made by the ancient Egyptians, who soaked wood in a solution of alum, and by the Romans, who soaked wood in a mixture of vinegar and alum. In 1638 paint containing clay or gypsum was used as a fire retardant in Italian theaters. Then in 1735 a patent was granted in England for a fire-retardant treatment containing alum, ferrous sulfate, and borax. Several of our best fire-retardant salts – ammonium phosphate, ammonium chloride, and borax – were actually discovered as long ago as 1821 by Joseph Louis Gay-Lussac.

In the interests of increased fire safety, building codes have become more exacting in recent years. New building materials that are less combustible than wood have also been developed. These changes have caused the wood industry to explore new fire-retardant treatments for wood and wood products that will improve protection without undesirable side effects.

The present annual output of fire-retardant-treated wood in North America is more than 140 000 m^3 (5 million cu ft). Indications are that this amount will be substantially increased in the near future.

How Wood Burns

The burning or combustion of wood is not a simple oxidation process but involves many complicated physical and chemical reactions. Although these reactions have been extensively studied, they are not yet fully understood. Any variation among species in the quantities of the major constituents (cellulose, hemicellulose, and lignin) does not greatly influence the products of the thermal decomposition of wood. More important are the conditions that exist during the thermal breakdown, including rate of heating, temperature, moisture content, and physical form. So far 213 products of the thermal decomposition of wood have been identified (Goos 1952).

Before wood can burn, it must be heated sufficiently to cause pyrolysis, or thermal decomposition. When the wood surface is heated to about 250°C (482°F), the only change is significant weight loss, which is caused by the evolution of adsorbed water. Between 250 and 400°C (482 and 752°F), active pyrolysis (MacKay 1967) takes place, and the wood is decomposed into flammable gases, tars, and a char residue. The proportion of these products varies widely, depending on pyrolysis conditions. Rapid heating, for example, produces more tar and flammable gases with less char than slow heating does.

If a source of ignition is present while the flammable gases are being

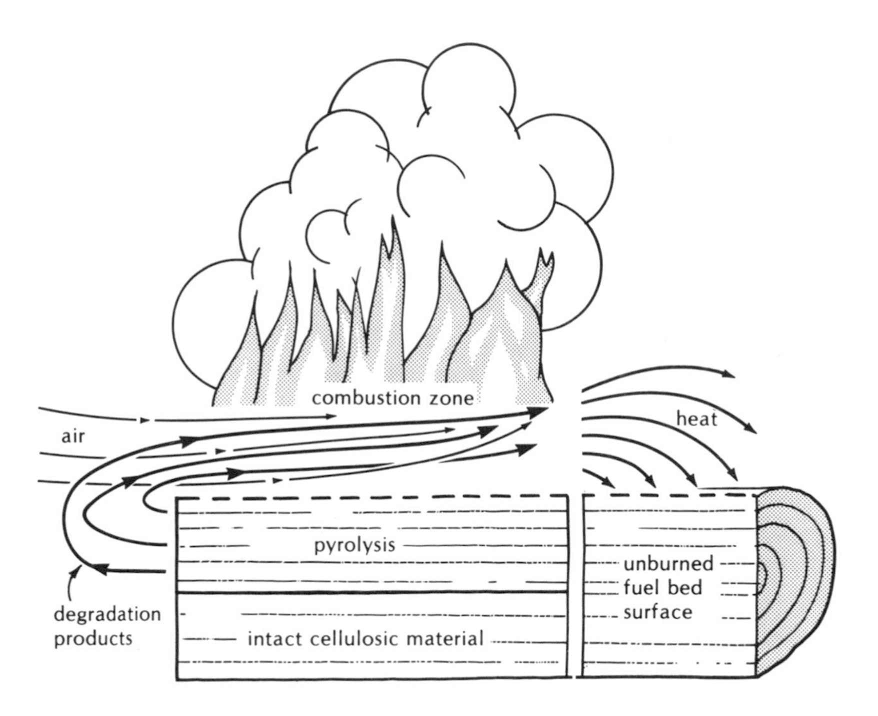

Figure 8.31 Schematic drawing of the sequence of combustion of wood.

evolved, the gases may catch fire. This is called pilot ignition and occurs only if the gases are mixed with air in appropriate proportions to form a combustible mixture. If heating is continued without pilot ignition, a point is reached at which spontaneous ignition will occur. While the flammable gases are being burned, the char that has formed cannot burn because of lack of air at the surface. When surface air becomes available, however, the char burns at a temperature of about 500°C (930°F) by means of glowing or visible combustion.

Wood is difficult to ignite in bulk and usually must be reduced to shavings or kindling size before a small but high-intensity heat source, such as a match (which has a flame temperature of about 450°C (840°F)), can ignite it. After ignition, the wood will continue to burn if the heat generated is sufficient to continue pyrolysis near the surface. Burning will then proceed if the heat can be conducted to the interior of the wood to bring about pyrolysis there. Since wood is a poor conductor of heat, the rate of burning from the surface inward is slow – generally 38 mm/h ($1\frac{1}{2}$ in./hr), although for low-density species the rate is somewhat faster and for high-density woods, somewhat slower. Thus, large timbers burn slowly, because of their bulk, and the fire resistance of heavy timber construction is superior to that of unprotected steel.

Steel begins to lose its strength at about 250°C (480°F), and at 550°C (1022°F) it has only about half its original strength (Richardson 1960). The fire resistance of wood and unprotected steel were compared in controlled roof tests (Forest Products Research Society 1960), using 102 mm × 356 mm (4 in. × 14 in.) wood joists and 356 mm (14 in.) steel-bar joists. The roofs were loaded to their design capacities of 146.5 kg/m² (30 lb/sq ft). After 13 minutes of fire exposure, the roof supported by the steel joists collapsed, while the roof supported by the wood joists remained structurally sound.

Figure 8.32 Wood panelling in the O'Keefe Centre in Toronto is pressure-treated with fire retardants. (Photo: Canadian Institute of Timber Construction)

Fire Retardants

Many inorganic compounds have been used as fire retardants for wood. The effectiveness of about 130 of these compounds, alone and in combinations, has been evaluated (Truax, Harrison, and Baechler 1956). The results show that diammonium phosphate is the best fire retardant, followed by monoammonium phosphate. Ammonium chloride, ammonium sulfate, borax, and zinc chloride are also effective. Commercial fire retardants used today, some of which are proprietary, are usually composed of a mixture of salts in order to achieve a balance of properties and to reduce costs. The main objection to inorganic fire retardants is their leachability, which prevents their use in exterior applications.

In recent years, much work has been done to develop leach-resistant fire retardants. Efforts have been concentrated on such amino-resin-forming compounds as urea, melamine, guanidine, and dicyandiamide in combination with phosphoric acid (Juneja 1972). These fire retardants are now available commercially and are produced as water-soluble monomers. The wood is impregnated with the monomers, which are then polymerized by heat to form insoluble polymers.

The methods of introducing fire retardants into wood are similar to those employed in wood preservation. However, loadings of more than 96 kg/m^3 (6 lb/cu ft), which are many times those needed for wood preservation, may be required for effective treatments. Retention of fire retardants, like that of preservatives, depends on the permeability of the species being treated.

If applied as directed, the fire-retardant paints and coatings available for wood will provide the fire protection prescribed by building codes. Because heavy applications are necessary, the use of such paints and coatings is expensive. Furthermore, they are not very washable. There are indications that the leach-resistant amino-resin fire retardants referred to earlier can be incorporated into existing finishing systems to give a moderate degree of fire retardancy at reasonable cost.

According to the most widely accepted theory, fire-retardant chemicals reduce the flaming combustion of wood by altering the pyrolysis reactions (Eichner 1966). It has been demonstrated that the presence of these chemicals causes a reduction in the amount of flammable gases and tars produced and a simultaneous increase in the production of char and water.

Certain fire-retardant chemicals lower the temperature at which the pyrolytic decomposition of wood occurs. Other fire retardants form a liquid, glassy, or foam layer on the wood surface that inhibits the escape of flammable products from the wood. This layer also blocks access of air to the wood surface and insulates it, preventing pyrolysis of the wood.

Properties of Fire-Retardant-Treated Wood

Fire-retardant treatments may affect hygroscopicity, strength, gluing, and paint-holding properties of wood.

Hygroscopicity

Many of the inorganic salts used in commercial fire retardants are hygroscopic, and at the retentions necessary for adequate fire protection, they affect the moisture content of the wood. At 80% relative humidity, the moisture content of treated wood may be double that of untreated wood (McKnight 1962).

Strength

The higher moisture content of treated wood causes a reduction in strength properties (Eichner 1966). The chemicals used, their retention levels, and the treating conditions also influence strength. For example, large amounts of borax make wood brittle. To compensate for the loss in strength, it has been recommended that the allowable design stresses for fire-retardant-treated lumber be reduced to 90% of those of untreated lumber (National Forest Products Association 1973).

Gluing Characteristics

In general, the treatment of wood with fire retardants adversely affects its gluing characteristics (Bergin 1962). The degree of interference varies, depending on the adhesive, the fire retardant employed, and the amount of fire-retardant chemicals present in the wood. The bond strength of some adhesives has been improved by increasing the glue spread, altering the curing schedule, treating the surface with chemicals, sanding the surface to remove crystals of fire retardant, and modifying the adhesive composition. Suitable adhesives for fire-retardant-treated wood include resorcinol, phenol-resorcinol, liquid or powdered phenolic, and melamine-formaldehyde resins.

Paint-Holding Properties

Wood treated with fire retardants can generally be painted without serious adverse effects (McKnight 1963). However, if heavy loadings of hygroscopic salts are present, giving the wood a high moisture content, then adhesion problems can arise. Most commercial formulations can be painted. Studies have shown that if white pine lumber and Douglas-fir plywood are treated with three commercial fire retardants and coated with five interior and five exterior paints they perform satisfactorily.

Measures of Fire Protection

Many test methods have been designed to evaluate the fire resistance of building materials and assemblies under actual fire conditions. These methods are used for research and development or for screening and range from simple laboratory tests to full-scale building fires. Reference is made here only to those methods that are generally used to evaluate wood products for compliance with North American building codes.

The test methods are usually classified in two groups: those that measure the surface-burning characteristics of building materials or assemblies and those that measure the fire endurance of these products. The surface-burning group of test methods is concerned with the propagation of flame over the surface of the material and related properties, such as fuel contributed and smoke generated. The fire-endurance test methods measure the length of time that building materials or assemblies will support their design loads and resist penetration by fire when subjected to standard temperature-time exposure conditions.

Most building codes in North America refer to the '25-foot-tunnel' test method for rating the surface-burning characteristics of building materials. Details of this test method are given in American Society for Testing and Materials (ASTM) Designation E84, Standard Method of Test for Surface Burning Characteristics of Building Materials. In this test, the specimen, which is 7.62 m × 51 cm (25 × 20 in.), is mounted so that it forms the top of the interior of the tunnel (Figure 8.33). Flames from a gas-fired burner, providing heat at a rate of about 24.4 W (5000 BTU/min), impinge on one end of the specimen. The maximum distance the flame advances during the test, or the time it takes the flame to travel the length of the tunnel, is noted. During the test the temperature and amount of smoke at the exit end of the tunnel may also be recorded. These data are used to calculate the flame spread, amount of fuel contributed, and smoke classifications, which are based on arbitrarily assigned values of 100 for select-grade red oak and 0 for

Figure 8.33 Tunnel furnace for fire hazard classification tests on materials and coatings.

asbestos-cement board. Well-treated fire-retardant lumber has a flame-spread index of less than 25. It will burn if a flame is applied to it, but burning ceases once the flame is removed.

The fire endurance of materials and assemblies such as walls, partitions, floors, roofs, and ceilings is defined by ASTM designation E119, Standard Methods of Fire Tests of Building Construction and Materials. In this test, one side of the assembly is heated according to a standard time-temperature curve. Endurance is defined as the period of resistance to this standard exposure that elapses before the first failure of the assembly is observed.

For door assemblies, the fire endurance is determined by use of ASTM designation E152, Standard Methods of Fire Tests of Door Assemblies. For window assemblies, ASTM designation E163, Standard Methods of Fire Tests of Window Assemblies, is used. In these tests, the doors or windows are mounted in their frames, using the required hardware, and then installed in a wall of the furnace. The fire endurance is determined by supplying heat to one side of the wall according to the standard time-temperature curve.

The fire resistance of roof coverings is evaluated by a special set of fire tests. In these tests, the fire-retardant characteristics of roof coverings are measured against fire originating outside the building. Roof decks are constructed and tested according to ASTM designation E108, Standard Methods of Fire Tests of Roof Coverings. Both before and after accelerated weathering, wood-shingle or shake decks are subjected to an intermittent-flame exposure test, a spread-of-flame test, a burning-brand test, and a flying-brand (burning embers) test (Figure 8.34). Roof coverings fall into three classifications based on these tests. A class A classification requires that the roof covering withstand severe fire exposure, a class B classification requires that the roof withstand

Figure 8.34 Flying-brand test on cedar shingles, to determine their fire resistance on exposure to flames from the burning brands.

moderate fire exposure, and a class C classification requires that the roof withstand light fire exposure.

DEGRADATION BY INSECTS AND MARINE BORERS

Degradation of wood and wood products by insects causes significant losses (Prebble and Gardiner 1958; McBride and Kinghorn 1960; Roff and Dobie 1968). Estimating monetary losses caused by wood-destroying insects is extremely difficult. Losses incurred include not only the product itself but the total replacement cost (labor, time, value of replaced products) as well. Although much of the damage, involving small sums of money, is scattered and never reported, it is estimated that these losses amount to millions of dollars annually.

This section describes the types of insect-caused defects or damage found in wood products, the insects responsible for the damage, and their control. Emphasis is placed on the most destructive insect families occurring in the Coleoptera (beetles), Hymenoptera (ants, bees, wasps), and Isoptera (termites) orders as well as on marine borers belonging to the class Crustacea (gribbles) and the phylum Mollusca (teredos).

Damage and Control of Insect Attacks on Unseasoned Wood

Pinholes

Small, round pinholes up to 3 mm ($\frac{1}{8}$ in.) in diameter with dark-stained walls are evidence of attack by ambrosia beetles (Coleoptera: Scolytidae, Platypodidae). Damage occurs when adult beetles bore into recently felled logs, unseasoned lumber, or, occasionally, standing green trees. Both hardwoods and softwoods are attacked, and the type of gallery (tunnel) and amount of damage caused vary with the attacking species.

Staining of gallery walls is caused by a fungus developed from spores carried into the galleries by the boring adult beetle. Occasionally the surrounding wood is stained as well with streaks running parallel to the grain. The adult beetles and larvae feed on the fungus, deriving little or no nourishment from the wood itself. The pinholes and staining (Figure 8.35) degrade the value of the products produced from the wood. Ambrosia beetles attack only green, unseasoned wood; when the wood is dried, the growth of the fungus ceases and the insects die. The striped ambrosia beetle, *Trypodendron lineatum* (Oliv.), and *Gnathotrichus sulcatus* LeConte are two ambrosia beetles commonly found in British Columbia that attack a variety of hosts, including Douglas-fir and western hemlock.

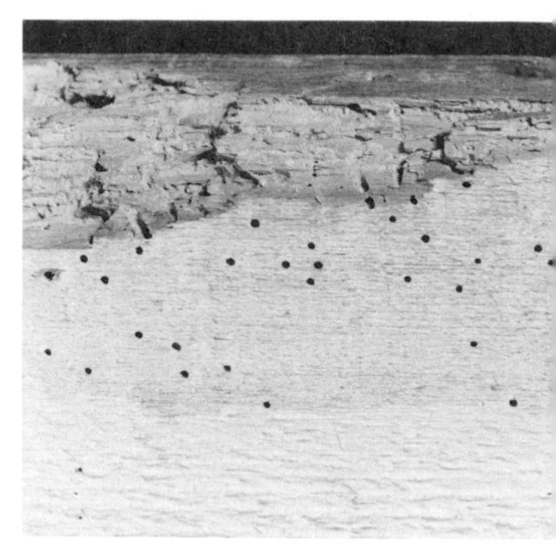

Figure 8.35 Pinhole damage in yellow birch caused by ambrosia beetles.

Wormholes

Large elliptical or circular holes 3–13 mm ($\frac{1}{8}$–$\frac{1}{2}$ in.) in diameter (and occasionally larger) are indications of attack by the larvae of round-headed (longhorn) wood borers (Coleoptera: Cerambycidae), flat-headed (metallic) wood borers (Coleptera: Buprestidae), or horntail wasps (Hymenoptera: Siricidae). The larvae excavate galleries throughout the wood in an irregular pattern. The galleries increase in size to accommodate the larvae's growth. Galleries of longhorn beetles are oval and are usually loosely packed with fibrous borings. Most species of metallic wood borers feed on or just under the bark, penetrating very short distances into the wood to pupate. The larvae of horntail wasps construct perfectly circular galleries filled with wood borings, which are tightly packed behind the larvae as they bore through the wood.

Recently felled timber and dying or dead trees of both hardwoods and softwoods are attacked by round-headed and flat-headed wood borers. Trees killed by fire are readily attacked by horntail wasps. Occasionally damage caused by these various insects occurs in air-seasoned products, usually because the material was infested before manufacture and the larvae have continued to develop. The golden buprestid, *Buprestis aurulenta* (L.), a wood borer, is long-lived, and the larvae may continue to excavate a network of galleries for a number of years after the infested log has been converted into lumber (Ross 1971).

Figure 8.36 Worm holes and discoloration of wood caused by sawyer beetles.

Wormholes facilitate the entrance of stain and decay fungi that cause discoloration of wood and accelerate wood rots. The deep tunnels and discoloration of wood (Figure 8.36) cause monetary losses by degrading the lumber.

Insect Control in Unseasoned Wood

Good management and prompt use of logs can greatly reduce losses caused by insects attacking recently felled timber or timber killed by fire. Logs left in the forest during the attack period suffer higher losses than those removed from the forests. Piling of logs at sawmills or storage areas reduces damage by longhorn and metallic wood borers by restricting egg deposition to the surface log layer. All wood borer attacks are prevented by removal of the bark, but severe checking may result if the logs are not ponded or processed shortly after debarking. Water misting or sprinkling of decked logs prevents attack by wood-boring beetles (Roff and Dobie 1968; Richmond and Nijholt 1972) and has the added advantage of preventing checking and development of staining fungi. Chemical control has been restricted because of public concern over the use of persistent chemicals, such as lindane. In some areas, however, lindane is still used to control wood borers, before egg laying begins (Ross and Downton 1966).

Damage and Control of Insect Attacks on Seasoned Wood

Powder Post

Powder post describes the final condition of wood infested by wood-boring beetles. The larvae feed inside the wood, riddling it with meandering tunnels and reducing the timber to an outer shell packed full of boring dust and excrement, called frass. Eventually the strength of the infested timber is so far depleted that structural members become seriously weakened, as indicated by sagging, shaking, or collapse. Often the surface of the infested material is dotted with small round pinholes, evidence that adult beetles have emerged (Figure 8.37). Associated with the exit holes are small piles of fine, sawdustlike dust ejected from the tunnels by the feeding larvae. The consistency of the boring dust may help identify the beetle causing the damage (Table 8.2).

Agents responsible for powder-post damage are beetles (Coleoptera) belonging to four families: Lyctidae (true powder-post beetles), Bostrychidae (shot-hole borers), Anobiidae (furniture and death-watch beetles), and Curculionidae (wood-boring weevils).

Powder-post beetles, or lyctids, feed on the carbohydrate content of wood. As a result, they attack only the sapwood of hardwoods, which contains sufficient starch (minimum of about 3%) and has vessels or pores large enough to accommodate the ovipositor of the female (Williams 1973a). Lyctids are usually found in sawmills, lumberyards,

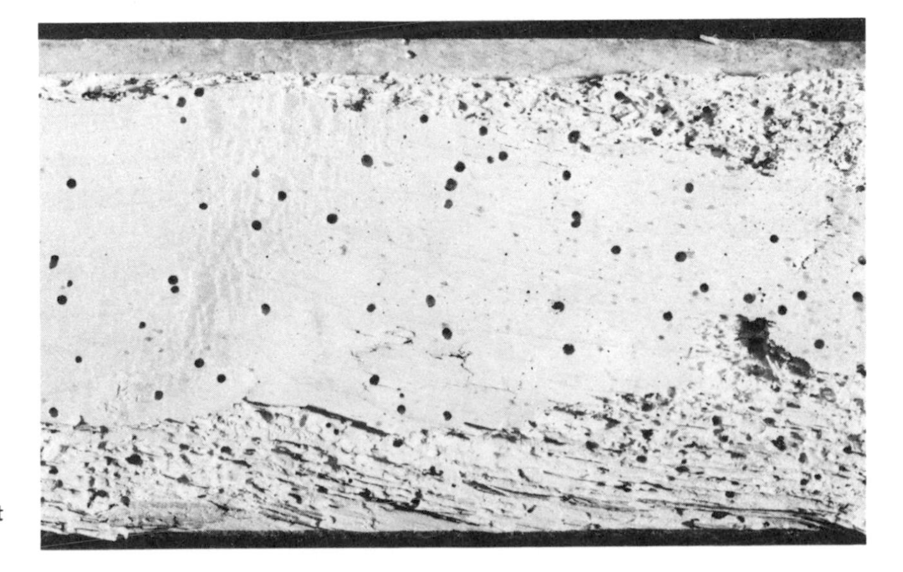

Figure 8.37 Powder-post damage in soft maple caused by *Lyctus* powder-post beetles.

Table 8.2 *Characteristics of insects causing degradation to wood products*

Insect	Damage agent	Wood attacked*	Boring dust	Galleries
Ambrosia beetles	Adult beetle	Unseasoned H & S; usually confined to sapwood of S. Do not reinfest.	White or cream-colored, accumulating on surface of wood.	Dark-stained walls, free of boring dust, uniform in size.
Longhorn beetles	Larvae	Unseasoned H & S; h & s. Do not reinfest.	Excelsiorlike.	Oval, winding irregularly through wood, increasing in diameter.
Metallic wood borers	Larvae	Unseasoned H & S; h & s. Do not reinfest.	Fine, powdery.	Tightly packed with boring dust, oval to flattened, not uniform in size.
Horntail wasps	Larvae	Unseasoned S; h & s; usually fire-killed trees. Do not reinfest.	Loose, fibrous shavings.	Tightly packed with boring dust to rear of tunnel; circular walls; winding, usually in a loop from s to h to s.
Powder-post beetles (*Lyctus*)	Larvae	Seasoned and partly seasoned; s of H with large pore sizes. Reinfest.	Fine talcumlike powder.	Loosely packed with boring dust; run in all directions, often merging into one another.
Bostrychids	Adults and larvae	Seasoned or partly seasoned. Usually s of H. Reinfest.	Coarser than *Lyctus*.	Larger than *Lyctus*, more circular in cross-section; tightly packed with bore dust.
Anobiids	Larvae	Seasoned H & S; usually confined to s. Reinfest.	Gritty, ellipsoidal pellets.	Packed with boring dust; run in all directions, often merging into one another.
Wood-boring weevils	Adults and larvae	Decayed wood or wood in very damp conditions, H & S. Reinfest.	Ellipsoidal pellets smaller than anobiids.	Packed with boring dust, often merging into one another; run in all directions.
Termites (subterranean)	Worker caste	Moist or decayed wood close to or in contact with ground; prefer earlywood of S.	Entirely consume wood.	Walls spotted with dark masses of earth and excrement; wood with honeycombed appearance.
Termites (dampwood)	Immature forms (nymphs)	Damp or decayed wood.	Entirely consume wood, excrete coarse, compacted pellets.	Wide and shallow, usually favoring soft rings.
Carpenter ants	Worker caste	Seasoned or unseasoned H & S; prefer softer earlywood of S.	Shredded wood particles or sawdust.	Clean, smooth walls; wood with honeycombed appearance.

*H = hardwood; S = softwood; h = heartwood; s = sapwood.

and manufacturing premises, where susceptible hardwoods having a moisture content of 8–25% are infested within the first year after processing. Infestations in houses almost always occur because wood, or occasionally furniture, containing eggs or larvae has been brought into the home.

Attacks by shot-hole borers, or bostrychids, on hardwoods are similar to those of lyctids and usually occur in similar situations. They produce coarser, more tightly packed frass and somewhat larger exit holes.

The anobiids (furniture and death-watch beetles) are able to digest cellulose by means of enzymes; thus, they will attack both the sapwood and the heartwood of all species. Factors favoring attack and increased larval development are high moisture content, low maximum temperature (not exceeding 22°C (72°F)), low resin content, and presence of fungal decay (British Forest Products Laboratory 1970). Infestations usually begin in damp environments such as crawl spaces or in ground-floor timbers. Although sound wood is also attacked, larvae grow more quickly in wood decayed by soft and white rots. Severe damage is often found in timber and joinery where old fungal decay is present.

Wood-boring weevils attack in situations similar to those conductive to anobiid attack except that they invade only damp, decayed wood. Exit holes are smaller and the frass is finer. Although infestations by these wood-boring insects are often confused with dry rot and other forms of fungal decay, damage by wood borers can be readily distinguished by the tunnels packed with boring dust within the wood and by circular flight holes visible on the surface.

Layering

Layering describes the damage caused by carpenter ants or termites in structural timbers, poles, and posts, as well as in living and dead trees. Both of these insects destroy the soft layers of wood and leave large, irregular galleries free of borings. If the walls of the galleries are clean and smooth, carpenter ants (Hymenoptera: Formicidae) are the destructive agents; walls spotted with dark masses of excrement identify the work of termites (Isoptera: Rhinotermitidae, Hodotermitidae).

Carpenter ants normally invade forest debris. However, nesting sites can be found in houses and in other structures between studs or roof rafters, behind panelling, baseboards, or door casings, and between layers of insulation. Carpenter ants are able to invade sound timbers (including cedar) as readily as deteriorating ones, and they prefer the earlywood of softwood species. Their activities give the infested material a laminated appearance (Figure 8.38). The wood is not eaten but is ejected from the excavated galleries as piles of shredded particles or sawdust. The excavated galleries are where eggs are laid, the grublike larvae are fed,and cocoons are stored. Carpenter ants eat various types of vegetable and animal matter, plant or insect secretions, and many accessible domestic foods, particularly sugar substances.

Signs of carpenter ant infestation are large, black foraging worker ants; piles of sawdust at the base of posts, along sills, or under exposed wood; a rustling sound when large numbers of ants move about behind walls or baseboards; and the appearance of winged ants.

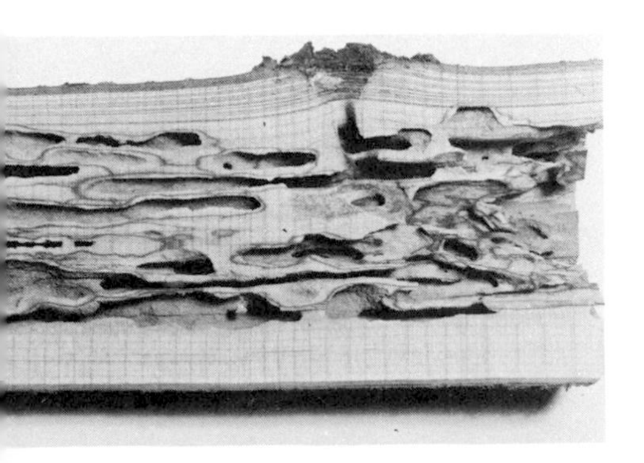

Figure 8.38 Damage in black cherry caused by carpenter ants.

Termites are responsible for the destruction of a wide range of human commodities and possessions, particularly wooden buildings or the wooden parts of buildings. In Canada two types of termites cause significant damage: the dampwood termite (Isoptera: Hodotermitidae), which is restricted to the coast of British Columbia and Vancouver Island, and the subterranean termite (Isoptera: Rhinotermitidae), which is found in southern Ontario as well as in southern British Columbia, including Vancouver Island (Gray 1969; Ruppel 1973).

Dampwood termites establish their colonies in damp, often decaying wood, preferring material with incipient brown rot or mechanical damage, which facilitates their entry and establishment. Once established, they can extend their activities into sound and even relatively dry wood. Infestations in buildings generally require wood-to-earth contact or wood that is continually wet. Dampwood termites do not tunnel through soil or build shelter tubes to new material.

Unlike dampwood termites, subterranean termites build their nests or colonies in moist, dead wood or cellulose debris buried in damp, warm soil. They construct flattened, earthen shelter tubes over the surface of foundation walls to reach susceptible material. The shelter tubes protect the termites from the drying effect of air; termites must maintain contact with the ground or some other source of moisture in order to survive. Ideal conditions are often found beneath buildings where crawl spaces are poorly ventilated; where scraps of lumber, form boards, grade stakes, stumps, or roots are left in the soil; or where the wood touches or is close to the ground. Cracks or voids in foundations and concrete floors make it easy for termites to reach wood that does not actually touch the soil. Termites also damage utility poles, fence posts, paper, fiberboard, and other wood products.

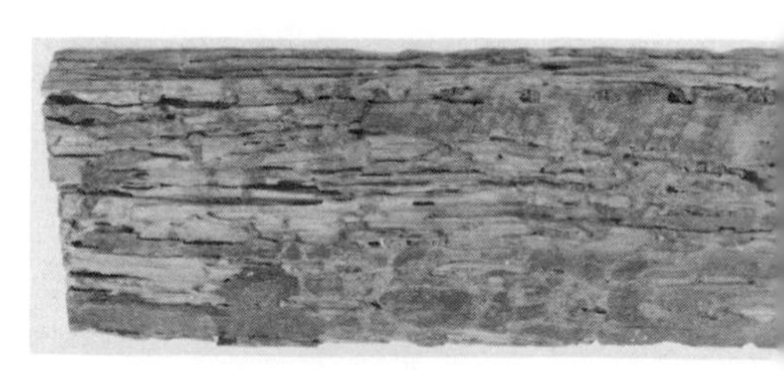

Figure 8.39 Section from structural timber destroyed by termites.

Telltale signs of the presence of termites are many. Shelter tubes, or mounds of darkish, fine grit appearing on the floor near baseboards, may indicate subterranean termites. Small brownish pellets collected in cracks, crevices, or openings in or around damp wood (i.e., between the plate and the foundation) or the accumulation of discarded wings may indicate dampwood termites. Any apparent weakening of structural timbers, particularly if they are damp, may signify termite attack. Tapping may produce a hollow sound, revealing evidence of termite galleries, or prodding with a sharp instrument may expose the laminated or honeycombed appearance of attacked wood.

Insect Control in Wood in Service

Much can be done to prevent attack in woodwork and furniture. Precautionary sanitation methods will help to prevent an attack, and regular inspections will detect any early signs of infestation. Remedial measures should be applied before an infestation has had time to spread or the insect has had opportunity to cause extensive damage.

Immediate action is needed to deal with insect infestations. Severe attacks affecting the stability of structural timbers require replacement of the affected material with preservative-treated wood. Minor infestations may be controlled by surface treatments with suitable contact insecticides (Williams and Johnston 1972; Williams 1973b). Penetration of the insecticide is greatly improved if it is injected into exit holes, with particular attention being given to cracks, crevices, and joints in woodwork and furniture. If carpenter ant nests cannot be located, an insecticide barrier along ant runways will weaken and kill the ants over a period of time (Smith 1957; Creelman 1972). Because some chemical treatments mar the finish of furniture and floors, it is advisable to determine the effect of the chemical on a small, unexposed portion of furniture before undertaking full treatment. Fumigation may be required if the extent of the infestation makes removal and replacement too expensive. Fumigation treatments provide only temporary control and do not prevent reinfestation.

The best time to provide protection against subterranean termites is during planning and construction of new buildings. The removal of all wood (stumps, roots, grade stakes, form boards, etc.) from the surface

Figure 8.40 Typical damage caused by marine borers (gribbles) to untreated piling in saltwater. Deep creosote impregnation by pressure treatment provides the most satisfactory protection against such attack. (Photo: Canadian Institute of Timber Construction)

layer of soil is a necessity in termite-infested areas. Infested soil must be treated, and if severe attack has taken place, a barrier of treated soil must be established adjacent to the foundation before any construction is started to bar termite access to the wood (Johnston, Smith and Beal 1972). In general, chemical control alone does not ensure permanent results; prevention of reinfestation is also necessary.

Within the building, crawl spaces must be well drained and well ventilated, and where the soil cannot be kept dry, a moisture barrier (a polyethylene sheet or roll roofing) should be in place. Any wood that may project through concrete floors or foundations (stair supports, posts, etc.) or is close to the ground should be pressure-treated by a standard process.

Marine Borer Attacks and Their Control

Marine structures and sawlogs stored in sea water are subject to rapid and extensive damage by wood-boring organisms, known as marine borers. Two distinct types of marine borers exist on the east and west coasts of Canada. Members of the wood-boring Crustacea, commonly called gribbles or pinworms, are equipped with several pairs of legs and are distantly related to the wood louse and crab. Members of the Mollusca, commonly referred to as shipworms, or teredos, are bivalves distantly related to clams and oysters. Water temperature, salinity, availability of oxygen, and source of food are factors affecting the incidence and prevalence of marine borers.

Shipworms, or teredos, become established when the free-swimming larvae settle on submerged timbers and bore small holes into the timber. Once in the wood, they increase rapidly in size using, the bivalved grinding shells on the head to increase the size of the burrow. At this stage they develop characteristic wormlike bodies. Contact with the surrounding sea water is maintained through two slender tubes, or siphons, through which the water is continuously pumped and from which oxygen and plankton are obtained. Shipworms have a life span of 18–24 months and grow to lengths of about 120 cm (4 ft), burrowing holes with diameters up to 2 cm (¾ in.) (Bramhall 1966). Under unfavorable conditions the shipworm retracts its siphons, plugs the burrow entrance, and remains virtually immune to all attempts to destroy it.

Damage done to sawlogs stored in sea water may result in extensive economic losses. Limited penetration by marine borers renders peeler or veneer logs valueless; deeper penetration results in loss of lumber and effects pulp recoveries and quality. In addition to losses in wood volume, losses in chip yield result from increased production of pin or unusable chips and dust by marine borers. When the pulp is made into paper, the presence of calcium carbonate crystals, originating from the grinding shells and the calcareous linings of burrows, weakens the paper (Murray et al. 1960).

Gribbles destroy marine piling and wharf timbers by boring interconnected burrows into the wood, extending up to 1.3 cm (½ in.) from the surface. The erosive action of the sea continually exposes fresh surfaces to attack, which is continuous and usually takes place over a relatively large area.

Measures for preserving timber against the attack of marine borers include covering the surface of the timber with an inert material that prevents the borers from gaining access to the wood and injecting the timber with a liquid poisonous to borers. Concrete or metal protective coatings are effective, provided they can be kept in good condition during service. Because surface treatment with preservatives offers little if any protection, deep creosote impregnation by pressure treatment in accordance with accepted standards is necessary (Canadian Standards Association 1974; American Wood Preserves Association 1974). The use of round piling is helpful since the band of sapwood readily absorbs the preservative. Any cutting or boring of the timber should take place before pressure impregnation, as these areas provide access points for marine borers.

Control of Insects in the Export Market

Stringent quarantine regulations governing the importation of wood and wood products have been initiated in several countries, notably Australia, New Zealand, and South Africa, to prevent the introduction of insects that could cause damage (Heath 1970; Tamblyn 1971). All wood, including wooden packaging material (crates, boxes, pallets or packing boards, dunnage), must be free of bark. Live insects in any part of the timber, or evidence of attack by wood borers, *Hylotrupes bajalus* (L.), or horntail wasps, *Sirex noctilio* (F.), results in quarantine impoundment for treatment at the importer's expense.

Some countries require a signed declaration that a preshipping inspection revealed all wooden packing material to be free of bark, visible signs of insects, and fungal attack. Others require an affidavit certifying that the shipment has been protected by a recommended treatment (fumigation with methyl bromide, kiln drying, steam/cold quench, hot borax immersion, or pressure impregnation with a suitable preservative).

Future Trends in Wood Entomology

Increasing public concern over the use of persistent chemicals such as lindane, and public health restrictions on insecticides that may contaminate domestic water supplies have restricted the use of chemical insecticides. These petroleum-based substances have also become more expensive because of the recent price increases of petroleum and petroleum-based products. Development of nonchemical methods of control has therefore received increasing attention and will continue to have high priority.

Promising results have been obtained from research on host selection by insects and on insect-produced chemical messengers. These chemicals may be pheromones, which convey messages to individuals of the same species (Borden and Stokkink 1973), or allomones, which are transpecific messengers. Feeding stimulants as well as feeding deterrents can be used to manipulate insect populations. Sex pheromones can be used to attract insects to traps, to disrupt mating by permeating the air with an attractive compound, or to monitor population fluctuations. Alarm pheromones of ants may be used to keep populations in a state of confusion or unsettlement (Ayre and Blum 1971). Attractant-insecticide and attractant-pathogen combinations may become effective termiticides (Esenther and Gray 1968; Lund 1969).

Biological controls are easier to apply, and are versatile and nonpollluting, but they require a major research effort before they can be widely used to control degradation of wood products by insects.

REFERENCES

American Wood Preservers Association. 1974. *Book of standards*. Washington, DC

Ashton, H.E. 1974. Exterior exposure study of stains and clear finishes. *Can. Paint and Finish.* 48: 2, 13

Atwell, E.A. 1948. *Red stain and pocket rot in jack pine: Their effect on strength and serviceability of the wood.* Can. For. Serv. Circ. 63. Ottawa

Ayre, G.L., and Blum, M.L. 1971. Attraction and alarm of ants (*Componotus* spp. – Hymenoptera: Formicidae) by pheremones. *Physiol. Zool.* 44: 77–83

Bardin, P.C. 1973. UV filler prepares particleboard for wood graining. *Ind. Finish.* 49: 9, 48

Bergin, E.G. 1962. *The gluability of fire-retardant treated birch veneer.* For. Prod. Res. Branch Rep. 191. Ottawa

Borden, J.H., and Stokkink, E. 1973. Laboratory investigations of secondary attraction in *Gnathotrichus sulcatus. Can. J. Zool.* 51: 469–73

Bramhall, G. 1966. *Marine borers and wooden piling in B.C. waters.* Dep. For. Publ. 1138. Ottawa

British Forest Products Laboratory. 1970. *The common furniture beetle.* Tech. Note No. 47. Princes Risborough, Aylesbury, Bucks.

Canadian Standards Association. 1974. *Wood preservation.* CSA standard 080-74. Toronto

Cech, M.Y. 1966. New treatments to prevent brown stain in white pine. *For. Prod. J.* 16 (11): 23–7

– 1968. *Dimensional stabilization of tree cross-sections with polyethylene glycol.* Dep. Fish. For. Publ. 1224. Ottawa

Collier, J.W. 1972. Review of wood finishing. *Ind. Finish. and Surface Coat.* 24 (285): 10

Cooney, D.G., and Emerson, R. 1964. *Thermophilic fungi.* London: Freeman

Cooper, P.A.: Graham, R.D.: and Lin, R.T. 1974. Factors influencing the movement of chloropicrin vapor in wood to control decay. *Wood Fiber* 6 (1): 81–90

Creelman, I.S. 1972. *Control of ants.* Dep. Agric. Publ. 1298. Ottawa

Cserjesi, A.J., and Roff, J.W. 1964. Retention of pentachlorophenol in lumber dipped in water solutions. *For. Prod. J.* 14 (8): 373–6

DaCosta, E.W.B., and Osborne, L.D. 1972. Laboratory evaluation of preservatives. VII. Effect of chemical structure on toxicity of organolead compounds to wood-destroying fungi. *Holzforsch.* 26 (3): 114–18

Eichner, H.W. 1966. Fire-retardant-treated wood. *J. Mater.* 1 (3): 625

Ellwood, E.L., and Ecklund, B.A. 1959. Bacterial attack of pine logs in pond storage. *For. Prod. J.* 9 (9): 283–92

Esenther, G.R., and Gray, D.E. 1968. Subterranean termite studies in southern Ontario. *Can. Entomol.* 100: 827–34

Forest Products Research Society. 1960. Wood outperforms steel. *For. Prod. J.* 10 (11): 617

Gjovik, L.R.; Roth, H.S.; and Davidson, H.L. 1972. *Treatment of Alaskan species by double diffusion and modified double diffusion methods.* U.S. Dep. Agric. For. Serv. Res. Pap. FPL 182. Madison, Wis.

Goldstein, I.S., and Loos, W.L. 1973. *Wood deterioration and its prevention by preservative treatments*, edited by D.D. Nicholas and W.L. Loos, Vol. 1, Chap. 10. Syracuse: Univ. of Syracuse Press

Goos, A.W. 1952. *The thermal decomposition of wood.* Wood chemistry, edited by L.E. Wise and E.C.Jahn, vol. 2. New York: Van Nostrand Reinhold

Gray, D.E. 1969. Lenzites trabea-invaded wood lure used as a subterranean termite survey tool in Ontario. *BWPA Termite Symposium*, pp. 117–27. London: Brit. Wood Pres. Assoc.

Hansen, A.T. 1963. Crawl spaces: How to avoid trouble with the foundation. *Canadian Builder* 12 (10): 63–4

Heath, J.E. 1970. Quarantine regulations on imported timber. *Forest and Timber (New South Wales)* 7: 13–14

Hill, R.R. 1972. Clear and natural finishes for wood. *Ind. Finish. and Surface Coat.* 24 (285): 20
Hudson, M. 1969. Chemical drying of southern pine wood. *For. Prod. J.* 19 (3): 21–4
Hulme, M.A., and Shields, J.K. 1972. Effect of primary fungal infection upon secondary colonization of birch bolts. *Mat. Org.* 7 (3): 177–88
– 1973. Treatments to reduce chip deterioration during storage. *Tappi* 56 (8): 81–90
Johnston, H.R.; Smith, V.K.; and Beal, R.H. 1972. *Subterranean termites, their prevention and control in buildings.* U.S. Dep. Agric. Home Garden Bull. No. 64. Washington, DC
Juneja, S.C. 1972. Urea-base fire retardant formulation and products. Can. Pat. 917,334
Kennedy, D.E., and Wakefield, W.E. 1948. *The strength of jack pine poles infected with pocket rot.* Can. For. Serv. Circ. 65. Ottawa
King, F.W., and McKnight, T.S. 1965. *A simple natural finish for exterior siding.* Dep. For. Publ. 1136. Ottawa
Krzyzewski, J. 1972. *Five year performance of slow release type groundline preservatives in creosoted pine poles.* East. For. Prod. Lab. Inf. Rep. OP-X-44. Ottawa
Leary, P.E. 1970. Interior coatings – Maintainability and durability. *For. Prod. J.* 20: 12, 36
Levinson, S.B. 1972a. Electrocoat, powder coat, radiate. Part II. *J. Paint Tech.* 44 (571): 28
– 1972b. Electrocoat, powder coat, radiate. Part III. *J. Paint Tech.* 44 (570): 38
Lund, A.E. 1969. Termite attractants and repellents. *BWPA Termite Symposium*, pp. 107–15. London: Brit. Wood Pres. Assoc.
MacKay, G.D.M. 1967. *Mechanism of thermal degradation of cellulose.* Dep. For. Publ. 1201. Otttawa
McBride, C.F., and Kinghorn, J.M. 1960. Lumber degrade caused by ambrosia beetles. *B.C. Lumberman* 44: 40–52
McKnight, T.S. 1962. *The hygroscopicity of wood treated with fire retarding compounds.* For. Prof. Res. Branch Rep. 190. Ottawa
– 1963. *Treatment of wood with fire retardants.* For. Prod. Res. Branch Publ. 1033. Ottawa
Miniutti, V.P. 1967. *Microscopic observations of ultraviolet irradiated and weathered softwood surfaces and clear coatings.* U.S. Dep. Agric. For. Serv. Res. Pap. FPL 74. Madison, Wis.
Murray, A.; Dowsley, A.H.; Walden, C.C.; and Allen, I.V.F. 1960. Losses in pulp yield and quality resulting from teredo (*Bankia setacea*) attack on logs stored in sea water. *Pulp Pap. Mag. Can.* 70 (10): T338–43
National Forest Products Association. 1973. *National design specifications for stress-grade lumber and its fastenings.* Washington, DC: NFPA
Palka, L.C. 1970. *Current trends in dimensional stabilization of wood.* West. For. Prod. Lab. Inf. Rep. VP-X-63. Vancouver
Prane, J.W. 1972. New polymers for coatings for the 70's. *Prog. in Organic Coat.* 1: 1, 3
Prebble, M.L., and Gardiner, L.M. 1958. Degrade and value loss in fire-killed pine in Mississagi area of Ontario. *For. Chron.* 34: 139–58
Richardson, H. 1960. The fire endurance of timber structures. *Dock. Harb. Auth.* May
Richmond, H.A., and Nijholt, W.W. 1972. *Water misting for log protection from ambrosia beetles in B.C.* Pacific For. Res. Centre Inf. Rep. BCP-4-72. Victoria
Roff, J.W. 1964. Hyphal characteristics of certain fungi in wood. *Mycologia* 56 (6): 799–804
Roff, J.W.; Cserjesi, A.J.; and Swann, G.W. 1974. *Prevention of sapstain and mould in packaged lumber.* Can. For. Serv. Publ. 1325. Ottawa
Roff, J.W., and Dobie, J. 1968. Beating the beetles: Technical report on sprinkling stored logs. *B.C. Lumberman.* 52: 60–71
Roff, J.W., and Whittaker, E.I. 1963. *Relative strength and decay resistance of red-stained lodgepole pine.* Dep. For. Publ. 1031. Ottawa
Rogers, B.S. 1973. Modern finishing methods. *Can. Paint and Finish.* 47: 6A
Ross, D.A. 1971. Bark beetles and wood borers in conifer logs in B.C. Forest Insect and Disease Survey, June Pest Report. Pacific Forestry Res. Centre, Victoria
Ross, D.A., and Downton, J.S. 1966. Protecting logs from long-horned wood borers with lindane emulsion. *For. Chron.* 42: 377–9

Rossell, S.E.; Abbot, E.G.M.; and Levy, J. 1973. A review of the literature relating to the presence, action and interaction of bacteria in wood. *J. Inst. Wood Sci.* 6 (2): 28–35

Ruppel, D.H. 1973. *Termites in B.C.* Pacific Forest Res. Centre, Forest Pest Leaflet No. 57. Victoria

Savory, J.G.; Pawsey, R.G.; and Lawrence, J.S. 1965. Prevention of blue stain in unpeeled scots pine logs. *Forestry* 38 (1): 59–81

Scheffer, T.C., and Clarke, J.N. 1967. On-site preservative treatments for exterior wood of buildings. *For. Prod. J.* 17 (12): 21–9

Scheffer, T.C., and Verrall, A.F. 1973. *Principles for protecting wood buildings from decay.* U.S. Dept. Agric. Forest Serv. Res. Paper FPL 190. Madison, Wis.

Sedziak, H.P.; Shields, J.K.; and Johnston, G.H. 1973. *Condition of timber foundation piles at Inuvik, N.W.T.* Can. For. Serv., Inf. Rep. OP-X-67. Ottawa

Sedziak, H.P.; Shields, J.K.; and Krzyzewski, J. 1970. Effectiveness of brush and dip preservative treatments for above-ground exterior exposure of wood. *Inter. Biodeterior. Bull.* 6 (4): 149–55

Sedziak, H.P.; Shields, J.K.; Krzyzewski, J.; King, F.W.; and McKnight, T.S. 1969. *Performance of preserved wood foundations and exterior cladding systems in experimental houses.* East. For. Prod. Lab. Inf. Rep. OP-X-17. Ottawa

Sedziak, H.P., and Unligil, H.H. 1973. *The use of preserved wood foundations in residential housing.* East. For. Prod. Lab. Inf. Rep. OP-X-79. Ottawa

Sell, J. 1968. Investigations on the infestation of untreated and surface treated wood by blue stain fungi. (In German.) *Holz Roh Werks* 26 (6): 215–22

Shields, J.K., and Atwell, E.A. 1963. Effect of a mold, *Trichoderma viride,* on decay of birch by four storage-rot fungi. *For. Prod. J.* 13 (7): 262–5

Smith, D.N. 1957. *Carpenter ant infestation and its control.* Dep. Agric. Publ. 1013. Ottawa

Smith, R.S. 1975. Automatic respiration analysis of the fungitoxic potential of wood preservatives, including an oxathiin. *For. Prod. J.* 25 (1): 48–53

Smith, R.S., and Afosu-Asiedu, A. 1972. Distribution of thermophilic and thermotolerant fungi in a spruce-pine chip pile. *Can. J. For. Res.* 2 (1): 16–26

Stamm, A.J. 1964. *Wood and cellulose science,* Chaps. 19 and 20. New York: Ronald Press

Tamblyn, N. 1971. *Termites and quarantine in Australia.* CSIRO For. Prod. News Letter No. 384. Melbourne

Tayelor, F.A., ed. 1974. *Architectural opportunities seminar.* West. For. Prod. Lab. Inf. Rep. VP-X-123. Vancouver

Thompson, W.S., ed. 1970. Pollution abatement and control in the wood preserving industry. Symposium sponsored by For. Prod. Util. Lab., Mississippi State Univ. and Southern Pressure Treaters Assoc. State College: Miss. State Univ.

Truax, T.R.; Harrison, C.A.; and Baechler, R.H. 1956. *Experiments in fireproofing wood.* U.S. Dep. Agric. For. Serv. Rep. 1118. Madison, Wis.

U.S. Forest Products Laboratory. 1961. *Comparative decay resistance of heartwood of different native species when used under conditions that favor decay.* U.S. Dep. Agric. For. Serv. Tech. Note 229. Madison, Wis.

– 1966. *Weathering of wood.* U.S. Dep. Agric. For. Serv. Res. Note FPL 135. Madison, Wis.

Unligil, H.H. 1972. Penetrability and strength of white spruce after ponding. *For. Prod. J.* 22 (a): 92–100

Verrall, A.F. 1962. Condensation in air-cooled buildings. *For. Prod. J.* 12 (11): 531–6

Verrall, A.F., and Mook, P.V. 1951. *Research on chemical control of fungi in green lumber, 1940–51.* U.S. Dep. Agric. Tech. Bull. 1046. Washington, DC

Whaley, J.H., Jr. 1972. *Furniture finishing textbook.* 2nd ed. Nashville, Tenn.: Production Publishing

Wilcox, N.W. 1970. Anatomical changes in wood cell walls attached by fungi and bacteria. *Bot. Rev.* 36 (1): 1–28

Williams, L.H. 1973a. Recognition and control of wood-destroying beetles. *Pest Control* 41 (2): 24–8

– 1973b. Identifying wood-destroying beetles. *Pest Control* 41 (5): 30–40

Williams, L.H., and Johnston, H.R. 1972. *Controlling wood-destroying beetles in buildings and furniture.* U.S. Dep. Agric. Leafl. No. 558. Washington, DC

9

M.N. CARROLL
Western Forest Products Laboratory,
Vancouver

Glues and Gluing

Figure 9.1 Dining room grouping constructed of cherry and maple solids with cherry face veneers. Gluing facilitates the matching of selected woods to produce such fine-quality furniture. (Photo: Kaufman of Collingwood)

The volume of glued wood products manufactured in Canada each year is exceeded only by the volume of lumber. Glued wood products range from commodity products to speciality items. Commodity products include plywood, particleboard, waferboard, fiberboard, and finger-jointed lumber, which are manufactured in millions of cubic metres each year (see Chapter 10). Specialty items include laminated timbers, furniture, cabinetry, toys, sporting goods, and many types of glued woodenware, which use a lesser amount of wood but still represent a high annual dollar value.

Gluing is one means by which the forest-products industries fill the gap between market needs and wood supply. If the trees are not long enough, shorter pieces are finger-jointed into longer pieces; if the trees are not wide enough, narrow pieces are edge-glued into wider material; if the trees are of low quality, the knots and other natural defects are cut out and the remaining pieces are glued together as required. The supply of beautiful woods is extended by peeling them into veneer and gluing them to core materials, such as particleboard. Particleboard itself is composed of mill residue particles glued together in sheet form. At the extreme end of the spectrum are the fiberboards. The original structure of the wood is here reduced to the elemental fibers, which are glued into a panel form that is highly uniform in structure.

Each year more of the products used in buildings are premanufactured with adhesives to reduce on-site labor. These products include sheathing, prehung doors, window units, stair sets, parquet flooring, ceiling and wall panelling of plywood, fiberboard or particleboard, roof sections in which cedar shingles are preglued to plywood sheathing for rapid installation of a finished roof, and factory-made counter and cabinet units.

The practice of gluing wood goes back many centuries, but the twentieth century has brought about two significant changes. First, gluing techniques have been adapted to high-speed mass production. This change means that gluing is no longer the simple art it used to be in the cabinetmaker's shop. In mass production, the gluing process has been speeded up tremendously, from hours to minutes or even seconds in some operations. To achieve this high-speed production with low-cost adhesives, careful control of all steps in the gluing process is required.

Second, highly durable adhesives have been developed to provide glue lines in wood products that are as strong and durable as the wood itself. The wood-products manufacturer depends almost entirely on the chemical industry for the technology of these adhesives, which is very complex.

To understand the modern approach to gluing wood, one must look at how adhesives function, the adhesives used in the wood industry, the techniques of mass production, and the durability of the final glued wood product.

HOW ADHESIVES FUNCTION

Gluing is basically an operation in which a liquid material is spread between the surfaces of the materials to be joined, and the two surfaces are held in intimate contact until the liquid has changed to a solid form. This change from a liquid to a solid is commonly referred to as setting, or hardening. For a nondemanding glue line in wood, the basic requirement is that the liquid wet the surfaces to be glued; for the more demanding glue lines, in which waterproofness and long-term resistance to weather exposure are required, the adhesive must penetrate into the cell structure of the surfaces being glued.

One must understand the various mechanisms by which a liquid glue sets to a solid in order to appreciate the processes used to glue wood. There are four basic mechanisms: solvent escape, hot melt, emulsion breaking, and molecular cross-linking. Some adhesives represent pure examples of these mechanisms; sophisticated adhesives may use two or even all of these mechanisms simultaneously in the gluing process.

The Solvent-Escape Mechanism

The simplest example of an adhesive depending on a solvent-escape mechanism is the ordinary postage stamp. The back of the stamp is coated with a dextrin gum that becomes liquified when it is wetted. After the stamp is pressed onto paper, the water diffuses into the paper, and the gum reverts to its solid form, or sets. A similar product is rubber cement, which is rubber dissolved in an organic solvent. When the cement is spread between two absorptive surfaces (paper or wood), it escapes by diffusion into the wood or paper, and a glue-line film of solid rubber is eventually formed. These are reversible mechanisms; if the glue line is exposed to the original solvent, the setting mechanism is reversed and the glue line is softened.

The Emulsion Mechanism

The simplest example of this mechanism is ordinary 'white' glue, an emulsion, or latex, of a plastic (most commonly polyvinyl acetate) in water. When this adhesive is spread between two pieces of wood, the water diffuses into the wood, leaving the little plastic particles behind. Under ordinary conditions, these plastic particles fuse together to form a continuous plastic film that becomes the glue line. This is the same mechanism that underlies latex, or water-base, paints. The film-forming process is irreversible.

The Hot-Melt Mechanism

In hot-melt gluing, a solid material is heated until it melts to form a liquid. This liquid is spread between the gluing surfaces, and the pieces are clamped together to form the glue line. When the liquid loses heat to the wood, the material reverts to its solid form. Roofing tar functions in this manner to glue the roofing felts to the roof sheathing and to each other. Hot-melt gluing is a reversible reaction; if the product is heated, the glue is softened again.

The Chemical-Setting Mechanism

A common example of this mechanism is found in epoxy adhesives, which are liquid chemicals that polymerize after mixing to form a hard, tough solid. The solid is not appreciably softened by solvents or by heat, so the glue lines formed from this type of adhesive are highly durable. All the highly durable resin adhesives used for wood set by chemical reaction. However, some of the chemically setting adhesives, notably urea-formaldehyde resins, are not durable, because the chemical-setting mechanism can be reversed if the glue line is exposed to heat and moisture.

ADHESIVES USED IN THE WOOD-PRODUCTS INDUSTRY

Adhesives are used for two main purposes in the wood-products industry. One is to glue wood to itself; the other, to glue wood to other materials. The other materials are usually overlays of some kind that are added to provide a protective or decorative surface. For example, vinyl plastic films with a wood-grain-print finish are applied to the surface of plywood or hardboard, and high-density overlays are applied to a wood-product base to form the highly resistant surface of a kitchen countertop.

The types of adhesives used fall into two general classes. One class is composed of the structural adhesives, which are rigid adhesives that form a glue bond that is harder and stronger than the wood itself. The other class is composed of the elastomeric adhesives, which form glue lines that are weaker than the wood and are usually softened by heat or moisture or both.

Structural adhesives are generally those that set by chemical reaction, whereas elastomerics are derived from natural or synthetic rubbers in the form of solvent-base solutions (contact cement), emulsions (water-base contact cements), or pressure-sensitive adhesives of the sort used on adhesive tapes. Elastomeric adhesives are adequate for holding things in place, such as ceramic tiles on a wall or vinyl tiles on a floor, or for joining floor sheathing to the joists to prevent a floor from squeaking.

Structural adhesives are subdivided into two classes: those with full

Figure 9.2 Kitchen cabinets and other household fixtures manufactured from glued wood products are familiar features of the modern home. (Photo: Council of Forest Industries of British Columbia)

durability and those with limited durability. The former set by an irreversible chemical reaction, and the latter set by a reversible reaction (that is, the chemical-setting mechanism is reversed when the glued product is exposed to weather or severe environments).

Resorcinol-Formaldehyde (RF)

Resorcinol-formaldehyde is made by prereacting resorcinol with a small amount of formaldehyde in aqueous solution to form a stable precursor called a novolak. At the time of use, the novolak is mixed with a polymeric form of formaldehyde called paraformaldehyde and a filler such as wood flour. The solid paraformaldehyde slowly depolymerizes to formaldehyde, which reacts with the novolak to form a cross-linked resorcinol-formaldehyde resin. A lower-cost variant of this is phenol-resorcinol formaldehyde (PRF), in which the novolak is formed from a mixture of phenol and resorcinol.

Both of these adhesives are classed as fully durable adhesives that set at room temperature, but the setting reaction may be accelerated by heating the glue line. Their most common use is for making laminated timbers and for finger-jointing lumber. A particular disadvantage of these adhesives is that they stain wood.

Phenol-Formaldehyde (PF)

Phenol reacts with formaldehyde much more slowly than resorcinol does, so the formulation of phenol-formaldehyde does not have to be handled in two stages like that of resorcinol-formaldehyde does. Instead the phenol is prereacted with the total amount of formaldehyde, and the combination is built up to a high molecular weight, using caustic soda (sodium hydroxide) to keep the resin molecules in solution in water. For plywood, the resin solution is extended with vegetable materials, such as wheat flour and ground corncobs. For particleboard, the liquid resin is used without extension.

Figure 9.3 Bridge constructed of glued-laminated timber on access road to a paper mill.

To obtain full setting of the resin in the glue line, high glue-line temperatures on the order of 110°C (230°F) are required. These temperatures are obtained by hot-pressing. A variant formulation of phenol-formaldehyde resin is used in manufacturing waferboard. This variant is a finely powdered form of a PF resin that is not soluble in water. In use, it is blended with the wafers and adheres to the surface of the wafers by mechanical adhesion. When the mat of wafers is hot-pressed, the heat melts the resin so that the small particles wet the adjoining surfaces. Thereafter a setting mechanism chemically cross-links the resin to a hard, infusible state. A combination of hot-melt and chemical-setting mechanisms is therefore involved. This type of resin can also be dissolved in alcohol and impregnated into a thin paper sheet that acts as a carrier, after which the alcohol solvent is removed. The dried sheets, known as a paper glue line, can be used in the manufacture of hot-pressed plywood when a nonaqueous glue line is needed. Paper glue line is sometimes used to apply plastic overlay films, such as Tedlar, to plywood by hot-pressing.

These nonaqueous PF adhesives are generally similar to the PF resins used by the plastic industry as molding compounds. Similar resins also form the base for medium- and high-density overlays (see Chapter 10).

PF resins are fully durable. The alkaline, water-based versions stain wood, but the non-water-base versions are nonstaining.

Melamine-Formaldehyde (MF)

MF resins are not as durable as the PF, PRF, or RF resins, but they are adequate for a large number of uses. The virtues of MF resin adhesives are that they form a colorless glue line, are nonstaining, and can be readily adapted to gluing processes that use radio-frequency heating to accelerate the cure of the glue line. They will form a glue bond at room temperature, but heating is required to develop their full durability.

MF resins are commonly made by reacting melamine with formaldehyde and spray-drying the resultant resin. For use, the dry, powdered resin is redispersed in water, and a glue mix is formed by adding fillers and extenders.

Urea-Formaldehyde (UF)

UF resins are the most inexpensive resin adhesives. They are colorless and nonstaining. If hardening agents such as the ammonium salts of strong acids (e.g., ammonium chloride, NH_4Cl) are added, UF resins can be made to cure very rapidly, either cold set or hot set. The largest use of UF resins is in manufacturing particleboard. They have limited durability and are suited only to products that are not exposed to weather.

MF and UF resins may be combined in any proportion to obtain reasonable durability at a lower cost than that of a pure MF resin. These combinations are sometimes referred to as fortified UF resins. Similarly, either MR or UF resins can be fortified by the addition of RF resin.

Casein

Casein glues are formulated from milk protein, using lime to bring the casein into aqueous solution. Because of their high alkali content, they stain wood. The glue sets primarily by loss of water to the wood, although a mild chemical-setting mechanism is also built into some formulations to improve water resistance. Because casein glue lines are softened by water, they are unsuited for structural wood products used in very humid environments. Their major use is in laminated timbers manufactured for use in dry conditions. Casein is the one water-base wood glue that will set at temperatures below 21°C (70°F).

White Glue

White glues are derived primarily from polyvinyl acetate (PVA) emulsions. They are the workhorse glues in furniture manufacturing, having almost entirely displaced animal glue and casein glue. They have numerous advantages, the most important of which is their indefinite pot life. The basic polyvinyl acetate resin is colorless and nonstaining. Special glues can be formulated with materials such as clay to improve the high-temperature strength, or dextrans to improve the speed of set. Special formulations, called quick-clamp adhesives, will set at room temperature in as little as 20 minutes.

Catalyzed PVA

Catalyzed PVA adhesives have similar working properties to those of the noncatalyzed adhesives, except a chemical-curing mechanism has been added to improve their resistance to heat and humidity. These adhesives set initially by breaking the emulsion and forming a plastic film. Thereafter the chemical reaction proceeds slowly at room temperature, so that a month or so may be required to complete setting. Heat speeds up the chemical-curing mechanism.

Hot Melts

Initially hot melts were mainly used in edge-banding panels for tables and case goods. Their particular virtue is that they can set in a matter of seconds. Recently they have come into use for applying veneers and other overlays to large surface areas. Hot melts are compounded from low-cost synthetic thermoplastics such as ethylene vinyl acetate.

Isocyanates
Isocyanate plastics are highly durable resins that are not commonly used in wood products because they are not water soluble. In recent years they have come into limited use in Europe in manufacturing a highly durable particleboard. Their particular advantage is that the chemical cross-linking mechanism by which they set can be catalyzed by moisture, even moisture picked up from the air.

Mastic Adhesives
Mastics are a class of elastomer adhesives (Vick 1971) that are usually based on rubber, dissolved in a solvent, with fillers such as clay added to make them into a thick paste. They set in a rubbery glue line. A more sophisticated mastic adhesive is based on a polyurethane polymer, which sets by a chemical reaction to a rubbery state when exposed to moisture.

An important asset of these elastomerics is their tolerance of a wide range of climatic conditions. They will successfully bond to wet or dry wood at temperatures ranging from below freezing to 40°C (104°F). They are nonstructural adhesives that find their main application in gluing on the building site: wall panelling to studs, floor sheathing to joists, ceramic tile to walls and floors, tiles and sheet flooring material to floors and some kinds of carpeting to concrete floors.

Figure 9.4 Cold-clamp gluing of laminated timbers. Gluing pressure is applied by tightening the nuts on the bolts that run between each pair of laminate layups. (Photo: Koppers International Canada Ltd.)

GLUING PROCESSES

Gluing processes involve two main operations: supplying sufficient pressure until the glue line has set, and usually, accelerating the set of the adhesive by applying heat. Gluing processes also fall into two general categories: assembly gluing, which applied to face-gluing, edge-gluing, and end-gluing of lumber; and platen pressing, which is used for panel products such as plywood and particleboard.

Cold-Clamp Gluing
The oldest form of gluing uses clamps to maintain pressure on the glue lines while the adhesive sets. This technique is still used commercially in manufacturing laminated timbers (Figure 9.4). Ordinarily no means of accelerating the set of the adhesive is used, and the clamps are left on for periods ranging from almost eight hours to overnight. For timbers made of heavy hardwoods such as oak, however, the entire assembly may be placed in a kiln and heated to temperatures of 65–70°C (150–160°F). Heating is used not to speed up the setting of the glue line but to advance the degree of chemical cross-linking in the PRF resin (cure) in order to improve the strength and durability of the glue line. The cold-clamp technique is particularly suitable for the manufacture of curved beams, which may vary in thickness within the length of the beam, because a beam is held tightly in place until the glue line has attained sufficient strength to resist any tendency of the lamination to spring apart when the beam is removed from the clamps.

Figure 9.5 Cold-press gluing of flush doors in a single-opening hydraulic press.

Cold-Press Gluing

The most common use of cold platen pressing is in the manufacture of flush doors, in which a thin plywood or hardboard skin is glued to a wooden frame. The adhesive is applied to the frame, skins are applied to both faces, and a stack of these assemblies is built up. The stack is placed in a hydraulic platen press, and gluing pressure is applied (Figure 9.5). The adhesives used for this process are of the quick-clamp type that will set to a sufficiently strong bond in about 30 minutes. These quick-clamp adhesives are ordinarily either a form of casein glue or a white glue.

Hot-Press Gluing

The use of the multiopening hot press in the gluing of softwood plywood is described in Chapter 10. Similar presses are used in the manufacture of waferboard, particleboard, and the fiberboards. These are larger presses that apply pressures of 2000–3500 kPa (300–500 psi) and operate at platen temperatures of 150–200°C (300–400°F). Their function is not only to supply heat to accelerate the cure of the adhesive binder but also to densify the particle or fiber mats to a relative density at which useful strength properties are developed. Radio-frequency heating is sometimes used to speed the rate of heating of the particleboard and to shorten the press time required for thick (>9 mm [3/8 in.]) boards.

Single-opening hot presses are also used for gluing particleboard. They are generally large presses (2.4 m × 17 m [8 ft × 56 ft]) that operate in conjunction with a heated prepress of similar size that serves to partially densify and to preheat the mat. The mat is transported through these presses on steel bands.

Continuous Hot-Press Gluing

Continuous hot presses are used to make particleboard and laminated veneer lumber (LVL). In the most common of these presses, the lower platen is solid and the upper 'platen' is formed by a continuous band of jointed platen segments that pass through the press area and are returned overhead. During the return, the platen segments are heated. As they pass through the press area, hydraulic pressure is applied by hydraulic cylinders acting through rollers, and the stored heat in the platen segments is transferred to the mat. Again, the mat (or the LVL layup) is conveyed through the press by steel bands.

Panel-flo Gluing

Panel-flo is an ingenious means of edge-gluing lumber in thicknesses up to about 10 cm (4 in.). Strips of lumber, prespread with glue on the edges, are advanced into a platen press. Side pressure is applied to develop the gluing pressure and the platen press is closed. The platen press serves two functions. First, it prevents the layup from buckling under the side pressure, and second, it supplies heat. The glue used is one that sets at room temperature. The heat from the platens penetrates into the top and bottom of each vertical glue line and rapidly sets the glue to a depth of about 3 cm (1/8 in.). Then the pressure is released and the assembly is removed from the press; the rest of the glue line sets in due time. If the press is fed a continuous layup, with the end joints

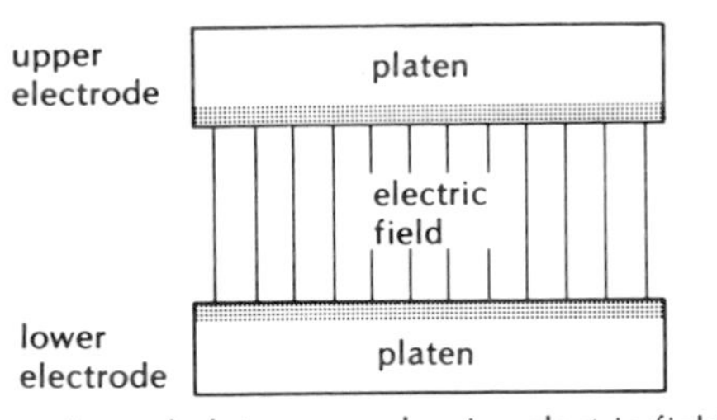

RF heated platen press showing electric field lines between electrodes

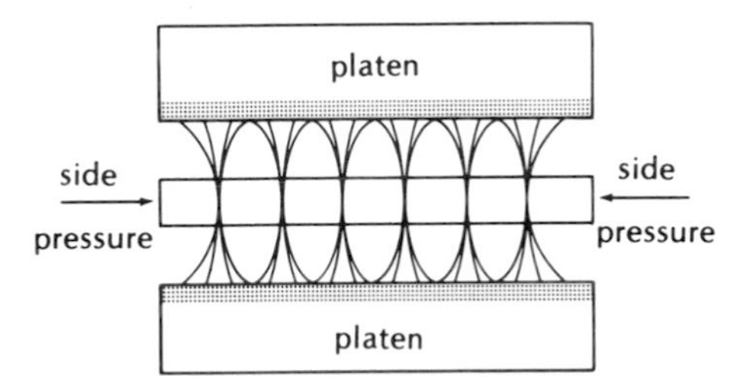

concentration of electric field in conductive glue lines of edge-glued lumber core (glue lines parallel to field)

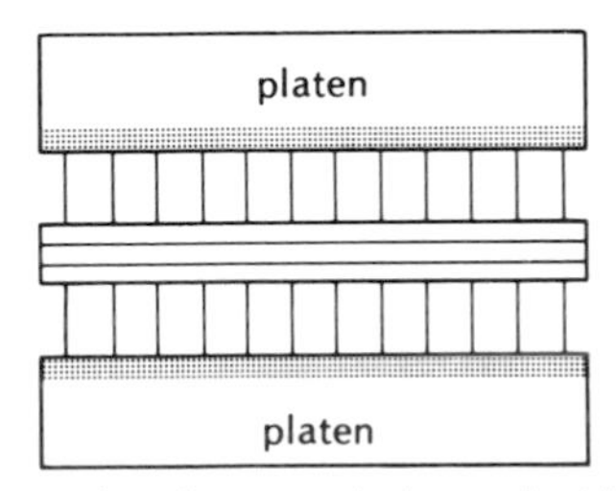

dielectric loss heating of plywood with glue lines running perpendicular to electric field

Figure 9.6 Schematic representation of parallel (*centre*) and perpendicular (*below*) modes of radio-frequency heating. In practice, the upper and lower electrodes are in contact with the wood.

staggered, a continuous length of edge-glued lumber is produced. However, the gluing process is discontinuous, operating on about a seven-minute pressing cycle.

Radio-frequency Gluing

Radio-frequency (RF) heating is a technique that can be adapted to almost any gluing procedure. Details of the technique are given in an excellent book by Pound (1974).

Basically when wood is placed in a radio-frequency electric field, the wood is heated more or less uniformly throughout its thickness. If glue lines are present, the setting of the adhesive is accelerated by the heat. This technique is used in manufacturing medium-density fiberboard in multiopening hot presses. It is also used to preheat particleboard mats before they enter a continuous hot press and to press curved plywood. The latter technique, in which the RF field heats the mass of the wood, is sometimes referred to as perpendicular RF heating.

A more efficient use of RF heating is in parallel heating, which is used for edge-gluing lumber and for finger-jointing. Electrodes are placed above and below an edge-gluing layup, so that the glue lines are parallel to the electric field between the electrodes. If the glue is formulated to make it conductive (for example, by adding common salt to a UF glue mix), most of the RF energy concentrates in the glue line. Figure 9.6 is a schematic representation of the parallel and perpendicular modes of RF heating.

The selectivity of the parallel RF heating technique depends on the relative energy loss in the glue line and the wood. The energy loss in the wood is a function of the moisture content of the wood. As the moisture content rises, more and more of the available energy is absorbed by the moisture in the wood, and the selectivity of heating the glue line is reduced. In addition, if the moisture content of the wood is high, the heating of the wood near the glue line can drive moisture from the wood into the glue. This moisture dilutes the glue and also impairs the process by which the glue diffuses into the wood to form a strong band, a process called flushing the glue line.

Crowder-Press Gluing

This technique is used for finger-jointing lumber and for edge-jointing veneers into a continuous sheet. The gluing pressure used in finger-jointing is developed by applying end pressure to the joint – that is, along the line of the lumber. In a continuous finger-jointing operation, the assembled finger joints (with glue applied) are passed through a series of crowder rolls that are driven at different speeds by a hydraulic drive. The differential speed drives the trailing portion of the joint into the leading portion and develops the necessary end pressure. The finger-joint lumber emerges from the crowder press as a continuous length. RF heating may be used to obtain a rapid set of the glue while the joint is still under pressure. In another technique, developed at the Western Forest Products Laboratory of Forintek Canada Corp., the machined fingers are predried and preheated in a kiln. When finger joints are assembled from these preheated sections, the heat stored in the wood sets the glue as the joint passes through the crowder press. These processes operate at speeds up to 30 joints per minutes. At these speeds, full cure of the glue line is not obtained in the pressing operation. However, sufficient set is obtained to enable the finger-jointed lumber to be handled, and full strength is developed in the joint shortly after manufacture.

Roll Laminating

Roll laminating is a high-speed technique for applying overlays or edge bands that depends on adhesives that bond rapidly. Gluing pressure is applied by a nip roll that provides momentary pressures. Three types of bonding can be distinguished; generally these are low-performance glue lines suitable for use in furniture and cabinetry.

1. For laminating flexible vinyl overlays to wood or a panel, a pressure-sensitive adhesive is preapplied to the vinyl sheet using a release sheet so that the vinyl sheet can be handled in roll form. The vinyl sheet is combined with the substrate (wood or panel) and passed through the nip roll. Not only can this technique be used to apply the film to the surface of a panel or a wood molding, but wraparound devices using this technique can carry the film around the edges and partially onto the back side.

2. Metal or high-density laminates can be applied to a panel core using contact cement. The contact cement is commonly sprayed onto both surfaces, and the excess solvent is 'flashed off,' or volatilized, in an oven. While the glue is still hot, the two parts are combined and passed through the nip roll, fusing the two adhesive films under pressure into a single glue line.

3. Surface laminates or edge bands can be applied to panels using hot-melt adhesives. The adhesive is melted and applied in the form of ribbons to the materials being combined just before they are passed through the nip roll. On passing through the nip roll, the ribbons of glue flatten into a thin film. This flattened film loses heat to the materials very rapidly and sets.

A special process has been developed for applying a vulcanized fiber overlay to lumber used as house siding. This product requires a durable glue line and the glue used is a catalyzed PVA in emulsion form. The emulsion is applied to the overlay sheet, which is passed by a bank of infrared heaters to evaporate the water and form a film. This plastic film leaves the heater section at a high temperature and is quite fluid. The overlay is then pressed onto the surface of the cold lumber and sets by almost instantaneous cooling, much like the film in roll laminating with hot-melt adhesives. Finally, the chemical cross-linking mechanism in the resin continues at the ambient temperature to increase the resistance of the glue line to softening by moisture and heat.

This technique is an example of a multiple-setting mechanism. The emulsion method is first used to form the glue-line film. This film is then fluidized by heat and the bond formed by a hot-melt technique. Both processes take place fast enough to prevent the chemical mechanism from advancing far enough to interfere with the gluing action. In the final stage, the chemical-setting mechanism cross-links the polymer to develop the necessary heat and moisture resistance for the lumber to withstand outdoor exposure.

Figure 9.7 A double-roll spreader typical of the sort used to apply glue to veneer or plywood.

Process Summary
The use of heat to accelerate the set of the glue line is an important factor in mass production. The glue line may be set by heating the entire wood product, as in hot-pressing of the panel products, or by using radio-frequency heating, or by a combination of both. For face- and edge-gluing, selective RF heating of the glue line can be used, as in the manufacture of decking, finger-jointed lumber, and straight laminated timbers. The panel-flo technique uses a glue that cures at room temperature and platen heating to 'tack' the glue lines together. This technique is used in manufacturing edge-glued decking, and one plant in the United States uses the technique to make finger joints. Finally, in the technique of preheat laminating, only the gluing surfaces are heated before the glue is applied, and the glue setting is accelerated by the heat stored in the surface layers of the wood.

The general trend in pressing techniques is toward continuous pressing. Such techniques are available for face gluing, edge-gluing, and end-gluing lumber; finger-jointing; edge-gluing veneer strips into a continuous sheet; and manufacturing particleboard and laminated veneer lumber. Roll laminating is used for applying thin overlay materials.

GLUE APPLICATION

The traditional means of spreading glue is the roll spreader. A film of adhesive is metered onto the roll spreader, which consists of a grooved rubber roll. The thickness of the adhesive film is set by a doctor roll. The spread of the adhesive can be varied, within limits, by adjusting the clearance between the doctor roll and the spreader roll. Higher or lower spread ranges can be obtained by using coarser or finer grooving on the spreader roll. Figure 9.7 shows a typical roll spreader of the sort used to apply glue to plywood. For finger joints, the spreader roll and the doctor roll must be machined to follow the contour of the fingers. Figure 9.8 shows a spreader for finger joints.

Spreader-roll equipment is simple in concept and operation. The glue is mixed in batches and fed to the spreader as needed from a storage tank. Many of the glue mixes used have a limited pot life because they are formulated with catalyst systems that function at room temperature. At the end of the pot life, the glue sets. When these mixes are used, the glue spreader and storage system must be washed clean every few hours. The need for this procedure emphasizes the advantages of water-base adhesives.

Figure 9.8 Typical glue spreader used to apply glue to the contours of a machined finger joint.

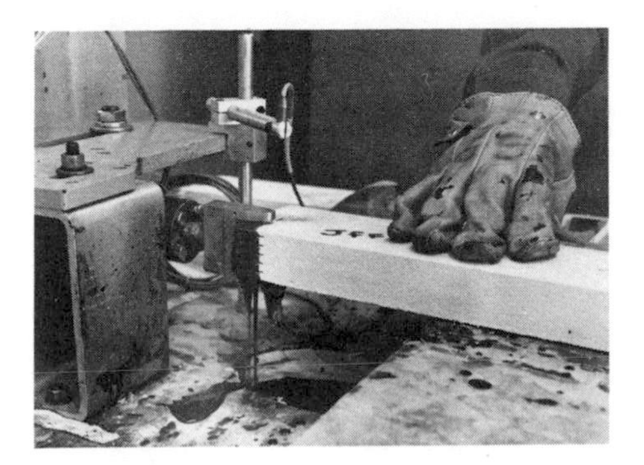

The curtain coater provides a more sophisticated method of spreading glue on veneers. In this method, the glue is pumped over a weir at a constant rate and falls from the edge of the weir in a continuous curtain. The surfaces to be spread pass through the curtain and receive a uniform coating of adhesive that can be varied by adjusting the pumping rate or the rate at which the wood passes through the curtain or both.

A variant of this technique is the extruder, which was developed for spreading glue onto lumber for laminated timbers. The extruder head is basically a piece of pipe with holes spaced at intervals along its length. The glue is pumped through these holes continuously and recycled. As the lumber is passed under the head, the glue is deposited on the surface in a series of parallel ribbons. When gluing pressure is applied, the ribbons are flattened into a thin film to form the glue line. Again, the spread rate is controlled by varying the speed of pumping or the rate at which the wood travels or both.

An extrusion technique is also used to apply hot-melt adhesives. The adhesive is melted and extruded onto the wood surface in the form of ribbons.

In the manufacture of particleboard, the adhesive binder is applied by spraying it onto the particles in a blender. In the manufacture of waferboard, the finely powdered adhesive is tumbled with the wafers in a blender, and the particles adhere to the wafers by mechanical adhesion. In the manufacture of fiberboards by the wet process, the fiber is handled in the form of a slurry in water. Alkaline phenol-formaldehyde resin is dissolved in this slurry, and alum is added to neutralize the caustic substance in the resin solution and to render the resin insoluble in water. The resin precipitate in the form of a fine emulsion and the individual resin particles are deposited on the fibers. In the manufacture of fiberboards by the dry process, the resin binder is metered into the attrition mill used to reduce the wood chips to fibrous form.

GLUE LINE STRENGTH AND DURABILITY

Selection of the lowest-cost adhesive and the most efficient gluing process is fundamental to the success of commercial gluing operations. For nondemanding products such as furniture, the choice can be largely controlled by speed and convenience in manufacture. For building products, the conditions of use are severe and the expected service life is measured in decades, so emphasis must be placed on obtaining highly durable glue lines.

The severity of the climate in which the product is to be used is another factor that must be considered in assessing the required durability. The descriptions given in this chapter and in Chapter 10 apply to the manufacture and use of products in the temperature zone, which includes most of Canada, Europe, and the United States. In tropical areas, more severe conditions of temperatures and humidity prevail. These conditions tend to exclude adhesives like PVA, casein, and UF, which are sensitive to heat and moisture, even from some products made for interior use.

Durability in a glue line permits it to retain its integrity, under the conditions in which the wood product is used, for a time period that is acceptable to the user. Durability requires an adhesive that resists softening by heat or moisture as well as chemical or biological attack. It also requires sufficient control of the manufacturing technique to ensure that the inherent durability of the adhesive is achieved in every part of the glue line. This goal is not easy to attain under mass production conditions with water-base adhesives.

Conditions of End Use

The two most important classes of end-use conditions are exterior, in which the glued wood product is exposed to the weather, and interior,

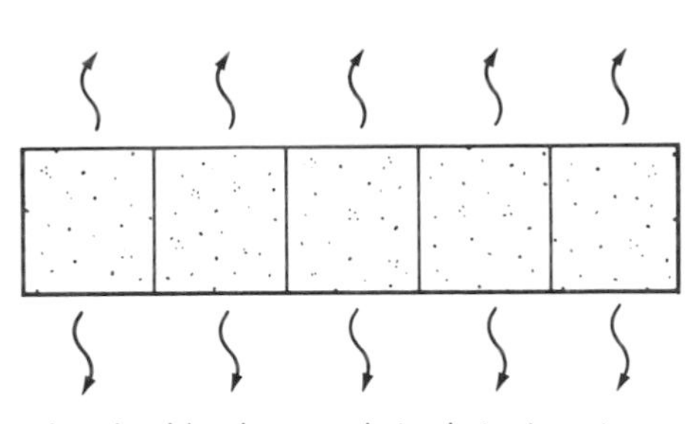

edge-glued lumber panels (end view): moisture leaving the surfaces

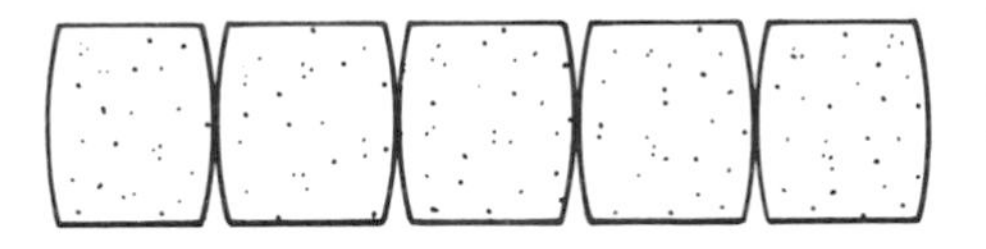

drying stresses relieved by splitting at glue lines which are weaker than the wood (delamination)

Figure 9.9 Schematic representation of the delamination in the glue lines of edge-glued lumber as a result of drying shrinkage on the two faces.

in which the product is protected from the weather. It should be remembered that end-use conditions include not only the final conditions of use but also the conditions that the product encounters during storage, shipment, and construction. In many cases these conditions are more severe than those the product will encounter in its final use. Fine furniture, for example, may be stored in a damp, unheated warehouse, shipped in a closed truck that can reach very low (−30°C [−20°F]) or very high (40°C [104°F]) temperatures, and kept on a showroom floor in the winter in an atmosphere of very low relative humidity. As another example, plywood or particleboard floor sheathing used in site-built housing may be directly exposed to weather, either in storage on site or after installation in the building for as long as six months. After the house is finished, both the floor sheathing and the fine furniture are required to *function* in indoor climatic conditions that are within the comfort range of the people living in the house.

The term interior exposure does not convey a precise picture of a set of exposure conditions. Interior ordinarily means that the wood-product is protected from rain or snow and that the moisture content of the wood product does not ordinarily rise above 16%. Exceptions are found in special-use indoor areas, such as buildings covering ice-skating rinks and swimming pools, in which the relative humidity is generally high.

How Glue Lines Fail

Glue lines fail primarily by delamination. Delamination is the splitting open of the glue line by the same forces that produce checking in wood during seasoning (see Chapter 7). When pieces of wood are subjected to drying, the surface layers dry first; the resulting tensile stresses built up in the surface layers of the wood can exceed the tensile strength of the wood perpendicular to the grain, and the wood splits open, or checks. If the tensile strength of the glue line perpendicular to the grain is significantly lower than that of the wood, the splitting takes place entirely in the glue line and is called delamination (Figure 9.9). These drying stresses are created when the humidity drops and are intensified if the surface is exposed to radiant heat, such as sunlight, or to convection heating from a space heater in a building. Not only do these outside heat sources cause moisture to evaporate from the surface of the wood, they also tend to drive moisture from the surface layers of the wood into the cooler interior of the wood. Figure 9.10 shows 15 cm (6 in.) square sections of laminated oak that have been exposed to the weather for two years. The one on the left, made with PRF adhesive, shows only the checking that develops in the end grain of the wood. The one on the right, made with a less durable adhesive, shows some delamination in almost every glue line.

Figure 9.10 Sections of oak laminated beams 15 cm (6 in.) square made in a laboratory and exposed to the weather for two years. The one on the left, made with phenol-resorcinol-formaldehyde (PRF) adhesive, shows checking in the wood. The section on the right, made with a less durable adhesive, shows some delamination in every glue line.

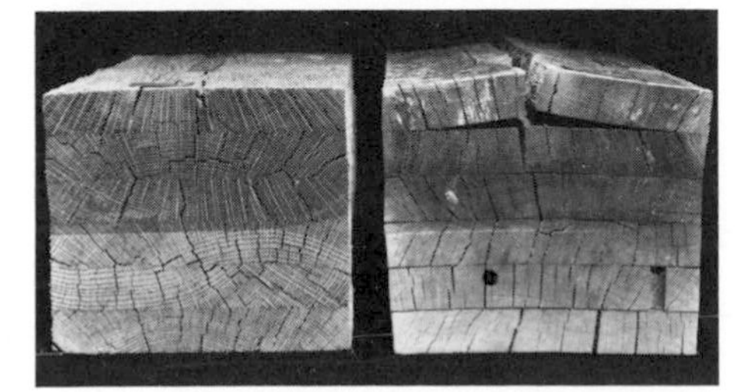

In outdoor exposure, drying stresses reach their maximum, particularly if the product is exposed to direct sunlight. As a result of the exposure conditions, the glue line may be (1) softened by heat (sunlight); (2) softened by moisture (rain, dew, condensation of water vapor diffusing through the walls of a building); or (3) gradually destroyed by fungal attack (casein glue), by hydrolysis (UF and MF glues), or by oxidation (glues based on natural rubber). By any of these mechanisms, the overall glue-line strength may be reduced so that the drying stresses required to produce delamination are even smaller than those required to produce checking in the wood.

Adhesives derived from resorcinol and/or phenol, by reaction with formaldehyde, are not softened by heat or moisture, are immune to biological attack, and are not attacked by any chemicals that would not also destroy the wood. However, while the adhesive itself has full durability, the glue lines will not be durable if the conditions of manufacture are not strictly controlled. More precisely, some *portions* of the glue lines formed in a wood product such as plywood or a laminated timber will not be durable. The two most common causes of this loss of durability are:

1. A portion of the glue line formed is weaker than the wood, even in the dry state. The drying stresses then concentrate in the glue line and produce delamination, which at the very least is unsightly (Gaudert and Carroll 1974). The glue line may weaken if it partially dries out before it is placed under pressure. This condition is known as dryout.

2. A portion of the glue line formed may be as strong as the wood, but the full chemical reaction has not occurred. This condition is called undercure. Undercured glue lines are softened by water and heat. This softening lowers the strength of the glue line to the point at which drying stresses produce delamination.

The Role of Wood Failure in Durability

The basic requirement for full durability is that the glue line be as strong as the wood and that it remain as strong as the wood throughout the lifetime of the wood product. In short-term tests full durability can be assessed by means of cyclic delamination procedures that expose samples of the product to alternate wetting and drying conditions. Such tests are found in the hardwood-plywood standard CSA 0115-1967 and in the laminated-timber standard CSA 0122-1977 (Canadian Standards Association 1967, 1977a).

An alternative procedure is used for softwood plywood and is described in CSA 0121-1978 and CSA 0151-1978 (Canadian Standards Association 1978a, 1978b). For plywood, small sections of the glue line are sheared open after they have been thoroughly saturated with water. If the wet strength of the glue line is higher than that of the wood, the failure will take place in the wood on either side of the glue line. If the glue line is weaker, the failure will take place either in the glue line itself (cohesive failure) or at the interface between the glue line and the wood (adhesive failure). Figure 9.11 shows typical block shear test specimens and plywood shear test specimens in which wood failure ranges from 0 to 100%.

Wood failure is a powerful analytical tool because it is applicable to wood of any strength (wood of different species, for example). In addition, if the glue line of the test sample fails at some point other than in the wood, examination of the type of failure will tell what went wrong in the manufacturing process so that process corrections can be made.

In exterior plywood, the two most common causes of difficulty are dryout and undercure. Both may be detected by a simple wet test rated according to wood failure. In softwood plywood made in Canada, the

Figure 9.11 Typical block shear and plywood shear test specimens before and after testing to expose the glue line. *Left*: no wood failure; only the glue is visible on the surfaces of the broken specimens. *Center*: specimens show a mixture of glue and wood failure. *Right*: 100% wood failure; no glue is visible.

long-term durability of the glue line is ensured by restricting the adhesive used to one based on phenol-formaldehyde.

GLUE LINE QUALITY CONTROL

Maintaining glue-line quality, or bonding quality, in a manufacturing operation requires control of wood temperature, wood moisture content, and glue spread. In most operations, control of assembly time, or the time that the glue is in contact with the wood before the final gluing pressure is applied, is also required. If the allowable assembly time is exceeded, the glue dries out and will no longer function properly. Frequently a minimum assembly time is also required in products made with water-base (or solvent-base) adhesives. A minimum time allows the excess moisture in the glue line to escape into the wood before heat is applied; otherwise "blisters" will form after hot-pressing.

Assessment of bonding quality in solid-wood products usually depends on measurement of wood failure or measurement of delamination resistance. In the reconstituted wood products, continuous glue lines are not formed, and alternative means have to be used.

Softwood Plywood

In softwood plywoods, product quality is standardized by controlling the quality of the veneers (veneer grade), the thickness of the veneers, the plywood construction (numbers of plies and thickness of each ply), and bonding quality.

As already noted, bonding quality of the finished plywood is assessed by means of a tension shear test made on samples. The CSA standard requirement is that these shear tests show an average wood failure of at least 80%. Additional restrictions are placed on the number of tests showing wood failure below 60% and below 30%. Massive comparison studies have been made in which plywood was exposed outdoors on a test fence for many years (Perkins 1950). These studies showed that if the average wood failure measured in the wet shear test was above 80%, the amount of delamination of the plywood on the test fence was negligible.

Most Canadian softwood plywood is made in British Columbia. About 85% of this plywood is made under third-party certification by the Council of Forest Industries of BC (COFI), and the procedures used by COFI are monitored by the Western Forest Products Laboratory of Forintek Canada Corp. The standards for this product are CSA 0121-1978 and CSA 0151-1978.

Hardwood Plywood

Bonding quality in interior-grade hardwood plywood as described in CSA 0115-1967 is measured by a cyclic delamination test in which small sections of plywood are alternately soaked in water and dried at 49°C (120°F).

Bonding quality in exterior hardwood plywood is measured by a test similar to that already described for softwood plywood, using wood failure as the criterion.

Laminated Timbers

Interior-type laminated timbers are made in Canada with either casein glue or UF glue. The test procedures measure both strength (block shear) and wood failure, and bonding quality is estimated from the wood-failure readings. Exterior-type laminated timbers are made with PRF adhesive, and the test procedures include a cyclic wetting and drying procedure that simulates outdoor exposure. Figure 9.10 shows the type of specimen used in this cyclic delamination test procedure.

Third-party certification is not used in all laminated timber manufacture. Certified manufacturers are certified by the Laminated Timber Administrative Board of the Canadian Standards Association. The standards for this product are CSA 0122-1977 and CSA 0177-1977 (Canadian Standards Association 1977a, 1977b).

Waferboard

Waferboard, as well as all particleboards and fiberboards, is made by 'spot-welding' the particles together with finely powdered or liquid adhesive. Wood failure cannot be used as an estimate of bonding quality in waferboard, because a continuous glue line is not formed. Instead samples of the finished waferboard are boiled for two hours as a convenient means of getting them wet. After boiling, the waferboard swells considerably in thickness as it recovers some of the compression applied in the hot-pressing process (see Figure 9.12). This swelling lowers the density of the board and consequently the strength of the board. The bonding quality is measured with CSA standard procedures by testing a waferboard specimen for bending strength after it has been boiled and cooled. Generally any errors in the formation of the glue bonds lead to greater swelling and lower strength.

Waferboard production is not certified in Canada, but a substantial amount of waferboard is sold in the United States and certified as meeting the requirements of a grade in the United States specification CS 236-66. Thus, any plant selling waferboard in the United States is, in effect, making a certified product. The product standard for waferboard is CSA 0188.2M-1978 (Canadian Standards Association 1978c).

Figure 9.12 A specimen of 9 mm waferboard before and after boiling in water. The boiled specimen (*right*) has increased in thickness by about 25%.

Medium-Density Fiberboard

Bonding quality in the exterior grade of medium-density fiberboard is assessed by a six-cycle delamination procedure in which the specimens are boiled, frozen, and oven dried in a cyclic manner. The specification requirements in Canadian General Standards Board CGSB-11-GP-3-1972 not only require a minimum retention of bending strength but also limit the thickness swelling to a maximum of 15% and specify 'no delamination or disintegration.' This requirement is the one Canadian specification for an exterior product that does not in any way specify the type of adhesive required. The implication is that any adhesive used in making fiberboard that survives the six-cycle accelerated aging test procedure has sufficient inherent durability to allow the product to be used as exterior siding.

SUMMARY OF GLUE LINE DURABILITY

The choice of adhesive for the necessary degree of durability is fundamentally governed by the severity of the environment to which the wood product will be exposed and the lifetime expected from the product. However, other factors influence the choice as well. A manufacturer of a commodity wood product for the general building trade has no control over the way the product is used. As a result, the softwood-plywood manufacturers in Canada have chosen to make all softwood plywood with PF glue lines, even though a substantial proportion may be used in places where the inherent durability of a PF glue line is not needed.

By contrast, laminated timbers are not a commodity product. Because they are generally made for a specific purpose in building construction, the end use is known. Thus, there is no difficulty in marketing an interior-grade and an exterior-grade product.

This distinction illustrates the general concept that although the required durability can be defined for a *specific* end use, the wood-products manufacturers often make products that have several end uses. In the practical sense, the durability required in the product is that required by the most severe of those end uses, unless the marketing can be strictly controlled.

Another factor worth mentioning is that in wood products made with fully durable glue lines, it is the glue line that is fully durable, not the wood product. The wood is still subject to stain and decay, plywood will still face check, phenolic-bonded particleboard and waferboard will still swell permanently in thickness and the surface will roughen, all as a result of exposure to extreme weather conditions. All of these dificulties can be avoided by appropriate modification of the wood product – for example, by preservative treatment, overlays to prevent face checking, and special manufacturing conditions for particleboard that eliminate swelling and surface roughening. For the vast majority of uses, however, these special treatments are unnecessary.

NEW PRODUCTS

The major developments in glued products taking place at the time of writing derive from two circumstances. One is the diminishing quality and availability of timber for manufacture into conventional lumber and plywood. The other is the long-term threat to the supply of phenol and resorcinol, both of which are now derived from petroleum.

The restricted supply of quality timber is leading to emphasis in research and development on techniques for deriving new building products from wood residues and forest residues. In the interior of British Columbia, the generally small log size available results in the conversion of only about a third of the log volume into usable lumber in a sawmill. The remainder is used as chips, sawdust, planer shavings, edge trim, and short pieces of lumber. The alternatives are to derive more building products from these mill residues or to develop techniques for creating building products from the parts of the forest that are not usable for lumber and plywood. Both of these approaches mean an increased demand for adhesives. Waferboard represents the most efficient use of adhesive in a reconstituted board, using only 2–3% by weight of a phenol-formaldehyde resin. Yet this level represents nearly three times the amount required to produce an equivalent amount of softwood plywood. Composite plywood, made by combining veneer faces with a strandwood core, requires about four times as much resin binder as all-veneer plywood does. Medium-density fiberboard , made to replace cedar and redwood sidings, may use 8–10% resin binder by weight.

The ever-growing demand for adhesives based on phenol or resorcinol elicits two responses. One reaction is that any shortfall in these substances derived from petroleum could be met by using coal tar as the source material. The other reaction is the desire of the forest-products industry to become self-sufficient in adhesives through the development of comparable adhesives from tree sources rather than chemicals.

The two most available sources of substitutes are the bark tannins and the degraded lignins available from the kraft and sulfite pulping processes. In South Africa a substitute adhesive based on wattle tannin has been commercialized (Saayman and Oatley 1976). This adhesive is not practical in Canada because it is higher in cost than the PF glues now in use. In the United States it is becoming common practice to extend the phenol-formaldehyde in plywood adhesives with small amounts of lignins from pulping liquors.

Figure 9.13 Checking a waferboard specimen bonded with spent sulfite liquor adhesive.

In Canada the Eastern Forest Products Laboratory of Forintek Canada Corp. has developed a technique, not yet commercialized, for bonding waferboard with spent sulfite lignin. A companion study explored the properties of a plywood glue mixture in which spent sulfite liquor is reacted with polyethyleneimine and with furfuraldehyde, which is derived from vegetable sources such as oat hulls (Clermont and Manery 1977). The adhesive was found acceptable during tests for bonding plywood.

Scientists at the Western Forest Products Laboratory studied plywood adhesives made from hemlock-bark extracts as early as 1952 (Maclean and Gardner 1952). Because of the PF resin from petrochemical sources was cheaper, the work was discontinued; it is now being revived, however.

Any new adhesive proposed for use in manufacturing glued wood construction products must have sufficient durability to ensure that wood products will remain serviceable for 50–100 years. This requirement has given rise to worldwide research on the use of accelerated aging techniques. In Canada, as in most countries, expertise has been concentrated in forest products laboratories. The work largely involves comparison of glued wood products exposed outdoors (for periods up to 30 years) with the results of accelerated aging tests on similar specimens. The program is based on studies of the basic chemical nature of the adhesives, the manner in which they do or might fail, and the glue-line hazards to which the wood products are exposed during use.

There is no doubt that the demand for adhesives for wood products will increase at an accelerating rate as the techniques of gluing are used more and more to bridge the widening gap between supply and demand of conventional wood-based building materials.

REFERENCES

Canadian Standards Association. 1967. *Hardwood plywood*. CSA standard 0115-1967. Rexdale, Ontario

– 1977a. *Structural glued-laminated timber*. CSA standard 0122-1977. Rexdale

– 1977b. *Qualification code for manufacturers of structural glued-laminated timber*. CSA standard 0177-1977. Rexdale

– 1978a. *Douglas-fir plywood*. CSA standard 0121-1978. Rexdale

– 1978b. *Canadian softwood plywood*. CSA standard 0151-1978. Rexdale

– 1978c. *Waferboard*. CSA standard 0188.2M-1978. Rexdale

Clermont, L.P., and Manery, N. 1977. *Lignin polyethyleneimine adhesives for wood*. East. For. Prod. Lab. Inf. Rep. OPX-177E. Ottawa

Gaudert, P., and Carroll, M.N. 1974. *The significance of tension normal stresses on the durability of glue lines in laminated timber*. East. For. Prod. Lab. Rep. OPX-94E. Ottawa

MacLean, H., and Gardiner, J.A.F. 1952. Bark extracts in adhesives. *Pulp Paper Mag. Can.* 53 (8): 111–14

Perkins, N.S. 1950. Predicting exterior plywood performance. *For. Prod. Res. Soc. Proc.* 4: 352–63

Pound, J. 1974. *Radio frequency heating in the timber industry*. London: Spon

Saayman, H.M. and Oatley, J.A. 1976. Wood adhesives from wattle bark extract. *Forest Prod. J.* 26 (12) 27–33

Vick, C.B. 1971. Elastomeric adhesives for field-gluing plywood floors. *For. Prod. J.* 21 (8): 38–44

Figure 10.1 Panel products are used extensively in modern housing construction.

10

M.N. CARROLL
Western Forest Products Laboratory,
Vancouver

Panel Products

The art of building consists largely of enclosing space. In a house the living space is enclosed to provide protection against the elements. Inside the house individual rooms are formed to provide privacy. Within the rooms are built-in cabinets, clothes closets, linen closets, and various kinds of furniture, such as wardrobes, buffets, desks, and liquor cabinets. All provide a means of storing various household goods so that they are readily accessible but out of sight.

The common structural feature of the house and its components is that they consist of a frame surrounded by skins, known as panel products, or sometimes just skins without any frame. Panel products are premanufactured by efficient mass-production techniques. Because the panel material is available in large sheets, the assembly of structures is greatly simplified and accelerated.

Equally important are the sophisticated finishes of great variety, durability, and beauty, which can be produced by modern production-line processes. These characteristics are achieved by such techniques as grain printing, embossing, applying specialty overlays, and finishing with high-performance plastic finishes baked in ovens or cured with ultraviolet radiation.

As an illustration of the modern demand for panel products, consider a simple, conventional one-story wood house of 150 m^2 floor space (about 1600 sq ft.) The exterior-wall and roof sheathing for such a structure requires some 400 m^2 (4300 sq ft) of roof and wall sheathing, 54 m^2 (580 sq ft) of soffits, and 110 m^2 (1100 sq ft) of exterior cladding. The inside of the house shell requires 150 m^2 (1600 sq ft) of floor sheathing and 250 m^2 (2700 sq ft) of inside ceiling and wall finish. Added to this is the finish for all the partition walls. Thus, more than 1000 m^2 (10 800 sq ft) of skin materials is required, not counting what is used for cabinets and furniture.

Before this century, these house areas were covered with lumber, piece by piece. Even when plaster finish was used, it was customary to cover the surface with wood lath to provide a key for the plaster. Today these areas are quickly covered with panel products in sheet form. The most common size is 1200 × 2400 mm (formerly 4 × 8 ft), but some panel products are used in sizes up to 1500 × 4200 mm (formerly 5 × 14 ft) as floors in mobile homes. These panel products are plywood, waferboard, particleboard, hardboard, insulation board, medium-density fiberboard (MDF), and gypsumboard.

When lumber is used to cover large areas, allowance must be made for linear expansion. In the cross-grain direction of wood, the shrinking and swelling with change in moisture content can amount to several percent; in the grain direction, shrinking and swelling are negligible. When lumber is used as cladding, it is customarily laid with some kind of lap joint or batten joint to allow for linear expansion and contraction, particularly when the wood is used on the outside of a house. When panel products are used, only minimal expansion joints are required between the panels because the inherent dimensional movement of the panels is very small – about $\frac{1}{10}$th to $\frac{1}{20}$th that in the cross-grain direction of lumber. However, because the architectural styles of houses were established in the days when they were built of lumber, panel products used on finish surfaces (not sheathing) tend to simulate the appearance of a surface finished in lumber. Thus, V grooves are machined into one type of interior wall panelling so that the finished wall will have the appearance of a wall panelled with random-width lumber planks; the V grooves correspond to the laplike expansion gaps that were used between the vertical planks of lumber and also provide a convenient means of concealing the joints between panels.

Decorative panel products not only simulate the appearance of lumber structures but often simulate the surface appearance of wood as well. For example, some interior wall panels include printed photographic reproductions of the grain patterns of such attractive wood species as walnut and rosewood. Another example is medium-density fiberboard (MDF) siding that is embossed during manufacture to provide a surface texture that simulates the surface of rough-sawn lumber. These simulation techniques illustrate our continuing interest in the beauty and textures of wood.

In general, panel products provide convenience not obtainable from lumber. They can also provide properties not readily available in sawn lumber. All panel products are characterized by resistance to warping and splitting and good dimensional stability. Some provide smooth, defect-free, and grain-free surfaces that will not check and that will take and hold paint finishes for exterior use. Plywood is the choice whenever high strength is the dominant property required. The reconstituted products, such as fiberboards and particleboards, fill many needs where strength is not the major requirement. They also allow the greatest use of our forest resources; they can use residue from sawmills and plywood mills, which usually convert not more than half the log into plywood or lumber.

The changing character of the Canadian forests is shifting the use of wood towards increasing amounts of reconstituted wood products. As a result, portions of the forest that were formerly left behind as unsuitable for lumber or plywood are now being used. An example of this is the use of Canadian poplars for waferboard. A large proportion of the poplars are unsuitable for efficient conversion into lumber or plywood because trees below 70 years old tend to be too small and crooked and trees over 70 years old tend to be badly decayed. None of these factors is a hindrance in the manufacture of waferboard, however, because the entire stem is usable for this product (Carroll 1976).

THE PANEL PRODUCTS

Plywood

Plywood is a wood panel structure created by gluing thin layers, or plies, of wood together with the grains of successive plies at right angles to each other. The purpose is to construct a product that overcomes the deficiencies of lumber in sawn form. One such deficiency is the high swelling movement in the cross-grain direction. A second is the tendency of wood to split along the grain. A third is the natural occurrence of knots. Knots and the tendency of wood to split along the grain are not detrimental to the use of lumber as framing material, but they deter its use in covering large areas.

The term plywood not only refers to plywood itself but is applied loosely to a wide range of products that use the plywood concept of cross-plying. At one extreme is the pin block in a grand piano, which is made by cross-plying three or more layers of a strong wood such as maple, each layer being some 6–12.5 mm ($\frac{1}{4}$–$\frac{1}{2}$ in.) thick. The main property required in this use is resistance to splitting. At the other extreme is a type of wall finish made by peeling very thin sheets, or veneers (0.08 mm [0.003 in.] thick), and cross-plying three of them to obtain a thin sheet that can be applied like wallpaper. In this cross-plied form, the plywood has the resistance to splitting that is needed for it to be handled in such a thin sheet. At the same time the lateral expansion that results from moisture swelling is controlled because the tendency of the wood in one ply to expand an undesirable amount in the cross-grain direction is restrained by the resistance of the adjacent plies. The swelling still takes place, but it is absorbed internally in each ply by compression of the porous wood structure.

Plywood dissipates the strength-reducing characteristics of wood, such as knots. If a log is cut into thin sheets, or veneers, and several sheets are recombined into a panel, the original knots in the wood occur at random in each ply, and their strength-reducing effect is diluted. For example, in roof sheathing, 9.5 mm ($\frac{3}{8}$ in.) plywood has replaced the 19 mm ($\frac{3}{4}$ in.) lumber sheathing formerly used. In the sheathing grades, 9.5 mm lumber would be too vulnerable to breaking during handling and to splitting during installation.

Used without qualification, the term plywood, generally refers to products made solely from veneers. All-veneer plywoods are the basis of building construction. For purely structural uses, such as sheathing, plywood is made mainly from softwood species and poplar in thicknesses of 6–31 mm ($\frac{1}{4}$–$1\frac{1}{4}$ in.), using veneers in the thickness range of 2.5–4.8 mm (0.1–0.188 in.). In Canada all such softwood plywood is made with a completely waterproof glue derived from phenol and formaldehyde (see Chapter 9). In North America the total production of such plywood is now about 1.9 million m² (2 billion sq ft) per year.

For interior panelling, thin (3 mm) three-ply plywood is made from the decorative, high-density hardwoods, such as yellow birch, although a large proportion of decorative plywoods are imports of lauan plywood prefinished with wood-grain patterns and textures. Such plywood wall panellings are the standard interior finish for mobile homes. Even thinner plywoods are made for 'skins' on flush doors.

A form of plywood called lumber-core plywood is used in the furniture industry. In this product, also known as blockboard, small pieces of lumber are end- and edge-glued to form a lumber panel. The surfaces of this lumber core are finished with thin (approx. 1 mm, or 0.025 in.) decorative veneers of handsome species such as birch, walnut, or rosewood. The most common form is a five-ply structure in which the lumber core is first cross-plied with veneers, called cross-bands, from a low-density species such as poplar or basswood, and the decorative veneers are applied to the surface so that the grain of these face plies is parallel to the grain of the core (Figure 10.2). This cross-ply construction provides dimensional stability. The cross-bands mask any surface irregularities in the lumber core to prevent them from 'telegraphing' to the very thin face veneers. The evolution of this product is described further under the heading 'Particleboard.'

Figure 10.2 A panel of lumber-core plywood cut away to show the lumber core, cross-band veneer, and surface veneer.

The Manufacture of Plywood

Veneer for plywood is made by peeling or slicing bolts of wood into thin sheets. Slicing is a technique in which a bolt of wood is clamped into dogs, or steel, teethlike projections, and a reciprocating knife moves progressively across the bolt, slicing off thin sheets of veneer the width of the bolt. These relatively narrow strips of veneer are dried, jointed, and edge-glued to form large sheets. The slicing technique is generally used only on decorative hardwoods, in which the grain patterns of the wood can be combined into repetitive patterns of great beauty. Slicing is also used to obtain highly figured veneers from crotches and burls.

The veneers for softwood plywood and some hardwood plywood are produced in a veneer lathe. A debarked section of the tree some 2.5 m (8½ ft) long is centered on the spindles of the lathe. These tree sections are referred to as veneer bolts or peeler bolts. The bolt is rotated at high speed, and a veneer knife is advanced into the side of the bolt at a speed proportional to the rotation speed. This action produces a continuous ribbon of veneer that is cut ('peeled') from the bolt in a spiral fashion (Figure 10.3). The peeling is continued until the bolt is reduced to a core about 101 mm (4 in.) in diameter. The core is then ejected from the lathe, the knife carriage is retracted, and a new peeler bolt is positioned. In plywood mills operating on small logs (averaging 280 mm [11 in.] in diameter), automatic charging equipment enables the lathe to peel as many as eight bolts per minute.

As it comes from the lathe, the veneer is clipped by a high-speed clipper to yield veneer sheets of the desired size (generally 1300 × 2500 mm [4¼ × 8½ ft]). If gross defects are present, they are clipped out, leaving pieces of veneer less than 1300 mm (4¼ ft) wide. If these underwidth pieces are of high quality, they may be edge-glued back into full sheet form to be used as outer plies. If not, they may be cut in

Figure 10.3 Peeling softwood veneer. (Photo: Council of Forest Industries of British Columbia)

Figure 10.4 Laying up construction plywood. (Photo: Council of Forest Industries of British Columbia)

half lengthwise to be used for the cores or cross-bands of the plywood. After clipping, the veneer is dried to 3–5% moisture content by being continuously passed through a multilevel hot-air dryer.

For five-ply plywood, the construction consists of a center veneer, two cross-bands, and a face and a back ply. For three-ply plywood, the construction consists of face and back plies and one cross-band. The glue is applied to both sides of the cross-band veneers (Figure 10.4), which are relatively narrow pieces 305–680 mm (12–27 in.) wide. The plywood layup is assembled by placing the back veneer and laying the cross-band veneer strips (with glue applied) crosswise, with the edges of the strips butted, until the accumulated width of the cross-band strips is equal to the length of the face sheet. Then the center veneer is laid with its grain parallel to the grain of the face sheet. Another layer of glue-coated cross-band strips is laid, and finally the face veneer is applied. For greater thickness the process is extended to create additional plies, but always with an odd number of alternating plies. A type of three-ply plywood is made with four layers of veneer. In this structure the two center veneers are laid parallel to each other so that they function as one ply that is twice as thick as the veneer. These layups may be prepressed in an unheated platen press to consolidate them into bonded structures that can be handled by automatic equipment.

The next stage in manufacture is hot-pressing, in which the layups are loaded into a multiopening hydraulic hot press (Figure 10.5). The thicker layups are loaded one per opening, whereas thinner layups may be loaded two per opening. Heat and pressure are applied by the hydraulic press, and the glue lines are brought to a temperature at which the glue sets.

After pressing, the panels are trimmed to size. Unsanded panels may be sold as is. Sanded panels are dimensioned in belt sanders. Surface defects in the finished panels, such as splits or open knotholes, may be filled with veneer inlays (patches) or a synthetic patching material (a fast-setting plastic) (Hancock 1975). Smaller defects such as wormholes are filled with plastic wood.

The various grades of softwood plywood and their uses are described in Table 10.1. This table is taken from the *Plywood Construction Manual*, which provides a wealth of information on the manufacture and use of softwood plywood (COFI 1976).

Hardwood plywood is manufactured in a similar manner, from either sliced or peeled veneers. The major difference is that the cross-plies are spread with adhesive and go through the glue spreader as full-size sheets rather than being handled in the form of small pieces. The reason is that in forming a core or cross-bands from narrow pieces of veneer butted together, gaps are unavoidably created between the pieces. These gaps are called core gaps. In softwood plywood these narrow gaps are easily bridged by the relatively thick face veneers and are not visible in the face of the panel. In hardwood plywood the thin face veneers cannot bridge such gaps, so solid core has to be used. Almost all hardwood plywood used in buildings is prefinished (*Woodworking and Furniture Digest* 1974).

For concrete form work, a market in which softwood plywood has essentially no competition from other wood panel products, coatings

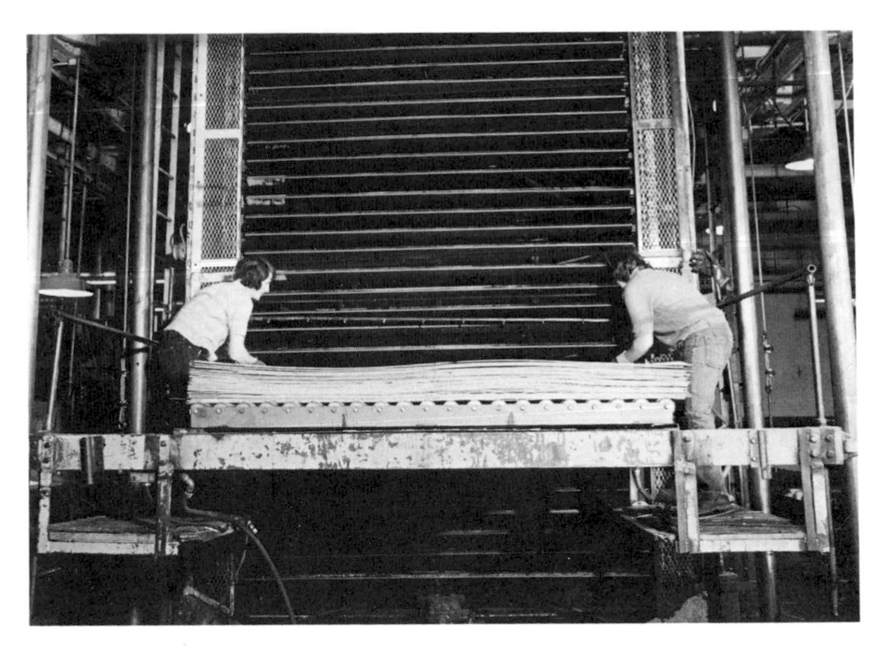

Figure 10.5 A multiopening hydraulic hot press used for pressing plywood. (Photo: Council of Forest Industries of British Columbia)

and overlays are commonly used to increase the resistance of the surfaces to abrasion and damage by the alkali in the concrete so that the panels may be reused many times. High-density overlay (HDO) is a paper-reinforced coating of phenol-formaldehyde resin that is hot-pressed onto the surface of the plywood. Medium-density overlay (MDO) is a paper overlay in which high wet strength and durability are achieved by the addition of substantial quantities of phenol-formaldehyde resin. MDO was originally developed as a paint-base overlay, but it is now used as a base for concrete form coatings as well.

Waferboard

Waferboard is a sheet material made from wafers which resemble small pieces of veneer. The wafers are 37–75 mm (1.5–3 in.) along the grain, an average of 0.5–1 mm (0.020–0.025 in.) thick, and of random width. Waferboard is a type of particleboard, since the wafers are a type of wood particle, and is made by a technology similar to that by which particleboards (q.v.) are made. However, it is a quite distinctive product, in both its properties and uses.

Figure 10.6 Veneer ribbons for waferboard produced by a disc flaker.

The Manufacture of Waferboard

The wafers are cut from 0.6 m (2 ft) long, or longer, roundwood bolts, mostly from poplar, by waferizing machines. One such machine has a special type of serrated knife mounted on the face of a disc. As the disc rotates at high speed, the bolts are pressed laterally (across the grain) into the knives. The cutting action is similar to that of a veneer slicer, except that the serrated knives yield ribbons of veneer instead of sheets (Figure 10.6). In subsequent drying and handling, these ribbons split randomly along the grain of the wood to form wafers.

The dried wafers are tumbled in a blender with a finely powdered phenol-formaldehyde resin in order to distribute the resin particles over the surfaces of the wafers. The wafers are then formed into a low-density mattress, or mat, by a sifting process so that the overlap-

Table 10.1 *Softwood plywood grades and uses*

Grade*	Specify by	Veneer grades			Characteristics	Typical applications
		Face	Inner plies	Back		
Good two sides (G2S)	CSA 0121	A	C	A	Sanded. Best appearance both faces. May contain neat wood patches, inlays, or synthetic patching material.	Furniture, cabinet doors, partitions, shelving, concrete forms, opaque paint finishes.
Good one side (G1S)	CSA 0121	A	C	C	Sanded. Best appearance one side only. May contain neat wood patches, inlays, or synthetic patching material.	Where appearance of one exposed surface is important. Floor underlayment where sanded surface desired. Concrete forms.
Select sheathing tight face (SEL TF)	CSA 0121 or	B†	C	C	Unsanded. Uniform surface.	Underlayment and combined subfloor and underlayment. Hoarding. Construction use where sanded material is not required. Concrete forms where smooth surface not necessary.
Select sheathing (SELECT)	CSA 0151	B	C	C	Unsanded. Uniform surface with minor open splits.	
Sheathing (SHG)	CSA 0121 or CSA 0151	C	C	C	Unsanded. Face may contain limited-size knots, knotholes, and other minor defects.	Roof, wall, and floor sheathing. Hoarding. Construction use where sanded material or uniform surface not required.
High-density overlaid 60/60 (HDO 60/60)	CSA 0121 or CSA 0151	B†	C	B†	Smooth, resin-fiber overlaid surface. Further finishing not required.	Bins, tanks, boats, furniture, signs, displays, forms for architectural concrete.
Medium-density overlaid (specify one or both sides) (MDO1S. MDO2S)	CSA 0121 or CSA 0151	C†	C	C†	Smooth, resin-fiber overlaid surface. Best paint base.	Siding, soffits, panelling, built-in fitments, signs, any use requiring superior paint surface.

*All grades, including overlays, are bonded with waterproof phenolic resin glue.
†Permissible openings filled.

ping wafers are generally oriented in the horizontal plane, but with the grain angle of the wafers randomly aligned in the plane. Finally, the mat is hot-pressed at high temperatures and high pressures – about 210°C (410°F) and 2.7 MPa (400 psi) – to form a waterproof bond between the particles. In the press the mats are consolidated to a relative density of 0.65, which is considerably higher than the 0.4 relative density of poplar, from which waferboard is made.

Because of the way the wafers are randomly aligned, the dimensional stability of waferboard is equal in all directions in the plane of the panel. The 'cross-ply' effect is still operating, but on a random basis rather than on a regular basis. The strength properties are also approximately equal in all directions in the plane of the panel.

The finished product has a lower strength-to-weight ratio than that of softwood plywood. For applications such as roof sheathing, it also has a lower strength-to-thickness ratio than that of plywood, so it is used in a greater thickness. For example, in Canada, 9.5 mm ($\frac{3}{8}$ in.) plywood is replaced by 11.1 mm ($\frac{7}{16}$ in.) waferboard for the same use.

The Canadian production of waferboard in 1978 was about one-third that of softwood plywood (on a volume basis), and almost three-quarters of that production has come into existence since 1972. Because of the vast Canadian reserves of poplar, Canada has a unique market position in waferboard and produces about 90% of the world production.

Particleboard

The terminology in particleboard is a bit confusing, and the reader is referred to other sources for details (Carroll 1966; Maloney 1977). The term particleboard is used here to encompass the range of products made primarily from wood residues from sawmills and plywood mills. These products differ from waferboard in that the wood particles do not have uniform length or thickness but are a random mixture of sizes. The main requirement is that the particles, of whatever size, have a reasonable length-to-thickness ratio. Cubical particles, such as sawdust, are undesirable because they will not orient with the grain of the wood in the particles lying in the horizontal plane of the panel. Sawdust may be used, however, if it is passed through a refiner; the rubbing action of the refiner fractures the sawdust particles along the grain, yielding fine splinters with an adequate length-to-thickness ratio.

The Manufacture of Particleboard

The manufacture of particleboard is roughly similar to that of waferboard. The mixture of particles is dried, and liquid glue (normally urea-formaldehyde resin solution) is sprayed onto the particles in a blender so that the atomized droplets of resin are distributed over their surface. The particles are then air-sifted into a mat and hot-pressed to obtain the desired density and to set the adhesive binder.

The major development in particleboard manufacture took place in the furniture industry, when particleboard replaced the lumber core in lumber-core plywood. The product for this use is variously referred to as core board, furniture board, or industrial board. Particleboard has also supplanted lumber almost entirely in the manufacture of furniture panels, but it is not limited to furniture markets. At one time 50% of North American particleboard production was marketed as underlayment. Underlayment particleboard is applied over a rough plywood or lumber subfloor to provide a smooth finish over which thin, resilient finish flooring can be laid. In the last ten years, the underlayment share of the market has diminished, but has been taken up by decking, a strong particleboard used in mobile homes as a combination subfloor and underlayment. Because mobile homes are built under cover, the floor is not exposed to weather during construction. Thus, a lower-cost particleboard bonded with a non-weather-resistant urea-formaldehyde resin is often used.

In furniture, replacement of the lumber core by particleboard took place in three stages. In the first stage particleboard was made from a homogeneous mixture of various particle sizes ranging from fine to coarse. Although the board was sanded, the varying size and orientation of the particles on the surface meant that the surface would tend to roughen when moistened (e.g., by the water in the glue used to apply the veneers to it). The five-ply construction with cross-bands alleviated this problem. Compared with a lumber core, however, this core had two deficiencies.

1. The screw-holding power of the particleboard core was substantially lower than that of wood. To some extent this deficiency could be improved by increasing the relative density of the board from the usual 0.7 to as much as 0.95 (*Woodworking and Furniture Digest* 1978). Because increased relative density gave a very heavy board, other means were sought to improve its screw-holding power. These included the use of longer screws, fastening of hinges with three or four screws instead of two, and filling the screw holes with glue before the screws were inserted.

2. The edges were coarse and could not be finished without some type of edge banding. For this purpose, strips of veneer were glued to the edges, or T-shaped plastic moldings were placed in slots cut into the edge of the board.

The second development in particleboard came in the 1950s with the introduction of a so-called wind-sifting process. In this process the mixture of particles was sifted to form the mat and passed through a cross-flowing airstream. The finer particles were carried laterally to a distance roughly inversely proportional to their size (Figure 10.7). In this manner the very coarse particles could be placed in the center of the board thickness, and the finest particles were laid on the surface. This arrangement produced a 'fine surface' or 'graduated construction board.'

The immediate advantage to furniture-core construction was that the need for cross-band veneers was eliminated. The standard veneered furniture panel then became a three-layer construction, and the decorative veneers were simply glued to the particleboard core. Further, the surface quality was now good enough that resin-paper overlays or wood-grain-printed vinyl overlays could be used to replace

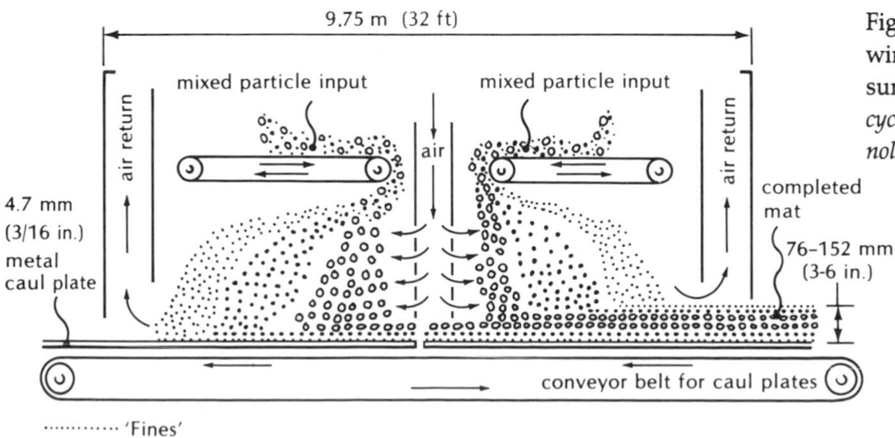

Figure 10.7 Schematic drawing of the wind-sifting process for forming fine-surface particleboard. (Adapted from *Encyclopedia of Polymer Science and Technology*, Vol. 4 [New York: Wiley, 1966])

Figure 10.8 Particleboard core with lumber lipping, covered with a decorative veneer.

the surface veneers in lower-quality furniture. In some cases, direct grain printing could be applied to the surface of the particleboard core.

The problem of the coarse edge still remained. For mitered corner joints, the coarse edge was overcome by an ingenious technique developed to form the joints. A flexible decorative vinyl film was glued to one side of the core panel. Ninety-degree V grooves were then machined into the core so as to just cut through the core and not through the vinyl. The V grooves could then be folded together to form a mitered corner joint, with the vinyl film acting as a hinge. If two of these 'V-fold' joints were placed adjacent to each other, they could be used to turn the edge of the board back on itself to present a completely finished edge.

For overhanging edges, such as a table top or dresser top, in which molded edges were desired, the problem of the coarse edge could only be overcome by lipping – surrounding the particleboard core with a wooden frame that could be machined and readily finished (Figure 10.8).

Medium-Density Fiberboard (MDF)

The final development in particleboard came in the 1960s with medium-density fiberboard, or MDF. For this product the wood residue is steamed and passed through an attrition disc mill, or refiner, and reduced to a fibrous form. Liquid resin binder is distributed over the fiber in approximately 10% of the weight of the board. With careful pressing techniques, the board can be made nearly uniform in density throughout its thickness. In contrast, conventional core boards tend to have very high-density surface layers on a low-density core layer.

The cost of manufacturing this product is about 10% higher than that of conventional fiberboards, mainly because of the energy required to reduce the wood to fibrous form. This cost is partially offset, however, by the use made of the residues from heavy hardwoods that would make an inferior particleboard.

The need for veneers or lipping is eliminated in MDF. The edges can be molded in the traditional patterns and finished quite simply. The surface can be embossed and wood-grain printed. The product is no longer a core board but a board that stands by itself as a panel product.

Thinboard

Thinboard is a type of particleboard, made from wood residues, that is distinguished by the pressing technique used. The conventional multiopening hot presses used for particleboard are uneconomical for particleboards less than about 6 mm ($\frac{1}{4}$ in.) thick. To overcome this problem, thinboard is made on a continuous rotary-drum press (Figure 10.9). Originally the material for this board was the same as that used in the thicker particleboards. In 1974, however, a plant in California began to make thinboard from wood fiber.

The Fiberboards

What are commonly referred to as the fiberboards belong to an earlier age (before 1945) when the wet-process manufacturing technique was dominant. In the wet-process technique, the wood is reduced to a coarse, fibrous form. The dry-process technique, used for products such as medium-density fiberboard, reduces the wood to fiber and

Figure 10.9 Manufacturing thinboard on a rotary hot press.

forms the basic mat by a dry-sifting process. In contrast, the wet-process fiberboards carry the fiber in an aqueous suspension and form the mat, now called a wet lap, by draining the fiber suspension on a screen. This wet lap is then placed in a hot press, which has a screen, like a flyscreen, on the bottom side, permitting escape of the water that is squeezed out of the lap by hydraulic pressure in the press. The screen also permits escape of the steam generated in the final stages of the hot-pressing operation. The resulting product is called screen-back hardboard, or smooth-one-side (S1S) hardboard.

The characteristic of fiberboard made by the dry process is that the moisture is removed before pressing and no screen is required. The result is a product called S2S (smooth-two-sides) hardboard. To add to the confusion of terminology, one can sand off the screen marks on the lower side of an S1S wet-process board and create an S2S board. Another way of creating an S2S product is to form a low-density fiberboard by the wet process, dry it, and repress it to high density under such extremes of temperature and pressure as 220°C (428°F) and 125 MPa (1800 psi). Under these conditions the wood lignin is softened so that it acts a binder for the new board.

Generally the distinction between the dry process (in which the fiber is dried before pressing) and the wet process (in which the water is removed in the hot-pressing operation) lies in the amount of auxiliary binder required. A wet-process fiberboard depends greatly on the papermaker's bond – the adherence of fibers to one another by chemical and physical processes as water is removed and drying takes place in a paper sheet – to develop strength in the board; the dry process depends on added resin binder. Thus, a fiberboard with a relative density of 0.9 made by the wet process may require 0.5% PF resin binder. The same board made by the dry process may require 3% PF resin binder because there is no contribution from the papermaker's bond.

Wet-process fiberboards can be made in a range of densities. At one

Figure 10.10 Composite plywood made with softwood veneers on an aligned-strand particleboard core.

end is the low-density insulation board panel that requires some 15% of a binder, usually starch, to reach a useful level of strength properties. In the middle are the medium-density fiberboards (around 0.7 relative density) that require about 8–10% resin. These fiberboards are similar to the MDF boards made by the dry process already described except that they have the screen pattern on the reverse side. For these MDF panels, the wet process has a slight advantage for making siding, because the bulk of the board can be formed from lower-quality fiber sources, including bark, and a thin surface layer of higher-quality fiber with a high resin content can be applied to develop a strong, weather-resistant surface.

At the high end of the density scale are the wet-process hardboards (strictly speaking, hard-pressed fiberboards) that use as little as 0.5% of an added phenol-formaldehyde resin binder. This board can also be made with lower-quality (i.e., lower-cost) fiber in the lower part and a special surface layer suitable for embossing or printing.

Composites

Composites are combinations of veneer faces on a particleboard core. One of these, in which thin decorative veneers are glued to a particleboard core, has already been described. This is described as a hardwood plywood in the Canadian Standards Association Standard 0115m. The term is somewhat of a misnomer because the board is not made entirely of veneer and does not use the principle of cross-plying. Indeed, it makes no significant difference in the end product whether the grain direction of the veneer runs parallel or at right angles to the long direction of the particleboard core. It would be more descriptive to refer to the board as an overlaid particleboard.

Another significant composite is a product not yet made in Canada and manufactured by one firm in the United States. The core is a special type of board in which wood veneer ribbons are made in a manner similar to that described for waferboard except that they are deliberately fractured to produce long, narrow particles called strands. When the dry-process mat is formed, the strands automatically orient themselves in the horizontal plane, but they are also aligned preferentially in one direction in the mat (Elmendorf 1969). When the mat is hot-pressed, the resulting board has anisotropic properties; the strength and stiffness in the direction of alignment are two to three times those in the cross-direction. At the same time, the dimensional movement in the cross-direction, resulting from moisture change, is higher than the movement in the direction in which the particles are aligned.

Figure 10.11 Aspen waferboard used as wall sheathing on residential buildings.

The composite is formed by applying softwood veneers to the core, with the face grain of the face and back veneers at right angles to the direction in which the particles are aligned. This product is referred to as composite plywood (Figure 10.10). Its strength and dimensional stability are similar to those of an all-veneer plywood of the same thickness (McKeen, Snodgrass, and Saunders 1975).

The attractiveness of this product is that various thicknesses can be achieved in the three-ply construction merely by varying the thickness of the core. Like the core of waferboard, the core of composite plywood can be made from roundwood that is unsuited for manufacture into lumber or plywood. The product is of major interest to existing softwood-plywood mills that are having difficulty getting an adequate supply of suitable peeler logs.

Figure 10.12 Plywood cladding simulating a reverse batten-lumber cladding style. (Photo: United States Plywood Corporation)

Gypsumboard
Although gypsumboard is not an all-wood product, it is included here because it rounds out the picture of panel products. Gypsumboard is simply a thick layer of gypsum plaster produced in panel form. Because plaster itself is weak, paper sheets are glued to the front and back of the plaster to provide sufficient bending strength for handling. The main virtues of the product are low cost, fire resistance, and the provision of a smooth, unbroken wall surface.

HOW PANEL PRODUCTS ARE USED

In this section it will be seen that the various panel products compete with one another and with lumber for the same uses. The choice of a particular product is based predominantly on cost, although aesthetics and maintenance costs must also be considered.

Roof Sheathing
Roof sheathing serves three main purposes. It ties a roof structure together to resist wind loads and snow loads, it supplies a nailing base for shingles, and it provides a working platform for those working on the roof. The most commonly used products for roof sheathing in Canada are waferboard and the sheathing grades of softwood plywood and poplar plywood.

Wall Sheathing
Wall sheathing is used on the outside of stud-wall framework to give the structure rigidity and racking resistance against wind loads (Figure 10.11). It can also provide a nailing base for subsequently applied finish cladding. The choice of panel products depends largely on economics, including shipping costs. In eastern Canada, waferboard is competitive with softwood plywood shipped from British Columbia. In British Columbia, plywood has the edge because of the cost of shipping waferboard from eastern Canada. A third panel product used in some areas is 'black joe,' an insulating fiberboard panel that has been impregnated with asphalt; this board does not provide a nailing base for cladding.

Figure 10.13 Rough-sawn lumber cladding. (Photo: Council of Forest Industries of British Columbia)

Finish Cladding
Finish cladding must be decorative and at the same time it must withstand the ravages of weather. The most common form of plywood cladding is the grooved panel shown in Figure 10.12. This is a five-ply panel, with dadoes cut down through the first two plies. The edges of the panel are rabbetted or grooved to form a lap joint that is indistinguishable in appearance from the dadoes. Plywood exposed to the weather will develop face checks, even when painted, so the most appropriate finish is one that simulates rough-sawn lumber and is suitable for staining (Figure 10.13). If the plywood is made with a medium-density overlay (MDO), its smooth surface will not check and will be entirely suited for painting.

Figure 10.14 A warehouse finished with waferboard. (Photo: MacMillan Bloedel Ltd.)

Plain waferboard is also acceptable as an exterior finish in some types of building (Figure 10.14), such as service stations and agricultural buildings (barns, granaries, and so on). The waferboard is finished by either staining or painting. In residential housing, waferboard is used more as accent panels than as the main finish.

The medium-density fiberboards, made by either the dry or the wet process, offer a great variety of finishes, since many different patterns and textures can be obtained by using embossing plates in the hot-pressing operation. Figure 10.15 shows plain-surface, medium-density fiberboard siding simulating the traditional lap siding. Figure 10.16 shows an MDF that has been embossed to give a surface texture resembling stucco plaster.

Interior Finish

The most common finish for interior walls, partition walls, and ceilings is gypsumboard, which provides a neutral base for decoration with paints, texture paints, and wallpaper. The notable exception is mobile homes, since plasterboard is too heavy and too fragile to withstand the flexing of the structure that develops during road transport. For mobile homes the traditional interior finish is a decorative hardwood plywood. In residential housing, V-grooved hardwood plywood is also used for decorative walls, or as the entire wall finish in special rooms such as dens, recreation rooms, and family rooms.

Hardwood plywood for these uses competes with hardboards in a wide variety of finish patterns (Figure 10.17). The embossing techniques available to the fiberboard process also make it possible to simulate other types of wall finishes, such as stone (Figure 10.19) and ceramic tile (Figure 10.18).

Figure 10.15 A traditional house with medium-density fiberboard (MDF) used as horizontal lap siding. (Photo: Masonite Canada Ltd.)

Floors

Before panel products were available, floors were constructed with lumber sheathing as the subfloor and with either wood-strip finish flooring, parquet tiles, or ceramic or slate tiles on top. As plywood became available, lumber sheathing was replaced by plywood sheathing.

The development of thin, resilient flooring materials made from modern vinyl plastics necessitated a very smooth surface on the subfloor. Such smoothness could be obtained by using sanded plywood with a defect-free surface. However, when fine-surface particleboard became available, it was widely used as underlayment. The result was a two-layer floor in which the lumber or plywood sheathing formed the structural subfloor, and underlayment-grade particleboard laid on top of this provided a base for the resilient flooring. This design is particularly adaptable to site-built housing, in which the subfloor acts as a working platform during construction of the house and is exposed to weather, dirt, and damage. After the house is closed in, the particleboard underlayment is added, and the finish floor is applied immediately.

In the past decade an increasing number of housing units have been built under factory conditions. As a result, structural particleboard or sanded waferboard is used as the whole floor, with either resilient flooring or wall-to-wall carpet laid directly on it.

Figure 10.16 MDF panels embossed and finished to simulate a stucco finish. (Photo: Abitibi Paper Company Ltd.)

Cabinets

Kitchen and bathroom cabinets are an integral part of modern homes and are usually premanufactured in a factory and installed during the construction of the house. The quality of the cabinetry covers a wide range. The top-quality lines use techniques similar to those used in the manufacture of fine furniture – fine-quality veneer surface finish, drawers and cupboards lined with high-density overlay, panel doors, and natural finishes that emphasize the beauty of the wood. At the other end of the scale, cabinetry may be built largely of particleboard and have a painted finish or a wood-grain vinyl sheet applied as the finish.

In between these extremes are many varieties of designs that use plywood, particleboard, hardboard, and medium-density fiberboard in various combinations. The main requirement in kitchen cabinetry is a surface that is durable and permits easy removal of grease and dirt.

The kitchen countertop in these units is usually made of particleboard with a high-density plastic sheet glued on to provide a working surface.

Figure 10.17 Three competing types of hardwood plywood for interior wall panelling: wood-grain print on hardboard (*bottom*), wood grain printed and embossed on imported lauan plywood (*center*), and printed melamine sheet overlaid on 3 mm ($\frac{1}{8}$ in.) thinboard (*top*).

Doors

Wood entrance doors are of two basic types. The first is the panelled door, the design used before panel products were available. The second type is the flush door, which was originally a solid slab of edge-glued wood. Because of the potential problems created by linear expansion in the cross-direction of the slab, a type of plywood construction is used in slab doors in which cross-piles of thin lumber are concealed within the structure of the door. Because panel products with good dimensional stability are now available, the flush door has become dominant.

Figure 10.18 Hardboard finished to simulate ceramic tile. (Photo: Masonite Canada Ltd.)

In modern practice, a flush door consists of a stile and rail frame of wood, thin skins covering the frame, and something to fill the gap between the skins.

Interior doors are generally of the hollow core type, in which spacing strips of wood are used to stiffen the skins. The skins are most commonly thin plywood (2.5 mm [1/10 in.]) to provide either a natural or painted finish, plain hardboard (usually painted), or hardboard with a prefinished wood-grain pattern.

In solid-core doors the gap between the skins is filled with a solid material. This core may be end- and edge-glued solid wood, a low-density (0.45 relative density) particleboard, or a somewhat heavier extruded particleboard (0.72 relative density) (Figure 10.20). The current emphasis on insulation is leading to the use of more sophisticated core materials, such as styrene and urethane foam, for exterior doors.

For industrial doors, and especially those requiring long-term fire resistance, metal faces are commonly used.

Concrete Formwork

Panel products have greatly simplified the process of erecting and disassembling concrete formwork, and the panels may be reused many times (Figure 10.21). With rare exceptions, Douglas-fir plywood is the panel product used for formwork because of its high strength-to-weight ratio. One exception is a type of construction in which insulating fiberboard several centimetres thick is used to make the form on which a concrete roof is poured. After the concrete has set, the fiberboard is left in position to act as insulation.

Preserved Wood Foundations

The walls forming the basement of a house have traditionally been made of masonry, either poured concrete or concrete blocks. Since about 1970 a new type of building construction has come into use in

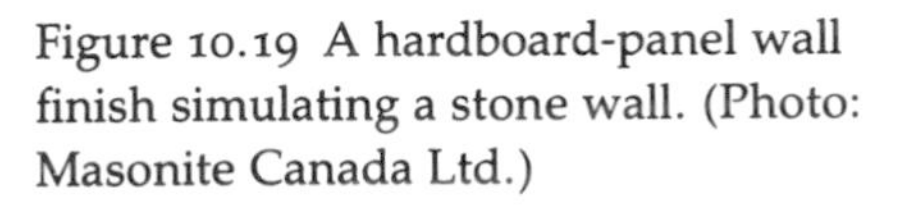

Figure 10.19 A hardboard-panel wall finish simulating a stone wall. (Photo: Masonite Canada Ltd.)

Figure 10.20 Imitation panelled doors made with molded hardboard skins finished in wood-grain pattern. (Photo: Weyerhaeuser Canada Ltd.)

Figure 10.21 Concrete formwork made with plywood. (Photo: Council of Forest Industries of British Columbia)

which the conventional stud and sheathing outside wall structure is extended below the soil level to form the basement (Canadian Wood Council 1976). Because contact with the soil would rapidly rot ordinary wood, the part below ground level is constructed from preservative-treated lumber and plywood (Figure 10.22). Preservative treatment roughly doubles the cost of the materials, but the several advantages gained offset the increased cost. Among these are:

1. The foundation can be erected in freezing weather, in which concrete does not set.

2. The stud space can be insulated directly, and the inside panelling provides a finish wall surface. In a basement made of masonry, furring strips are applied to the walls and floors to create space for insulation and to provide a nailing base for applying the finish wall panels. Among other things, furring strips sacrifice valuable floor space and headroom.

Other Uses

This listing of the uses of panel products is far from complete. Other uses include fencing, crating and packaging, soffits in housing, boat hulls and decks, produce boxes (Figure 10.23), doghouses, and bird-houses. These examples illustrate the characteristics of the various panel products and their interchangeability. In many other uses, the panels are cut into small sections. For example, particleboard stair treads are cut from 32 mm (1.25 in.) thick particleboard panels, and insulating board is cut into tile form for use as a sound-absorbing wall or ceiling finish.

Figure 10.22 Preserved wood foundation built with plywood. (Photo: Council of Forest Industries of British Columbia)

Figure 10.23 Plywood pallet bins used for airlifting cranberries. (Photo: Council of Forest Industries of British Columbia)

THE MARKETPLACE

The Canadian marketplace cannot be viewed in isolation. Canada uses and exports softwood plywood and waferboard but imports other panel products, notably lauan plywood and medium-density fiberboard. MDF is not yet made in Canada, because the Canadian market is not large enough to make a plant of economical size (approximately 300 tonnes per day) feasible. The composite plywood described earlier is not made or used in Canada at the time of this writing, but it is expected to come into production and use in the near future.

Softwood plywood is the major panel product produced in Canada, most of it in British Columbia. Waferboard is made largely in eastern Canada; although poplars are available in the west, they are also available in Ontario, and a plant in Ontario does not have the burden of shipping its product thousands of kilometres to reach the major markets in eastern Canada and the northeasten United States.

As the market expands and the supply of timber available for plywood diminishes, one can expect a larger share of the panel-product market to go to waferboard, particleboard, and composites. The growth in the panel products industry described in this chapter has taken place in the last thirty years, giving striking evidence of the inherent demand for panel products. Such products as thinboard, MDF, waferboard, and composite plywood have become commodity trade items only in the last five to ten years.

REFERENCES

Canadian Wood Council. 1976. *Construction guide for preserved wood foundations*. Ottawa

Carroll, M.N. 1966. Composition board. In *Encyclopedia of Polymer Technology*, Vol. 4. New York: Wiley

– 1976. Growth of waferboard in Canada. *Forest Prod. J.* 26 (11): 26–30

COFI. 1976. *Plywood construction manual*. 3rd ed. Council of Forest Industries of BC, Vancouver

Elmendorf, A.E. 1969. Orienting Wood Strands. U.S. Patent 3,478,861

Hancock, W.V. 1975. *Evaluation procedures for synthetic patching materials for repairs to plywood panels*. West. For. Prod. Lab. Inf. Rep. VP-X-131. Vancouver

Kirk, R.E., and Othmer, D.F. 1966. Particleboard. In *Encyclopedia of chemical technology*. New York: Wiley

Maloney, T.M. 1977. *Modern particleboard and dry process fiberboard manufacturing*. San Francisco: Miller Freeman

McKeen, H.B.; Snodgrass, J.D.; and Saunders, R.J. 1975. Commerical development of composite plywood. *Forest Prod. J.* 25 (9): 63–8

Woodworking and Furniture Digest. 1974. Decorator panelling from lauan. 76 (4): 45–7

– 1978. High density particleboard cabinetry. 80 (1): 22–4

11

R.J. PAYNE
Council of Forest Industries of British Columbia

Houses and Structures

If one excludes pulp and paper, housing is the forest industry's largest market for wood products. About 200 000 housing units are built in Canada each year, and this volume is expected to be maintained or exceeded in the years to come. Over 80% of the houses in North America are timber-frame structures, and in 1970 an average residential house required 17.6 m^3 (10 840 fbm [foot board measure]) of lumber and 4.7 m^3 (5358 sq ft) of wood-based panel products (9.5 mm [3/8 in.] basis). Projections to the year 2000 show that although the quantity of lumber per house will remain almost constant, the amount of wood-based panel products will rise by 10% as greater use is made of these products for roof, wall, and floor sheathing.

As we become increasingly urbanized, the pressure to use available land is rising exponentially. Already almost 77% of Canada's population lives in cities of over 100 000 inhabitants, and population movement to urban areas is increasing annually. The number of households in Canada is also steadily increasing, largely as a result of the rise in marriages. It is estimated that the number of marriages has increased by one-third between 1971 and 1981. Nearly all these new families require housing accommodation, and if present trends continue, over half will want single-family timber-frame dwellings.

The growth in the demand for housing will generate a significant increase in the demand for wood products. Canada is fortunate in possessing a rich, renewable forest resource capable of providing these much-needed building products.

Like the cost of other manufactured building products, the cost of forest products has increased over the years. However, despite increased costs, timber-frame construction is still the most economical method of house building, and lumber and other wood products account for only a minor and decreasing portion of the total cost of housing. The major contributors to rising housing costs are land, high interest rates, and services.

Figure 11.1 Large, single-family homes near Ottawa. Extensive use of wood siding enhances the beauty of these houses. (Photo: Canada Mortgage and Housing Corporation)

NATIONAL BUILDING CODE

Public regulation of the design and construction of buildings is now a generally accepted part of the overall building process. Early building laws were concerned only with the prevention of structural failure. Today, three primary requirements – structural sufficiency, fire safety, and protection against health hazards – are the basis of most laws that deal with the construction and use of buildings.

From their inception, building regulations have been designed to protect the public. Traditionally they were prepared by local authorities and based to a great extent on experience within the community. In general this led to the development of a heterogeneous assortment of local ordinances, which followed no consistent order, differed in many matters of detail, and were frequently so specific in their acceptances that the introduction of new materials or methods of construction was extremely inconvenient. Under such codes it was difficult and often impossible to use wood or wood-based products to their full structural and economic potential.

The National Building Code of Canada was first published in 1941. Although only advisory in nature, it was written so that it could be adopted by any municipality in Canada as its own building bylaw. Today this national set of building regulations is in use in one form or another, through voluntary adoption, by building authorities representing over 75% of the population of Canada. In addition, several provincial governments have adopted the national code as a mandatory code for all building within their provincial boundaries.

The nationwide acceptance of the National Building Code, and its supporting standards, directly contributes to the efficiency and economy of timber-frame structures in various ways. For example:

1. It facilitates the work of designers and builders who operate in more

than one area by obviating the need to design different residential structures for different regions.

2. It facilitates the production and marketing of manufactured products throughout the nation and permits the use of standard products that are readily available at lower costs than those of custom-made products.

3. It prevents the use of inferior products and poor workmanship and thus provides a degree of consumer protection.

4. It simplifies the task of building inspectors, mortgage lenders, and insurance underwriters by providing a common standard for structural and financial assessment.

5. It facilitates the acceptance of new, often more economical and more efficient methods and materials of construction.

Chapter 15 contains a broader discussion of the application of the National Building Code to the wood industry.

CONSTRUCTION OF BUILDINGS

The buildings of Canada, particularly our houses, provide a valuable historical record of our society, our attitudes, our values, and our lifestyles. The homes we have built, from the early shelters of the Indians and primitive log cabins of the first settlers to today's high-density row houses and apartments, are of inestimable value to the social historian.

Perhaps the single-family house on its own lot best represents the dream home of the average Canadian family. The rows of houses in the suburban developments that ring our urban centers testify that this dream has been realized for many. That this dream is not attainable by every family, however, can be seen from the new forms of housing that are now taking shape. High-density developments such as condominiums, townhouses, and apartment complexes are becoming the rule rather than the exception. Many of the types of housing that are being built today were unknown a decade ago. No longer is housing being built just for the child-oriented family. Students, the elderly, the handicapped, and other social groups with special needs are all benefiting from today's approach to housing.

Materials of Construction

From earliest times humans have built their houses from materials close at hand, using the tools and technology available to them. Because the vast forests of Canada are this country's prime natural resource, it is natural that a tradition of wood construction has evolved. As a construction material, wood has many unique qualities – it is light yet strong, easily worked, and highly durable; it has high impact strength; and it is readily available.

The tree in its natural form first dictated the type of shelter that could be built with wood. As Canada was settled by Europeans who brought their construction tools and skills with them, the log gave way to the timber plank and ultimately to the wood building materials in current use.

Figure 11.2 Modern townhouse development in Vancouver. Western red cedar shingles are used as exterior-wall and roof finish. (Photo: Council of Forest Industries of British Columbia)

Canada's forests support a large, nationwide industrial complex that produces a wide variety of building materials. These products – which include lumber, siding, panelling, wood-based panel products, shingles, and shakes – are all manufactured primarily for the residential construction market.

Lumber of various species groups is used for studs, joists, beams, rafters, trusses, plates, and other structural members. Lumber boards are occasionally used as sheathing, but because of their strength, rigidity, and large panel size, plywood, waferboards, and particleboards are the preferred materials for sheathing floors, walls, and roofs. A wide variety of wood siding is used on residential construction as exterior finish, often in combination with other materials. New types of prefinished plywood are also being used for exterior-siding applications, often combining wall sheathing and siding in a single application. Western red cedar shingles and shakes are a traditional roofing material still popular today.

Wood is also used for doors, window frames, moldings, kitchen cupboards, and built-in closets, and many houses now being built feature wood panelling on walls and ceilings as a decorative design element.

Plywood panels are also used to construct the formwork for concrete foundations. A new form of foundation, called the preserved wood foundation system, uses preservative-treated plywood and lumber for the actual foundation walls.

The use of wood in residential construction extends beyond the structural frame and interior and exterior panelling, siding, and trim. It is used extensively in landscaping design to construct fences and screens, garden furniture, decks, and patios. Most forms of housing in Canada, whether low-, medium-, or high-density developments, place strong emphasis on the provision of private outdoor living areas.

Canadian builders have developed a sophisticated building technology based on wood construction, which makes them among the most efficient in the world. Not only are Canadian wood products exported to many countries, but our methods of construction have been exported along with them.

House Framing Systems

The most widely used method of residential construction, from single-family detached houses to row housing and three-story apartments, is the wood-frame system. As the name implies, in this system, the structural frame is constructed of dimension lumber, with members usually spaced 300–600 mm (12–24 in.) apart. The three main types of house framing techniques are platform (or western), balloon, and post and beam.

In the platform frame system, the floor joists of each story rest on the top plates of the story below (or on the foundation sill for the first story), and the bearing walls and partitions rest on the subfloor of each story (Figure 11.4). The chief advantage of this method is that the floor system, which is assembled independently from the walls, provides a working platform for the assembly and erection of the walls and partitions. This is by far the most widely used method of frame construction because of its speed and simplicity. It also lends itself to the use of prefabricated building techniques.

The balloon frame system differs from the platform frame system in that the studs used for the exterior walls are continuous, extending in one piece from the foundation wall to the top plate supporting the roof. This system is rarely used today because the long lengths of lumber that are required are not readily available across Canada; furthermore, the system is not compatible with prefabrication.

In post and beam framing (Figure 11.5) the posts and beams support the loads. Partition walls in this system are not load-bearing structures. The frame is composed of lumber decking, beams, and posts supported on a foundation. The floor and roof decks transfer loads to the beams, which in turn carry them to the posts and on down to the foundation. This method of construction, still popular to some extent today, dates back to some of the earliest buildings of Greece. Traditional Japanese homes, half-timbered Tudor houses, and early American and English colonial houses are examples of post and beam construction. By the middle of the nineteenth century, platform frame construction, a faster and more versatile system, began to replace this method.

Figure 11.3 Roof trusses prefabricated from dimension lumber.

The small builder, constructing only a few houses each year, relies to a great extent on the use of precut lumber and prefabricated components, which are purchased from specialized suppliers who deliver them to the job site ready to be installed. Among these components are roof trusses, windows that are ready glazed with hardware installed, prefinished kitchen cabinets, and prehung doors. The use of these factory-fabricated elements, which are readily available in most areas of Canada, is a great advantage to the builder because it helps lower on-site labor costs and increases productivity.

The majority of houses in Canada are constructed on site using the platform or western frame system, and incorporate many or all of the prefabricated components that are available. Some builders also prefabricate components on site using their own framing crews.

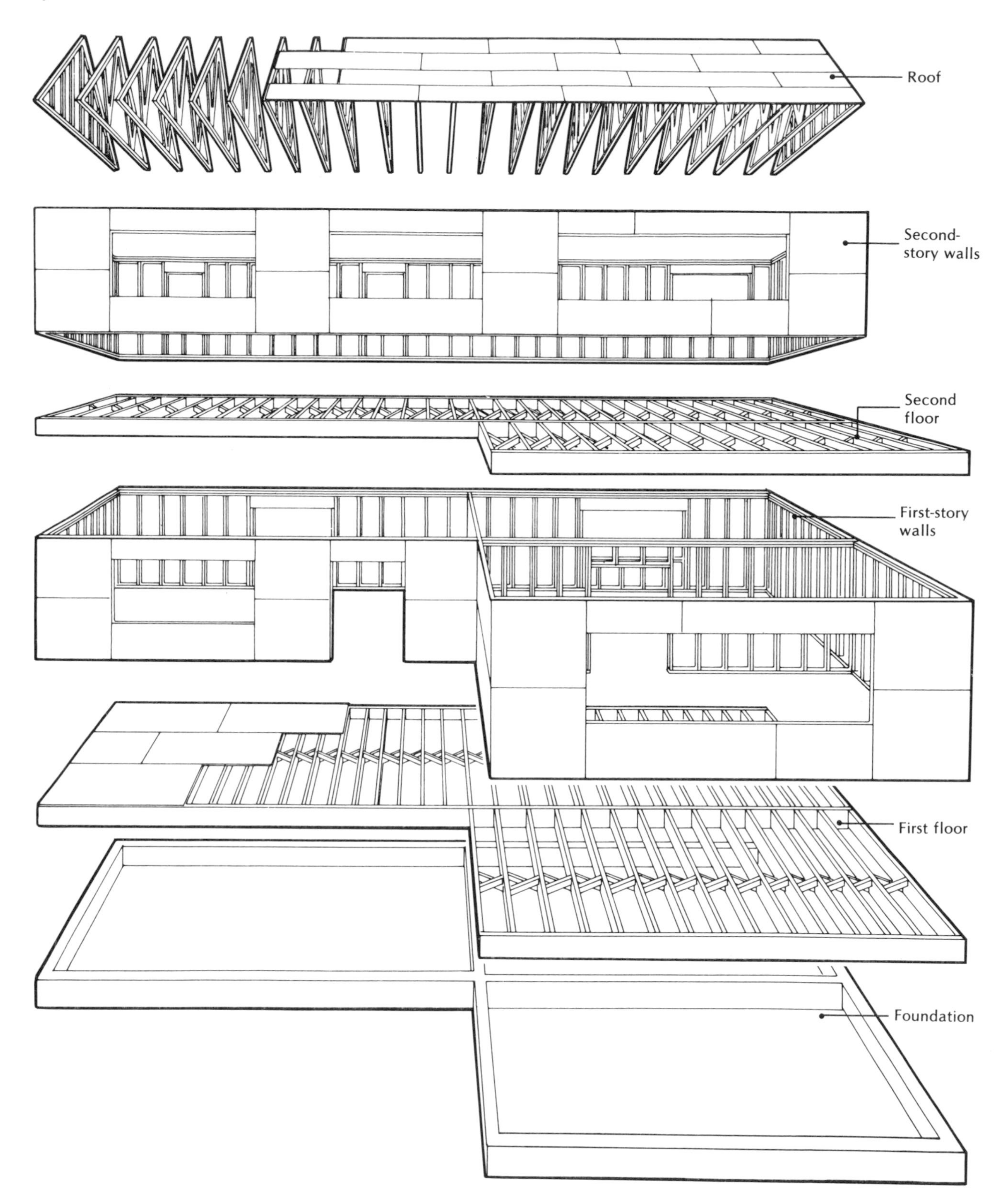

Figure 11.4 Elements of a platform-frame house.

Figure 11.5 Rectilinear post and beam house design. (Photo: Council of Forest Industries of British Columbia)

They often preassemble wall sections of lumber and either plywood or other panel products on the site, using the completed subfloor as a working platform. In some cases, when prefinished plywood sidings are used, the wall section can go into place with its exterior finish completely installed.

Figure 11.6 Factory assembly of windows for on-site house construction. (NFB Photothèque. Photo by Mike West)

Factory Fabrication

The prefabrication industry in Canada is becoming more and more significant, and a growing number of builders rely primarily on in-plant production facilities. These builders may construct a house in component parts, in sections, or in completely finished form, and then transport it to a prepared foundation for final assembly. Factory fabrication offers the advantages of controlled working conditions, a high degree of mechanization, and a lower labor-to-material ratio than on-site construction. However, it entails a large capital expenditure for plant and equipment facilities, which must be kept in continuous use if

Figure 11.7 Framing and sheathing large components of factory-fabricated houses (Photo: Council of Forest Industries of British Columbia)

the operation is to be competitive. The cost of transporting the house to its site must also be taken into account.

Mobile homes are also fabricated in the factory, where units are completely finished, often with furniture and appliances (Figure 11.7). Although these houses are designed to be transportable, they are used most frequently as permanent residences, sited in mobile-home parks or trailer camps.

Builders using prefabricated components are still using the wood-frame construction method, but they rely on plywood and wood-based panel products to a greater extent than the site builder does. The reason is that the strength and rigidity of these products allow them to withstand the racking loads in transport.

Roof Styles

The wood-frame construction system can be adapted to a wide variety of housing designs, from the flat roof of a post and beam house to the contemporary versions of the shed roof that are fashionable today. The roof line of a house, whether flat or sloped, on a single-family house or large row-house development, is an important consideration from both a design and a structural point of view. The roof line contributes most to the character and identity of a house.

The roof slope of a wood-frame house can take many forms (Figure 11.8), including the shed roof, gable roof, hip roof, gambrel (or mansard) roof, and the intersecting roof. All are used in residential construction in Canada, and all may be constructed using the wood-frame system. The roof system may incorporate trusses, rafters, or beams.

Rural Buildings

Some of the most interesting examples of wood construction in Canada are the old log cabins of the pioneers and that vanishing landmark, the two-story barn. The log cabin is a natural form of construction for a forested country such as Canada, and, indeed, log cabins are still being constructed.

The traditional two-story barn, built of wood and neatly painted, is a beautiful example of our rural past, but industrialization has changed the face of farm buildings just as it has changed the face of farming. As farms became larger, more specialized, and more mechanized, the need arose for larger and more efficient farm buildings. The most popular type of farm structure in use today is one with a clear-span interior that can be easily adapted to the particular requirements of the farmer, his equipment, crops, and livestock. In these buildings, which are often built with spans over 21 m (70 ft), the clear-span trussed roof relies on bearing supports only at the extremity of its span.

Many designs of agricultural buildings incorporate a high degree of prefabrication. A popular type of farm structure, which can be constructed by the farmer himself, is the rigid-frame building. The rigid frame is essentially an arch constructed from four straight pieces of lumber joined at the crown and haunches by plywood gussets. When sheathed with plywood, these structures develop great rigidity and have the advantage of being easily fabricated and erected on site. They offer a clear-span interior, and their use has spread beyond agricultural applications to churches, community halls, and recreation centers. The advantages of rigid-frame construction are its relative economy, simplicity of construction, ease of erection, and flexible interior area.

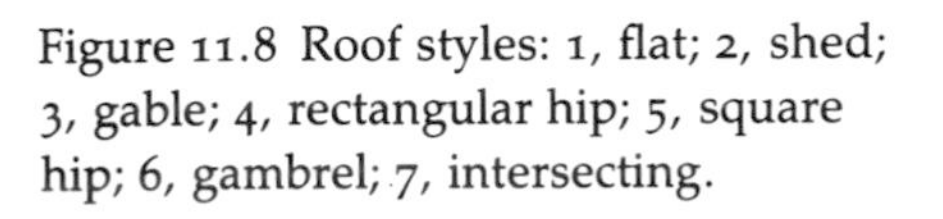

Figure 11.8 Roof styles: 1, flat; 2, shed; 3, gable; 4, rectangular hip; 5, square hip; 6, gambrel; 7, intersecting.

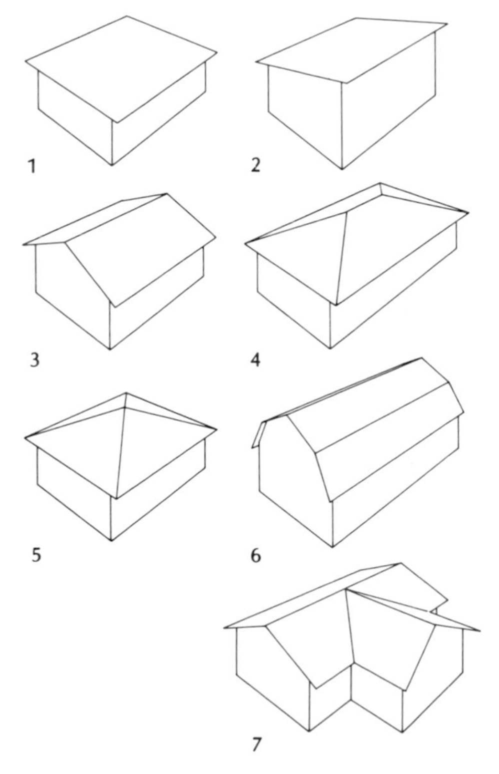

RECENT DEVELOPMENTS IN HOUSING CONSTRUCTION

Prefabrication and Industrialization

The industrialization of housing production has been evolving for centuries. More and more cutting, assembly, and finishing operations are being performed off site. Subassemblies such as roof trusses, prehung doors, kitchen cabinets, and even complete bathrooms are now being mass-produced as production techniques become more sophisticated.

The reasons for the movement from sites to factories are clear; mass production permits much more efficient use of labor and greater reliance on highly mechanized and specialized equipment in factories, labor costs are lower because in most cases employees working on assembly lines need not be as skilled as on-site workers, and management can schedule production operations with precision because production is independent of weather conditions. Better supervision of production operations and closer quality control are also possible, and significant savings in cost may be realized through the bulk purchase of many materials and components.

Modular Construction

A technique that is rapidly gaining support among house builders is the use of modules, which are complete or semicomplete sections of houses or apartments that are preassembled in a factory on production lines. In the final stages of this production, cabinets, appliances, and mechanical hookups are added as required. Interior walls are often

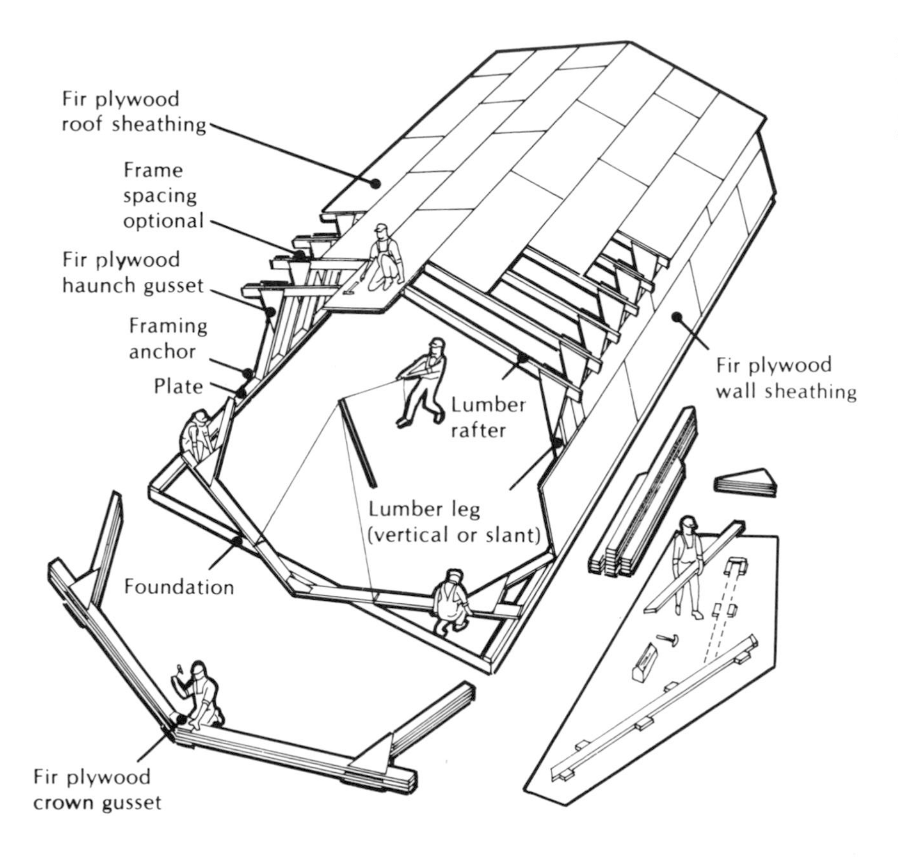

Figure 11.9 Rigid-frame construction. (Adapted from *Plywood in Farm Structures*, Council of Forest Industries of British Columbia, 1968)

Figure 11.10 Assembling prefabricated building modules. (Photo: Council of Forest Industries of British Columbia)

prefinished. The modules are wrapped and transported to the site, where they are quickly assembled.

Mobile Homes

Mobile homes, completely assembled and finished in factories, are the best example of industrialization in housing. In Canada mobile homes account for approximately 10% of all new housing sales – and sales are growing at the rate of 20% per year. The mobile-home industry uses nearly 1.5 million m^3 (825 million fbm) of lumber annually in North America, of which 50% is estimated to be Canadian. When travel trailers and motorized homes are added to this market, the total approaches 2 million m^3 (1 billion fbm) annually. Western and eastern white spruce are the preferred species, with the major supply coming from the northern interior of British Columbia.

Canadian mobile homes use an average of 3.25 m^3 (2000 fbm) of lumber and 743 m^2 (8000 sq ft) of wood-based panel products such as plywood, particleboard, and hardboard per unit. Because of its high strength-to-weight ratio, wood is well suited for the structural framework above the steel chassis. It is used for floor joists, wall studs, stringers, roof rafters, trusses, and cabinet framing. Because nearly all components for mobile homes are assembled in jigs, the physical size, quality, and condition of the wood must be uniform, providing dimensional stability and ease of assembly as well as reliable nailing and gluing properties.

A wide variety of plywood and wood-based panel products are used for structural floors, exterior-wall and roof sheathing, and interior-wall and cabinet finish. These products are used because they are light in weight and capable of being transported without damage.

Softwood plywoods predominate for structural skins, but the use of phenolic-bonded waferboard is growing. These products are usually fastened to lumber framing with glues and nails or staples.

Interior finish varies with the decor and style. The interior panelling of high- or custom-line models is often natural 6 mm ($\frac{1}{4}$ in.) Canadian hardwood plywood. The objective is to achieve luxury of appearance combined with minimum maintenance and low cost.

Figure 11.11 Preserved wood foundation for condominium housing project. (Photo: Council of Forest Industries of British Columbia)

Preserved Wood Foundations

The increasing cost of materials and labor for the construction of concrete or masonry foundations has given impetus to the development of a new construction technique – preserved wood foundations (Figure 11.11). The system is basically conventional frame construction using treated lumber and plywood, and the foundations can be constructed without the need for special subcontractors.

Preserved wood foundations used today are constructed in the same way as conventional wood-frame walls built above grade. They are stud walls with plywood sheathing, the most common stud size being 38 × 140 mm (2 × 6 in. nominal) with either 12.5 mm ($\frac{1}{2}$ in.) or 15.5 mm ($\frac{5}{8}$ in.) thick plywood sheathing, depending on stud spacing. All wood material in walls contacting soil and the 38 mm (2 in. nominal) dimension lumber footing plates are pressure-treated with wood preservatives. Copper chrome arsenate (CCA) or ammoniacal-copper-arsenate (ACA) water-borne salt treatments are preferred because of the clean, odor-free condition of the wood after treatment.

Preserved wood foundations offer many advantages and benefits. In addition to long life, reduced construction time, and year-round building capability, they provide larger below-grade living areas that are more convenient and consequently more economical to finish than conventional basements. Insulation of lower living areas is improved because the low thermal conductivity of wood keeps the temperature more stable, regardless of the season. As a result, less fuel is required to maintain a comfortable environment in the lower level. This factor is important because this area can then be used as additional living space, equivalent in utility and comfort to the remainder of the home.

HEAVY TIMBER CONSTRUCTION

Heavy timber construction is defined in the National Building Code of Canada as that type of combustible construction in which a degree of fire safety is attained by placing limitations on the minimum sizes of wood structural members that can be used and on the thickness and composition of wood floors and roofs. Good structural practice, such as avoidance of concealed spaces under floors and roofs, and use of approved fastenings, construction details, and adhesives for structural members are also required.

Heavy timber construction is one of the oldest methods of building in Canada. During the past 150 years, its superior performance has been demonstrated in thousands of buildings, a large number of which are still in use.

As originally conceived, the heavy timber building was a multistory structure designed and used primarily for industrial and storage purposes. Today its use has been expanded to include schools, churches, auditoriums, gymnasiums, supermarkets, and various other structures such as bridges and factory, mine, and marine installations.

In keeping with the trend to reduce manufacturing costs, since a single-story building is cheaper to build than a multi-story one, industrial buildings have spread outward rather than upward. The result is that many modern heavy timber buildings have large areas but are only one story high. However, where land is limited, the modern multistory heavy timber building has proved its worth.

If either solid-sawn or glued-laminated stress-graded lumber and timbers are used, precise structural design procedures can be applied to heavy timber construction, resulting in a completely engineered structure. With properly designed fastenings, such structures can be relied on to carry all anticipated loads safely.

Solid-Sawn Lumber

The field of heavy timber construction using solid-sawn lumber is greatly diversified. It covers many types of buildings as well as a number of engineered structures such as bridges, piers, wharves, towers, and trusses where members 140 × 140 mm (6 × 6 in. nominal) and larger are used.

Practical considerations such as the unavailability of large, long timbers and their unsuitability for curved members invariably limit the use of solid-sawn lumber for heavy timber construction. This limitation has led to the development of glued-laminated lumber.

Figure 11.12 The Nass River bridge in northern British Columbia. Solid-sawn and glued-laminated timber structural members were used in its construction. (Photo: British Columbia Forest Service)

Glued-Laminated Lumber

Structural members of glued-laminated lumber are manufactured from comparatively short lengths of dimension lumber up to 38 mm (2 in. nominal) in thickness, glued together with the grain of the laminations essentially parallel to the length of the members. The members may be straight with horizontal or vertical laminations. Curved members are horizontally laminated to permit bending of the laminations during gluing. Horizontally laminated members are the most widely used because they are adaptable to a wide range of structural applications.

The advantages of glued-laminated members are that their size and length are not limited by the sizes of available sawlogs, the influence of strength-reducing natural growth characteristics of wood can be controlled through selection, and architecturally pleasing designs using curved members can readily be obtained. However, glued-laminated members may be more expensive than solid-sawn members of equivalent capacity.

Figure 11.13 Curved glued-laminated arches provide a clear span for a school gymnasium. (Photo: Council of Forest Industries of British Columbia)

Fire Resistance

Heavy timber framing has excellent fire resistance, which has been demonstrated in many buildings over the years, and often provides much better protection than noncombustible materials (see 'How Wood Burns' in Chapter 8). Building codes recognize the superior performance of heavy timber framing by allowing larger floor areas for structures of this type. Fire-insurance companies give lower rates to buildings of heavy timber construction.

In those structures where aesthetics are an important consideration, the beauty of the exposed wood members combined with the excellent fire resistance of the heavy timber framing produces highly desirable results.

DESIGN OF TIMBER STRUCTURES

Because wood is a natural material, its physical and mechanical properties vary from piece to piece. This variation is well understood and affects both working stresses and design. For example, the proper design of a wood structure places high-grade material in areas of higher stress and the more inexpensive grades in areas of lower stress.

Allowable Unit Stress

Allowable unit stresses of wood vary both between species and within species. Certain species are naturally stronger than others. Building codes recognize the superior strength of Douglas-fir and other high-density woods, and accordingly these species are assigned higher allowable unit stresses.

The allowable unit stresses within a species are dependent on the grade of lumber sawn or cut from that species. The better grades are naturally allotted higher stresses. Lumber grades can be determined either mechanically (giving machine-stress-rated lumber) or visually. Because a relationship exists between strength and stiffness of wood, the strength of machine-graded lumber can be estimated by measuring its stiffness. The strength and stiffness of visually graded lumber is estimated by observing the frequency of natural variables such as knots. Establishing the grade is a part of the manufacturing process, and therefore the practical design of wooden elements is similar to design with other structural materials.

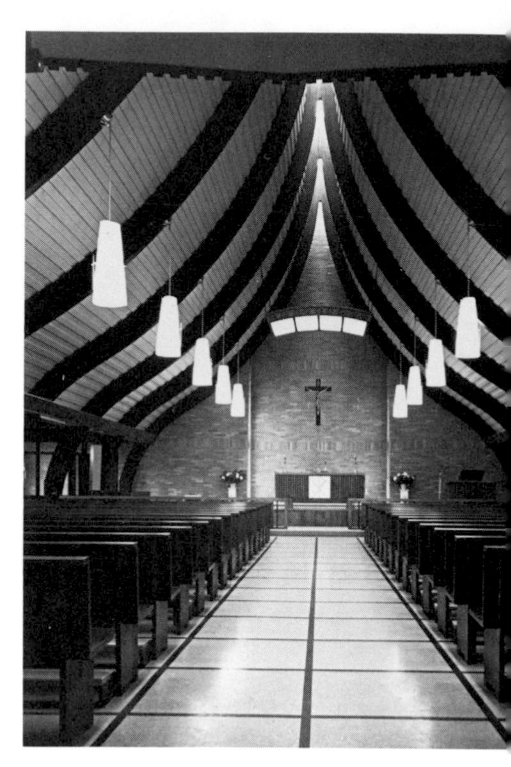

Figure 11.14 The beauty of exposed heavy-timber framing is used effectively in the interior design of this church. (Photo: Canadian Wood Council)

Wood is strongest in compression and tension parallel to the grain. Allowable unit stresses in compression parallel to the grain range from 1.5 to 5 times the allowable stresses perpendicular to the grain. Tension perpendicular to the grain in glued-laminated members is assigned an allowable unit stress amounting to about $\frac{1}{30}$ the same stress parallel to the grain.

The shear strength of wood is greatest perpendicular to the grain. In fact, when members are tested to destruction in the laboratory to determine their ultimate bending strength, shear failures are extremely rare. Permissible stresses for longitudinal shear are generally low: below 690 kPa (100 psi) for sawn timber and about 1030 kPa (150 psi) for glued-laminated members.

The compressive and tensile properties of wood parallel to the grain exhibit an unmatched efficiency in strength-to-weight ratio. Thus, wood is most efficient when used for members in which the main stresses are axial tension or compression. The most obvious examples are posts, piles, studs, and trusses. Design loads of 300–500 kN (34–55 tons) per pile are not uncommon, and bow-string wood trusses have been found to be one of the most economical ways of attaining roof spans greater than 30 m (100 ft).

Load Sharing

The strength and stiffness of wood can vary between two or more pieces that are visually similar. To guard against the potential hazards of such variations in the properties of wood, load sharing, a concept unique to wooden structures, is employed. Load sharing is based on the assumption that a series of wood members will support a given load

Figure 11.15 Heavy-timber construction in a large warehouse. (Photo: Koppers International Canada Ltd.)

if provision is made to compensate for their variation in strength. Thus, a load-sharing system consists of three or more essentially parallel members spaced at a maximum of 600 mm (24 in.) on center, arranged or connected so that they support the load together. Whenever this criterion for load sharing is satisfied, the allowable unit stresses of the system can be increased by 10%. Many of today's wood structural systems employ load sharing in one way or another. Trusses and plywood structural components, such as stressed skin panels, are typical examples.

Connections

Two main types of connections are used in wood assemblies: glued joints and mechanical fasteners. Each requires a different engineering approach, and each has certain advantages over the other. In general, adhesives are used to manufacture structural members, and mechanical fasteners are used to assemble structural members into a system. The major exception to this rule is nail gluing, in which both adhesives and mechanical fasteners are used to assemble systems such as residential floors.

Glued joints require careful control of moisture content, temperature, glue chemistry, and surface smoothness of the wood. These requirements dictate that the work be done in a shop or plant where skilled personnel are available and ambient conditions can be controlled. Two important examples of the use of glued joints are in the construction of glued-laminated beams and plywood panels.

In glued-laminated beams, end joints in the various laminations can be staggered along the length of the member, and lower grades of laminating stock can be used near the central plane of the member where the bending stresses will be lowest. The most noteworthy advantage of glued-laminated members is that they can be manufactured in larger sizes and longer lengths than sawn lumber members.

Plywood panels compensate for the possible inclusion during manufacture of weaker species by functioning as load-sharing systems. The major advantage of plywood is that the cross-lamination of the plies gives the panel greater shear capacity, greater width, and greater dimensional stability than sawn lumber.

Highly efficient glued systems can be constructed using the greater length of glued-laminated members in conjunction with the shear properties of plywood. Examples of these systems are plywood web beams, stressed skin panels, and folded plates. Production volumes of these plywood structural components are small, however, in comparison with the volumes of glued-laminated members and plywood panels.

Mechanical connections include common nails, bolted joints incorporating split-ring or shear-plate connectors, steel plates, metal truss plates, steel dowels, and glulam rivets. Mechanical fastenings are often more efficient when used in double shear. The use of multiple wood members also contributes to the efficiency of the joint.

Apart from the standard applications in housing and other light-frame construction, nails can also be used effectively in a variety of engineered structures. Nails are frequently used, for example, in combination with plywood gusset plates in the assembly of trusses. The advent of the power-nailing gun, hardened steel nails, and nail shanks coated with thermosetting glues has increased the efficiency with which trusses can be manufactured. Truss plates, which combine the function of gusset and nail in a steel plate from which nail-like teeth have been punched, have supplanted nailed plywood gusset plates in many instances. These plates are applied by a press, making high production rates possible. The cost of trusses using truss plates is relatively low, and they are in widespread use in residential construction.

Nails are also used in conjunction with sheet metal. Tests show that very high loads per nail can be achieved by driving nails through an assembly of lumber, plywood, and multiple-sheet metal plates. Applications of this system are to be found in nailed I-beams with lumber flanges and plywood webs.

An example of the use of specialized nails and steel plate is to be found in glulam rivets as specified in Canadian Standards Association Standard CSA-086, Code for the Engineering Design of Wood. In this system, steel side plates are prebored and attached to the members with hardened steel nails, or rivets, of elongated cross-section. The wedge-shaped rivet heads are hard driven into the plate to create a high degree of fixity between nailhead and plate. The broad face of the rivet should always be oriented parallel to the grain direction of the member. The result is very high strength and stiffness, enabling compact connections to be made between large glued-laminated members carrying high loads.

Bolted joints must be designed with consideration given to the angle of the load in relation to the direction of the wood grain. Hankinson's formula gives the relationship between permissible loads parallel, at an angle, and perpendicular to the grain:

$$N = PQ/(P \sin^2 \theta + Q \cos^2 \theta) \quad (1)$$

where N is the allowable load at angle θ to the grain direction, P is the allowable load parallel to the grain, Q is the allowable load perpendicular to the grain, and θ is the angle between the direction of the grain and the direction of the load.

Bolted joints can be extremely strong, especially when the bolts function in multiple shear, but such joints should be inspected periodically for correct tightness. Variations of the bolted joint are found in lag screws that thread into prebored holes.

Several proprietary wood connectors are in common use. Spike grids and toothed plate connectors, with teeth projecting from both surfaces, press into the wood members as they are drawn together. Shear plates, which fit into a circular groove precut into the wood, transfer load from the wood through the shear plate into the bolt. Split rings fit into grooves precut into two adjacent wood surfaces that are held together by a bolt. In the last case, the load is transferred directly from one wood member to another.

Split rings and shear plates both require relatively large amounts of wood to be removed during precutting; therefore, the load-carrying capacity of the remaining section must be considered. Generally, the design of mechanically fastened joints must take into consideration a variety of factors, such as end- and edge-spacing distances, moisture content, service conditions, and the effect of the number of connectors used.

Because the cost of fabricating and installing connections may amount to a large percentage of the total cost of a timber structure, it is important to engineer the details of a structure before designing the members.

Loads

Timber structures are economical because of their high strength-to-weight ratio and their capacity to handle increased loads of short duration, such as snow, wind, and vehicular bridge loads. Working stresses for short-term loads such as wind and earthquake are therefore increased by 33%. Furthermore, because of the low weight of timber structures, their design is not usually critical under earthquake loads. In fact, timber is particularly efficient in resisting shock or impact loads. Plate elements, such as timber decking or plywood wall and roof sheathing, functioning as diaphragms, are highly efficient in resisting wind and other lateral loads because the connections to the supporting framework are usually continuous along the length and width of the structure. Thus, although a single nail may have a low lateral load-bearing capacity, the cumulative strength of a series of nails is high when they are spaced 100–150 mm (4–6 in.) apart. It is not unusual to encounter large structures that rely on roof and wall diaphragms to provide wind bracing rather than on diagonal bracing within the framework.

Design Methods

As shown in the building codes, timber is generally assigned working stresses and allowable deflection, and, in practice, standard engineering design methods can be applied to timber structures.

Beams can be simple or continuous, and the usual bending and shear stresses are taken into account in addition to deflection. Recommendations for lateral bracing and support are given in the design codes. If beams are constructed with plywood webs and lumber flanges, the horizontal shear stress is used to determine the required width of the glue lines or the number of nails. When lumber is used on the flat, as in decking, bending stresses and deflections must be modified, depending on the selected spacing of joints.

Columns are designed according to slenderness and effective height in addition to cross-sectional area and permissible axial stress. In the usual situation where a beam must resist axial load in addition to bending load, or a column must withstand lateral load, the combined strength is checked by the standard interaction formula found in building codes.

Truss designs should allow for some rotational rigidity at the connections. Because it is difficult to identify the exact degree of rigidity, however, it is common practice to consider a truss to be pin-jointed and to make subsequent approximate allowances for connection rigidity. Residential roof trusses are generally designed on the basis of load tests conducted on complete truss assemblies. Because top chords of trusses are stressed in compression, adequate lateral support against buckling must be provided by the roof sheathing system. Web members subjected to compression may require lateral bracing. Timber roofs can be light in weight; thus, it is important that roof members be securely attached to the walls or other supports and that stress reversal be investigated in case the roof should be subject to uplift.

Plywood is unique because the grains of adjacent plies are perpendicular to each other. When plywood is bent in one direction, those plies parallel to the principal stresses contribute much more strength than the perpendicular plies do. Consequently the section properties of plywood, which relate to its stiffness and strength, are a combination of those properties both parallel and perpendicular to the grain. The section properties that should be used when designing with Douglas-fir plywood are given in Canadian Standards Association Standard CSA-O86.

From an engineering point of view, plywood is most usefully employed in resisting shear stresses in structures such as plywood web beams or diaphragm roof assemblies. In these cases shear is based on the full cross-section of the plywood.

CSA standard O86, *Code for the Engineering Design of Wood*, is the major Canadian standard for the design of timber buildings and other structures. It deals not only with plywood but also with lumber, glued-laminated timber, and connections in considerable detail, and it should be consulted for design of any wooden structures.

Limit States Design

Limit states design has recently been introduced into Canadian structural engineering as an alternative design procedure. Eventually

Figure 11.16 Large glued-laminated beams provide an unobstructed interior in a pulp and paper mill in Quebec. (Photo: Koppers International Canada Ltd.)

the limit states design will replace all other design procedures now in use.

In the past, in developing structural design standards, engineers have considered independently such structural materials as concrete, masonry, steel, aluminum, and wood. The general approach was to use common loading requirements but to apply a single overall factor of safety to the material strengths. Because of the diversity of materials, this approach resulted in a proliferation of complex structural standards using different factors of safety, but all having the same objective of providing a safe, functional, and economical structure.

The limit states design includes the basic structural requirements for the prevention of collapse, the prevention of excessive deflection or vibration, or other such serviceability requirements. But in this procedure, safety criteria that are more consistent with risk and consequences of failure are provided by replacing the existing, single overall factor of safety with partial factors of safety. These partial factors of safety are independent of the type of construction material and hence result in similar overall factors of safety for structures, regardless of the materials of construction.

A common limit states design format, independent of the structural material, is now being incorporated into the National Building Code of Canada.

Finite Element Methods

A recent development in engineering analysis is the finite element technique of analysis of stresses and deformation in structural elements. The technique has been applied to some extremely difficult problems, and useful results have been obtained. Applications to shear in plywood, glulam rivet connections, tension perpendicular to grain in pitched-tapered beams, and shear strength of glued-laminated beams have improved engineers' predictions about the behavior of these wood structures under load.

Many references in the literature refer to basic work by Zienkiewicz (1971), which is much beyond the scope of this chapter. Undoubtedly the finite element method will be used more extensively in the future as timber engineers strive to develop more efficient and economical designs based on detailed stress analysis.

BIBLIOGRAPHY

American Institute of Timber Construction. 1974. *Timber construction manual*. 2nd ed. New York: Wiley

Canadian Institute of Timber Construction. 1970. *Modern timber bridges*. Ottawa

Canadian Wood Council. 1975. *Construction guide for preserved wood foundations*. Ottawa

Central Mortgage and Housing Corporation. 1974a. *Canadian wood-frame house construction*. Ottawa

– 1974b. *Canadian housing statistics*. Econ. Stat. Div., Ottawa

Council of Forest Industries of British Columbia. 1968. *Plywood in farm structures*. Vancouver

Department of Industry, Trade, and Commerce. 1970. *The mobile home in Canada*. BEAM Program, Mater. Branch, Ottawa

Dickens, B.H. 1973. *Problems associated with the development of use of wood in construction and possible solutions – Legislative and code aspects*. Tech. Pap. No. 386, Div. Bldg. Res., Nat. Res. Counc., Ottawa

Foschi, R.O. 1969. Deflection of multilayer-sandwich beams with application to plywood panels. *Wood and Fiber* 5 (3): 182–91

Gowans, A. 1966. *Building Canada – An architectural history of Canadian life*. Don Mills, Ont.: Oxford Univ. Press

Gurfinkel, G. 1973. *Wood engineering*. New Orleans: South. For. Prod. Assoc.

Hutcheon, N.B. 1971. *Codes, standards and building research*. Tech. Pap. No. 357, Div. Bldg. Res., Nat. Res. Counc., Ottawa

Laminated Timber Institute of Canada. 1972. *Timber design manual*. Ottawa

National Association of Home Builders Research Foundation. 1969. *An industrial engineering study comparing direct costs of concrete block and treated wood foundations*. Washington, DC

National Forest Products Association. 1970. *Wood structural design data*. Washington, DC

National Research Council. 1975. *National building code of Canada*. Assoc. Comm. Nat. Bldg. Code, Ottawa

Payne, R.J. 1971. *Plywood construction manual*. Vancouver: Council For. Ind. BC

Smith, R.C. 1970. *Principles and practices of light construction*. Englewood Cliffs, NJ: Prentice-Hall

Timber Engineering Company. 1956. *Timber design and construction handbook*. Toronto: McGraw-Hill

Wass, A. 1973. *Methods and materials of residential construction*. Reston, Va.: Reston

Western Wood Products Association. 1973. *Western woods use book*. Portland, Oreg.

Zienkiewicz, O.C. 1971 *The finite element method in engineering science*. London: McGraw-Hill

12

J.E. KORHONEN
Ontario Forest Research Centre,
Maple

Other Uses and Processes

For practical and aesthetic reasons, wood is widely used throughout the world for a great variety of products. Its strength, durability, lightness, workability with hand tools, range of color and grain pattern, moderate cost, and other attributes make it the most important raw material on earth. Fortunately it is also a renewable resource, and trees make our environment beautiful as they grow into merchantable timber.

In this chapter both the processes and the products of some of the most important secondary wood-using industries are discussed. A comparison of the value added by the primary forest-based industries and by the secondary wood-using industries demonstrates their relative importance (Figure 12.1).

FURNITURE

Furniture is by far the most important single product of the secondary wood-using industries. Good furniture must satisfy many conflicting demands. It must be functional yet decorative; comfortable and aesthetically pleasing, both to the eye and to touch; and relatively light in weight yet strong and durable. Wood continues to be the material best suited to meet these requirements.

The manufacture of furniture brings together a number of skills and production techniques. Although the number and sequence of the various production functions may vary from plant to plant, the basic processes, illustrated in Figure 12.3, are the same.

There is a trend toward vertical integration within the furniture industry. In this arrangement a single firm operates its own sawmill, dry kiln, rough mill (cut-up plant), and final manufacturing plant or plants. Several firms have established sawmills, dry kilns, and rough mills close to the log source. Complexes of this type ship rough components ready for machining, assembly, and finishing to factories located close to large urban centers. Sometimes some of the high-grade lumber produced is sold to other furniture companies, and the lower grades of lumber are usually marketed for pallet boards and stringers. Furniture components are also known as hardwood dimension stock, or simply, dimension stock.

The Canadian furniture industry is centered in the eastern townships of Quebec and in southwestern Ontario, where the major concentrations of hardwoods were plentiful before the forests were harvested. Yellow birch, hard maple, and elm are the most important species, but beech, red oak, and many other species are also used. White pine, a coniferous species, was a favorite furniture wood of the early settlers because it is easy to work with hand tools. Its use declined with the establishment of well-equipped factories that could process hardwoods without problems, but recently there has been a strong renewed demand for white pine furniture.

Particleboard is used in large quantities by the furniture industry, and it has largely replaced lumber for corestock. It is usually overlaid with plastic and resin-impregnated paper veneer for commercial and low-cost household furniture, or with thin, high-value wood veneer for high-quality items. Veneers have been used in furniture manufacturing since early times, permitting optimum utilization of expensive species such as walnut, cherry, true mahogany, rosewood, and teak.

Particleboard provides considerable savings in both weight and construction costs without any sacrifice in quality. Made from both softwood and hardwood residues, it reduces the drain on our very limited resource of dense hardwoods, such as yellow birch. For many applications it is superior to lumber corestock because good-quality particleboard, properly overlaid, is often more dimensionally stable than solid wood.

In Europe much low- and medium-cost furniture, and commercial furniture, are manufactured from particleboard with no overlay at all. The board usually contains three layers of particles: the two surface layers consist of small particles, whereas the center layer is relatively coarse. The board surfaces are pressed and sanded to produce a smooth, hard surface. These large particleboard sheets are cut into furniture panels and finished, using liquid finishes applied by curtain-coating equipment.

Molded plastic carvings are used extensively in modern furniture manufacture because of the very high costs of woodcarvings made by hand or machine, but their popularity has declined recently. The reasons for the decline are the rapid rise in the price of plastics and a renewed interest in wood for aesthetic reasons. Woodcarving is discussed in more detail later in this chapter.

One unusual feature of the furniture industry is that by tradition many retailers are unwilling to identify the manufacturer of their furniture. They may even remove the manufacturer's labels. To protect their identity, many manufacturers of medium- and high-quality furniture burn their brand name, or mark, into the wood in locations such as the inside of a drawer side or the underside of a table.

Although its products are generally associated with furniture, the kitchen-cabinet industry operates quite independently of the furniture

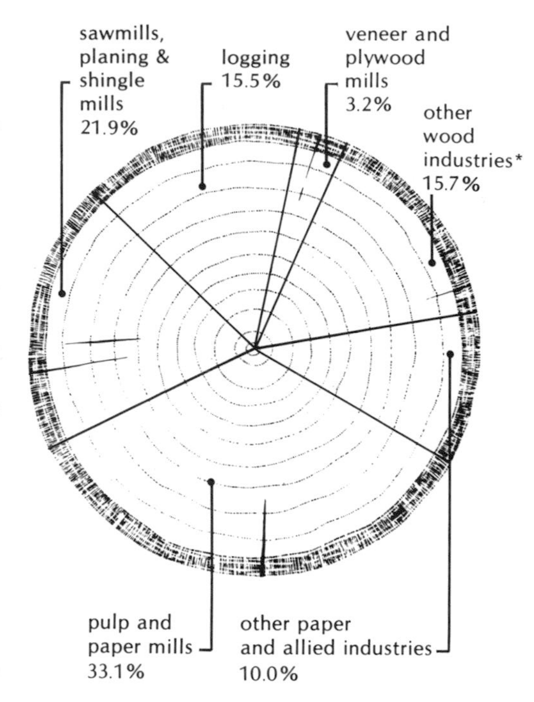

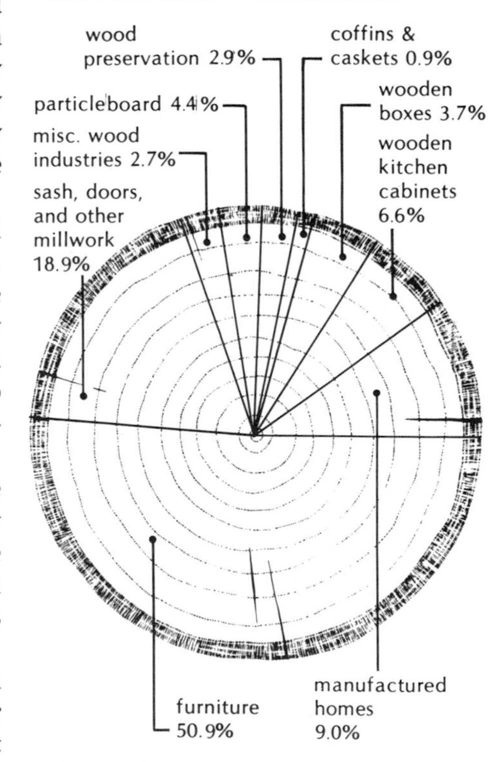

Figure 12.1 Value added (1978) in the forest-based and secondary wood-using industries.

Figure 12.2 Display of Upper Canada antique-style furniture constructed of pine with maple used for structural strength. (Photo: Simmons Limited)

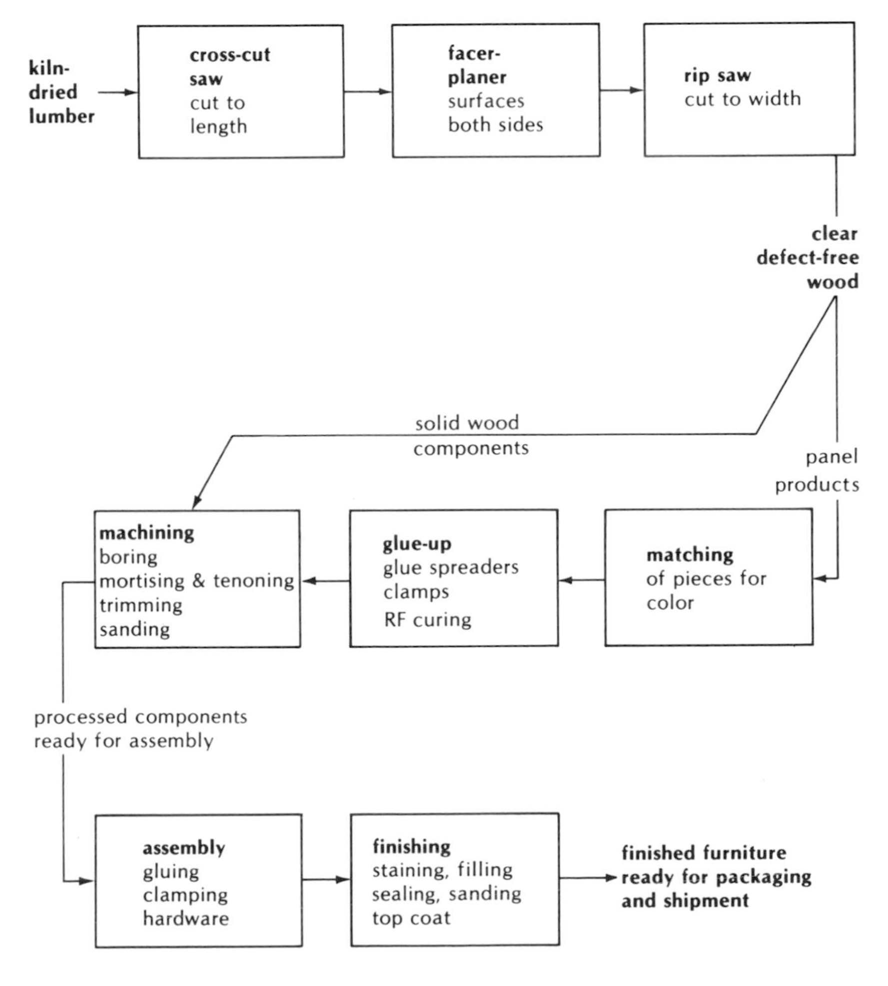

Figure 12.3 Flow diagram of basic production steps in manufacturing furniture solids.

industry. The manufacturing processes involved have many similarities, however.

COFFINS AND CASKETS

Wood is traditionally the preferred material for coffins and caskets. Manufacturing processes are similar for expensive coffins and furniture in that the wood is cut to size, assembled, upholstered, and finished. Because appearance is a prime consideration in both products, the same species are often used. In eastern Canada these are mainly red oak, white elm, hard maple, and yellow birch.

Less expensive coffins are made from wood species such as cedar, cottonwood, and white pine. The lumber is usually edge-glued, and then the coffin shell is covered with fabric.

Figure 12.5 Shaping furniture parts on a bandsaw. (NFB Photothèque. Photo by George Hunter)

WOODCARVING

For many years decorative woodcarvings used as embellishment on wood furniture have been produced by two methods: by hand carving with chisels and by multiple-spindle carving machines. Today, insufficient numbers of skilled craftsmen are available for hand carving, and the cost of this slow, careful work is prohibitive for all but the most expensive furniture.

Most decorative woodcarvings are produced on multiple-spindle carving machines controlled by a tracer that the operator moves over the surface of a prepared pattern. The tracer causes the carving spindles to move in unison over a number of wood panels and to duplicate exactly the contours of the pattern. Up to 16 wood panels can be carved simultaneously (Figure 12.6).

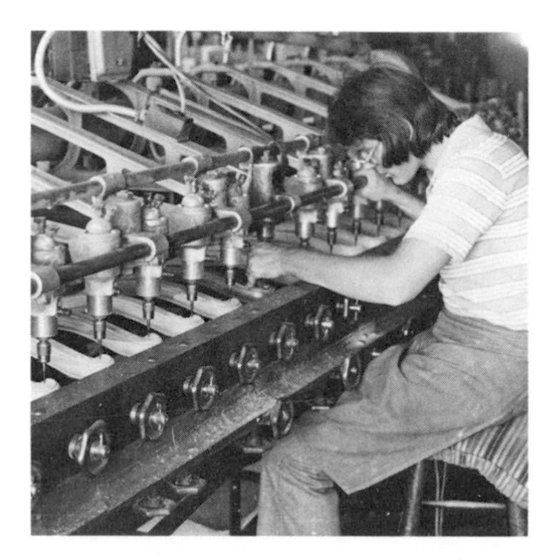

Figure 12.6 Multiple-spindle carving machine. (Photo: *Woodworking and Furniture Digest*)

The cost of producing carvings in this manner has probably been the highest of any machine operation in furniture manufacturing because of the high labor content per part, both in the set-up time required and in the operation of the machine. As a result, woodcarvings have largely been replaced by molded plastic carvings in the production of medium- and low-cost furniture. The rapidly rising cost of plastics, and an apparent aversion on the part of many consumers to plastic in a piece of furniture that is purchased as a symbol of high quality or affluence, are causing a reversal of this trend.

Woodcarving is now practiced almost exclusively by producers of high-quality household and church furniture and by hobbyists.

Figure 12.4 Ornate wood carving surrounding a window in the Senate Chamber, Ottawa. (NFB Photothèque. Photo by Chris Lund)

WOOD TURNINGS

Wood turnings include a great variety of products ranging from furniture parts and bowling pins to textile bobbins, baseball bats, ax handles, billiard cues, drawer knobs, and golf tees. They have only their method of manufacture in common – all are turned on a lathe, which rotates the wood against a cutting blade. Dowels, which are wooden pegs or rods with a circular cross-section, are commonly considered as turnings, even though they are produced by a dowel machine, in which a cutting blade rotates around the wood.

Aside from the strength and durability requirements of each

individual product, the foremost consideration in choosing a wood species for turning is the ease with which it can be machined to a smooth surface. Accordingly, the principal species used in wood turnings are white and yellow birch, hard maple, and beech. At one time large quantities of white pine were used for window-shade rollers, but this product has virtually disappeared from the market. More information on the machining properties of Canadian wood species is available in a publication by Cantin (1965). It should be noted, however, that Cantin used constant rotation-speed and knife-angle settings; some species that he listed as poor for turning can be turned satisfactorily with different settings.

Figure 12.7 Back-knife lathe.

Two basic lathe types are used for the manufacture of turnings. Relatively long, slender turnings are generally prepared on a long-knife type of lathe; the most frequently used is the back-knife lathe. This lathe has a milled-to-pattern knife that contacts the full length of the rapidly rotating work piece. It is restricted to turnings that are round in cross-section, including baseball bats, spools, bobbins, and billiard cues.

The manufacture of turnings that are not round in cross-section is accomplished with a lathe of the peripheral milling type, in which a rapidly rotating cutter head is brought to bear against the slowly turning work piece. A common example of this lathe is the automatic shaping lathe. Not only does the carriage motion of the automatic shaping lathe control the diameter of the completed work piece, but, through careful timing of its oscillation, it also can be used to produce turnings that are oval or polygonal in cross-section. Mediterranean-style table legs and pick handles are examples of turnings produced on the shaping lathe.

An additional degree of complexity is introduced when the required turning is asymmetrical both longitudinally and in cross-section. Ax handles, gunstocks, and shoe lasts are typical products of this type. They are machined on a copying lathe, which, in principle, is similar to the shaping lathe except that a follower, acting on a master pattern of the part to be turned, guides a narrow cutter head in both longitudinal and cross-feed directions. The copying lathe may be compared with the multiple-spindle carver discussed under 'Woodcarving.'

Figure 12.8 Worker placing ax heads on handles. (NFB Photothèque. Photo by Ted Grant)

Most turnings are sanded in an automatic machine that rotates the turnings slowly against ribbons of rapidly rotating sandpaper. Very small turnings, such as drawer knobs and golf tees, are tumble-sanded simply by placing large numbers of them in a rotating drum. Lacquer finish is also applied to golf tees, large wooden beads, and similar articles in tumble drums. Long cuttings that require a finish, such as furniture legs, are spray-coated with fillers, sealers, stains, and so on and finished either before or after assembly.

MILLWORK

Handrails, stair treads, window sashes and frames, interior and exterior doors, molding, and picture frames are some of the many products known as millwork. Cross-sections of typical examples of millwork that have been shaped on a molder are shown in Figure 12.9. Because of the high length-to-width ratio of the work pieces, in

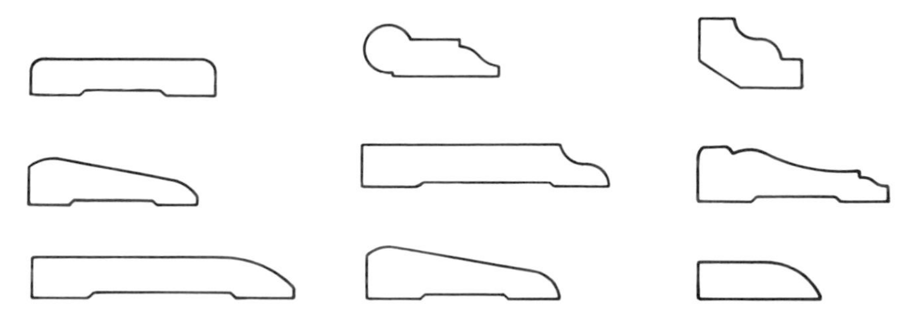

Figure 12.9 Typical cross-sections of millwork shaped on a molder.

millwork, the lumber is first ripped to width and then crosscut to length. This procedure is the reverse of the practice in furniture production, but it provides a higher recovery of long, clear components and thus permits greater use of available material.

In its simplest form, a molder has a top cutter head with two side heads behind it and finally a bottom cutter head. This arrangement allows lumber to be milled on all four sides in a single pass through the machine.

An important process in millwork is finger-jointing. In this process short pieces of clear lumber that would otherwise be discarded are end-glued into long, continuous strips that can then be crosscut to any desired length (Figure 12.10). Finger-jointing usually takes place after the lumber has been dried and ripped and the defects have been removed, but it now may take place when lumber is in the green condition as well (i.e., immediately after sawing). It is expected, however, that the latter practice will be restricted to softwood dimension lumber.

Softwoods – particularly white pine, Douglas-fir, and hemlock – are usually used for millwork, but soft-textured hardwoods such as poplar and aspen are often used for interior moldings and picture frames. Dimensional stability, ease of machining and finishing are the main criteria used in selecting material from these species for millwork.

Competing materials, particularly aluminum windows, have made strong inroads into the wood-millwork market, but both high heat loss and increased condensation are distinct disadvantages of this particular product. Improved finishing methods and the use of plastic overlays and inserted tracks now provide wood products with the ease of maintenance of aluminum but with none of its disadvantages.

BENTWOOD

Bentwood provides a means of producing both functional and aesthetic curved wood products. There are several ways of producing bentwood; it can be made by application of heat, moisture, and pressure; by gluing thin laminations of wood into the required shape; and by chemical treatments.

Bent solid wood, composed mainly of hardwoods with moisture contents of 20–25%, is used in the manufacture of furniture, canes, sporting goods, and curved boat members. The wood is plasticized by steam heating to 100°C (212°F); then it is held in the new shape until it has cooled and its moisture content has been reduced to the optimum level for its intended use. The bending process consists mainly of

A

B

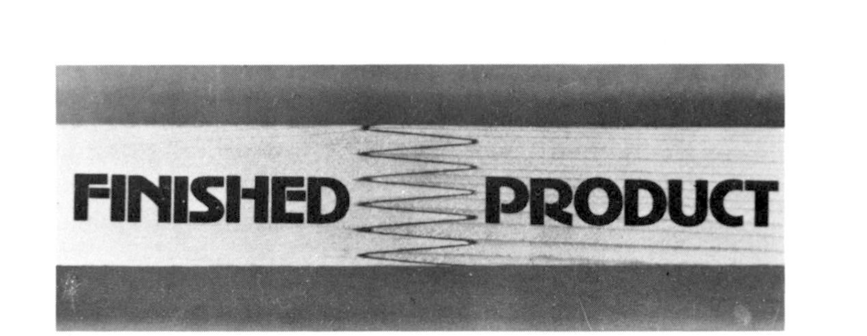

E

D

C

Figure 12.10 Some important steps in producing finger-jointed lumber: (A) machining the fingers, (B) spreading glue on the fingers, (C) feeding fingers into the assembly unit, (D) cutting continuous strip to length, and (E) finished product.

compressing the wood fibers on the inner side of the bend because only a very slight lengthening is tolerated by wood on the outer side before fracture occurs.

The lamination process for making curved wooden members is usually carried out by gluing the laminations while they are held in the shape required. This technique is used for a wide range of products – from golf-club heads to furniture parts to massive glued-laminated arches. The thickness of the various laminations is dictated by the degree of curvature needed and the moisture content of the laminations. Plywood made with exterior-type adhesives can also be curved after manufacture by processes similar to those used for bending solid wood, although it is not as stable as plywood laminated in the required shape.

Wood can be plasticized by several chemicals: liquid ammonia, urea, dimethylol urea, dimethyl sulfoxide, and low-molecular-weight phenol-formaldehyde resin. Limited commercial application has been made of wood bent with the use of urea and dimethylol urea.

Figure 12.11 Curved glued-laminated support for a staircase – a spectacular example of bentwood. (Photo: Canadian Wood Council)

SPORTING GOODS

Hockey Sticks

About 4 million hockey sticks are manufactured in Canada each year. The one-piece bentwood hockey stick of the early 1900s remained basically unchanged until the 1950s, when the three-piece stick was developed to take advantage of more efficient production methods. Various minor developments have included curving the blade for better forehand puck control and strengthening the blade by overlaying it with a fiberglass mesh.

Hockey sticks must be able to withstand tremendous impacts; a slapshot can send a puck hurtling at more than 159 km/h (100 mph). Only the toughest wood species, such as white ash, rock elm, and hickory, can be used in their construction. Sticks with laminated wood veneer handles have recently appeared on the market.

Baseball Bats

Baseball bats are produced from white ash dowels or squares that are kiln-dried and then turned to shape on a back-knife lathe. A flame heat treatment that is sometimes used before finishing produces attractive brown lines along the grain. Bats are made in various sizes and weights to suit individual requirements.

Bowling Pins

The two basic types of bowling pins are made of hard maple to provide maximum resistance to splitting, which may occur as a result of the repeated impact of the bowling ball.

Small-pin bowling is strictly a Canadian game and requires a relatively small pin turned from a one-piece blank. The large-pin game uses a much larger pin, which involves a more complicated manufacturing technique. A laminated blank, usually constructed of 19 pieces, provides greater economy because smaller pieces of high-quality hard maple can be used. Both types of pins are turned on a back-knife lathe.

Quality control is stringent, and particular attention is paid to the plastic finishing treatment in order to produce a durable pin. Some pins are made entirely in Canada, but most are made from imported blanks that are turned and finished in Canada.

Lacrosse Sticks

Lacrosse sticks are a rare example of a product that is still almost completely handcrafted. To withstand the rigors of lacrosse, a one-piece stick is necessary, and only hickory can provide the toughness and resilience required. A one-piece stick is produced by bending the wood to the required shape following a basic procedure that has been practiced for centuries. The blank is steamed to soften the wood and to render it pliable; then it is bent to shape and placed in a restraining form to allow the bend to set. Another bend is made later to ensure that the stick will hold its shape.

The only lacrosse-stick plant in Canada, operated by the Mohawk Indians near Cornwall, Ontario, produces about 85 000 lacrosse sticks annually and accounts for about 98% of the world's supply.

Figure 12.12 Bowling alley. The floor and pins are made of hard maple. (Photo: Canadian Wood Council)

Other Sports Products

Although the one-piece wooden downhill ski has almost disappeared, many plastic-coated multilaminated wooden skis are still manufactured. Wood is still a prime material for water skis, sleighs, and snowshoes, as well as for canoe paddles, pool-table frames, billiard cues, golf-club heads, and table-tennis paddles.

HARDWOOD FLOORING

Hardwood provides a flooring material unsurpassed in natural beauty and trouble-free durability. Compared with tile and wall-to-wall carpeting, hardwood flooring has both the lowest annual maintenance and the lowest long-term cost. While synthetic tiles and wall-to-wall carpets have recently outsold hardwood flooring, the preference is restricted to those installations where initial cost is the prime consideration. It has long been recognized that when ruggedness and durability take precedence, hardwood remains the preferred flooring material. Bowling alleys provide an example of extremely harsh conditions where hard maple flooring is the only acceptable material.

Despite the hardness that gives it great durability, hardwood flooring is resilient. In addition, by absorbing the body shocks produced in walking, hardwood flooring minimizes fatigue due to tired feet.

The two basic types of hardwood flooring are strip and parquet. One type of strip flooring consists of tongue and groove strips of uniform width that are laid continuously edge to edge and toenailed through the tongue; another type consists of thin, narrow strips face-nailed over softwood shim stock. Parquet flooring, which is laid much like floor tiles are laid, continuously repeats simple geometric patterns. It is prefabricated from thin, narrow strips of hardwood,

Figure 12.13 Parquet and strip hardwood flooring in a school gymnasium. (Photo: Canadian Wood Council)

which are glued to a veneer backing or tied together with thin wires embedded in the wood.

A third type of flooring in which blocks of hardwood are laid with the end grain uppermost, is sometimes used in foundries. This type of flooring not only provides a long-wearing surface but also minimizes the breakage of dropped castings.

Wood shrinks and swells less in the radial direction (along the rays) than it does in the tangential direction (parallel to the annual rings). Therefore, for better-quality flooring, edge-grain strips, with the annual rings perpendicular to the wide face of each strip, are preferred. Flat or tangential grain, with the annual rings parallel to the wide face, is less desirable. Quarter sawing to expose the rays is sometimes employed to produce edge-grain strips that are sawn almost exactly parallel to the rays. This practice is becoming rare, however, because of its high production cost. The appearance of oak flooring is greatly affected by grain orientation in the strips because oak is a ring-porous species. Grain orientation is less important from an aesthetic point of view for diffuse-porous species such as maple and yellow birch.

BOATS

Despite competition from plastics and aluminum, wood continues to be highly regarded by many boat builders and buyers alike. The preference for wood can be attributed to a variety of factors, the most important of which include high strength-to-weight ratio, ease of workability and fabrication, high impact and abrasion resistance, good screw-holding properties, and aesthetic appeal. Wood is a favored material for fishing and sailing craft of several tons displacement, and its replacement by plastics and aluminum has largely been restricted to pleasure craft under 7.6 m (25 ft) in length.

In the construction of canoes, eastern white and western red cedar are usually selected for the ribs because the wood of both species is light and strong and resistant to decay. Sitka spruce is a preferred material for the masts and booms of small sailing craft.

Larger wooden vessels consist basically of a skeletal frame and keel, usually of white or red oak overlaid with planking of various species, including oak, Douglas-fir, and red pine. Teak and mahogany are preferred for decking material because they resist saltwater damage, but their use has become limited because of high cost. Vessels constructed in this fashion usually give many years of trouble-free service; for example, the famed RCMP wooden patrol boat *St. Roch* sailed around the top of North America twice and was in service for many years.

Strength and natural resistance to decay are essential qualities of wood that is used in the construction of large boats. White oak was long the most favored species, but other species can be used if they are given a preservative treatment. Indeed, the great porosity of a species such as red oak, which is normally susceptible to decay because it readily soaks up water, can be turned to great advantage through preservative treatment, which renders it even more decay resistant than white oak.

A unique example of a ship for which wood was the only suitable material was the minesweeper that was widely used during the Second

Figure 12.14 Construction of a wooden boat by Lemont Hutt at Alberton, Prince Edward Island. (NFB Photothèque. Photo by Dunkin Cameron)

World War. Normally navy ships are of steel-hull construction, but minesweepers had to be fabricated of nonmagnetic materials in order not to attract magnetic mines. Inspection of such vessels after 15 years of service showed them to be in sound condition and still serviceable.

Shortly after the Second World War, marine-grade hardwood plywood came into prominence as a basic boat-building material. It has the strength and durability to resist the rigors of a heavy sea; yet it can be manufactured by the bentwood lamination process to the compound curves required for modern hull designs. Because of the very high strength requirements, veneers of only two species, coastal Douglas-fir and western larch, have been used in the fabrication of softwood marine-grade plywood panels. Recently, however, the use (and availability) of this plywood seems to have decreased.

Examples of boats that have been constructed from marine-grade plywood and are capable of enduring the harshest of conditions range from the famous PT boats of the Second World War to modern unlimited-class hydroplanes that reach speeds in excess of 518 km/h (200 mph) (Nash 1965).

Many modern hulls are of a composite nature – that is, a fiberglass skin overlaying plywood. Wood provides strength and aesthetic appeal, and fiberglass provides a more durable finish as well as contributing to the strength of the composite.

COOPERAGE

Figure 12.15 A cooper assembling barrels. Partly completed barrels are seen in the foregound. (NFB Photothèque. Photo by Ted Grant)

Cooperage is a collective term applied to a variety of wooden containers that consist basically of an assembly of staves bound together with hoops. The barrel is probably the oldest wooden container known and is still the principal form of cooperage in use.

In the past the barrel was almost the sole container for the storage and transport of liquids. So widespread was its use that the standard barrel (0.164 m^3 [36 gal]) is still a legal unit of measure.

Unlike the fabrication of most wood products, in which individual craftsmanship has been reduced to a series of simple, mechanically performed operations, barrel making remains an art. The barrel's structural integrity and watertightness depend on the cooper's skill in evenly fitting and raising staves of random width. The fitting of curved staves to form a double arch, with each stave resting against its neighbor, gives the barrel both its characteristic shape and its great strength. There are two classes of cooperage: tight cooperage, in which the barrel must not leak, and slack cooperage, where the barrel does not have to be watertight because it is used for holding nonliquid contents such as nails. The market for slack cooperage has practically disappeared because for many products the barrel has been replaced by metal and plastic drums.

Tight cooperage for the distillery and wine industries is usually made from white oak imported from the United States. Tight cooperage for fish, meat, and pickled products is usually made from maple or gum, and the barrel is lined with paraffin. For the aging of premium whiskeys, only high-quality barrels of white oak are acceptable. They are charred on the inside so that the resulting charcoal will absorb impurities during the aging process. Red oak is suitable only for slack cooperage.

BOXES AND CONTAINERS

In recent years wood has been in constant competition with corrugated cardboard, plastics, and metal as the first choice for packaging material. Formerly, thick-walled, solid wooden boxes and crates, sometimes meant for reuse, were the rule. Today their use is confined to long-distance transport of heavy goods such as automotive parts and machinery destined for export markets. These containers are generally constructed of softwoods, such as pine and spruce, with a load-bearing bottom of maple, elm, birch, or other hardwood.

The decline in use of the wooden container is due not only to higher raw material and labor costs but also to the modern demand for sturdy, lightweight containers that are cheap and expendable. For this reason the wood in apple boxes has been replaced by container-board, and the traditional basswood butter box and circular cheese box made of elm are no longer found. In spite of these encroachments on the use of wooden containers, wood is still widely used for many important packaging applications.

Large plywood bins are used to collect and transport fruit from the grower to the packing plant. Plywood is also being used in the large crates of containerized shipping operations. These crates have metal frames and are often constructed with sides of Douglas-fir plywood reinforced with a fiberglass coating. Ease of repair is the asset of this type of construction.

Another current use of wood in containers is in the wire-reinforced boxes used for shipping fruits and vegetables and some factory products. These thin-walled containers are reusable and can readily be collapsed for storage and shipment back to the farm or factory. They have considerably higher crushing resistance than comparable corrugated container-board products do, and they retain their stability on absorbing moisture. Hardwoods – often birch, soft maple, and poplar – are used in their construction.

PALLETS

A pallet is a low, sturdy platform on which materials can be piled to facilitate their storage and handling with forklift trucks. Because they permit mechanization and more compact storage, pallets are a vital part of handling systems in all segments of industry. So widespread is their use that 6½ million new pallets were manufactured and documented in 1978 by Statistics Canada, and many more were manufactured by the numerous small, unlisted producers scattered across the country.

The most common type of pallet in use is the four-way entry pallet, which permits the forks of a lift truck to be passed under a load from all four sides. This general-purpose pallet may be constructed of any strong wood species, but hardwoods are favored because of their ability to give years of service under repeated abuse. High resistance to nail withdrawal, impact, and abrasion are important considerations, and commonly used species include maple, beech, oak, and elm. The cost of raw material is of great importance in the highly competitive pallet industry, and less desirable species may be used when pallet standards are not followed. Plywood can be a suitable alternative for hardwood decks (load-bearing surfaces), however.

Most pallet plants also manufacture less expensive one-way pallets (not intended for reuse) and shipping skids such as those attached to the bottoms of new stoves and refrigerators.

Figure 12.16 Douglas-fir plywood pallet bins used for shipping fruit in British Columbia. (Photo: Council of Forest Industries of British Columbia)

SHINGLES AND SHAKES

The manufacture of shakes is undoubtedly one of Canada's oldest industries. Early shakes were split by hand, using a froe and mallet, after the log had been cut into blocks of the proper length. A second type of shake is manufactured by hand-splitting to almost double thickness and taper-sawing down the thickness to produce two tapered shakes with a split-shake upper surface and a flat-sawn underside. This shinglelike shake makes for easier laying. The essen-

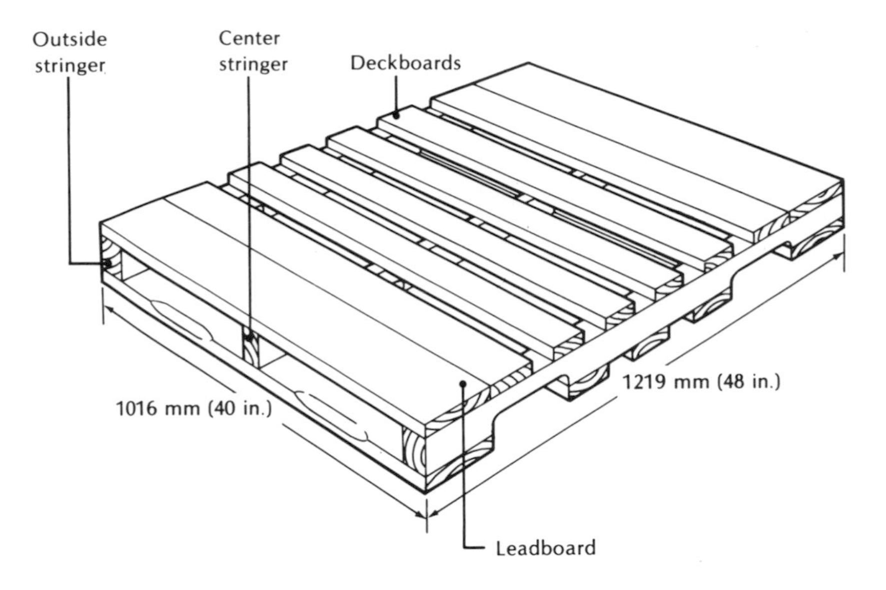

Figure 12.17 Typical four-way entry pallet of the notched stringer design.

tial difference between shakes and shingles is shown in Figure 12.18.

Today the sawing of shingles is carried out on large, upright circular saws by one man, who saws, edges, and grades the shingles. Shingle bolts, which are quarter cross-sections of logs cut to shingle length, are held and fed by a carriage that tips back and forth as successive tapered shingles are sawn by the large saw. Simultaneously the operator does the edging on a much smaller second saw.

Most of Canada's shingle production is centered in British Columbia, where the shingles are made from the heartwood of western red cedar logs. This species offers the advantages of very high durability, fine-textured wood, high thermal insulation, and large tree diameter. This last feature is particularly important because it enables a high proportion of edge-grained shingles to be cut. Edge-grained shingles are less prone to warp and split with lengthy exposure to weathering than are flat-grained shingles. A considerable volume of shingles is produced from eastern white cedar in Ontario, Quebec, and New Brunswick.

Figure 12.18 Types of shingles and shakes: (A) edge-grained, sawn, tapered shingle and (B) split shake (not tapered).

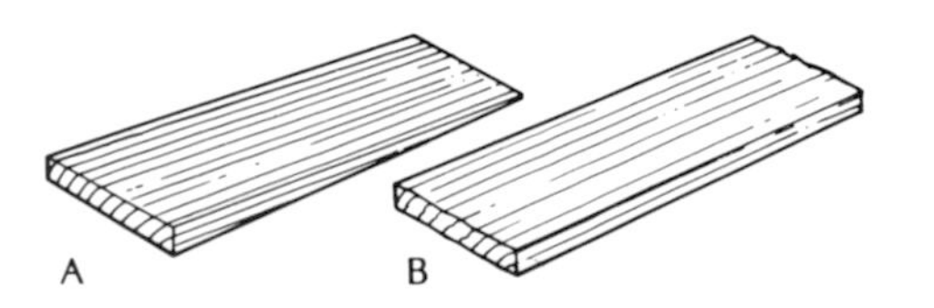

BROOMS, BRUSHES, AND MOPS

The broom, brush, and mop industry manufactures a wide variety of products: household, industrial, and curling brooms; household and industrial brushes; paintbrushes; personal brushes; and floor, dust, and dish mops. While the industry relies heavily on imported raw materials for the bristles, a significant amount of wood, which is not imported, is used in the manufacture of the handles; indeed, it is the largest raw material cost.

The corn broom has always dominated the Canadian broom market. Recently, however, there has been a shift toward brooms with bristles made of synthetics and other vegetable fibers. This trend reflects a desire for less expensive products, and the relatively soft, synthetic bristles are well suited for light sweeping on finished floors.

Handles for light household brooms are usually made from a variety of softwoods, while maple and birch are used for heavy industrial brooms. Broom handles are produced by passing long squares of wood

Figure 12.19 Western red cedar shingles are used as exterior-wall and roof finish on this house in southwestern Ontario. (Photo: Council of Forest Industries of British Columbia)

through a dowel machine, which is simply a rapidly rotating cutter head.

Curling brooms, the highest-priced brooms produced, are still made in the traditional way to withstand rigorous use – with corn bristles on a maple or birch handle.

Despite the use of plastics in lower-priced items, maple, beech, and birch wood are still important in the manufacture of high-quality household and personal brushes.

Mops of various types are produced in Canada. Various species, including red and white pine and aspen, are used in the manufacture of light-duty, wet, and dry mops. Maple and birch handles are often used in the heavier duty 'yacht-style' mops.

WOODEN MATCHES

Total daily match production in Canada provides some seven matches for every man, woman, and child. The majority are made of paper, but approximately 14% are wooden matches, which are of the 'strike-anywhere' type.

Match production is a highly automated process, and great efficiency has enabled retail match prices to remain unchanged for many years. Only square-splint wooden matches are produced now.

Match splints are manufactured from aspen-poplar veneer. The logs are cut into short bolts about 46 cm (18 in.) long and steamed (to soften the wood); then the bolts are fed into a small veneer lathe. The emerging sheet of veneer, which is 2.6 mm ($\frac{1}{10}$ in.) thick, is scored, or slit across the grain with knives, into strips 43 mm ($1\frac{11}{16}$ in.) long and guillotined along the grain into splints 2.6 mm ($\frac{1}{10}$ in.) wide. Defective splints and slivers are removed by perforated vibrating screens, and the good splints are arranged and packed into trays, dried, and then shipped to a dipping plant for final processing.

LAMINATED VENEER LUMBER

Lower-grade logs are used in this product, which consists of rotary-peeled thick veneer laminated into long, wide planks that are usually sawn into structural dimension lumber (Bohlen 1972). The conversion process from logs to laminated veneer lumber (LVL) is similar to that of plywood manufacture.

A six-ply product of exterior-board quality, 1.2 m (4 ft) wide, made from 6 mm ($\frac{1}{4}$ in.) veneer with all plies running in the same direction and with regularly spaced butt joints, has been developed in Canada and used commercially for roof panels, 1.2 m (4 ft) wide by 12.2 m (40 ft) long. Limited amounts of structural dimension lumber have also been manufactured from LVL and tested (Bohlen 1975).

REFERENCES

Bohlen, J.C. 1972. Laminated veneer lumber – Development and economics. *For. Prod. J.* 22 (1): 18–26

– 1975. Shear strength of laminated veneer lumber. *For. Prod. J.* 22 (2): 16–23

Cantin, M. 1965. *The machining properties of 16 Eastern Canadian woods.* Dep. For. Publ. 1111. Ottawa

Nash, N. 1965. Marine-grade plywood takes to the water. *Wood Work. Dig.* Oct., pp. 45–6

Statistics Canada. 1978. Cat. 35-209. *Wooden Box Factories.* Ottawa

BIBLIOGRAPHY

Baldwin, W.C. 1968. Pallets from low grade hardwoods. *For Prod. J.* 18 (3): 11–13

Clark, W.M. 1965. *Veneering and wood bending in the furniture industry.* New York: Pergamon Press

Highley, T.L.; Scheffer, T.C.; and Selbo, M.L. 1971. Wood minesweepers are sound after 15 years of service. *For. Prod. J.* 21 (5): 46–8

Industry, Trade, and Commerce Canada. 1974. *A review of the broom, brush and mop industry in Canada.* Wood Prod. Branch, Ottawa

Koch, P. 1964. *Wood machining processes.* New York: Ronald Press

Martens, D.G. 1971. *A comparison of flooring costs under residential conditions.* U.S. Dep. Agric. For. Serv. Res. Pap. NE-200. Washington, DC

Panshin, A.J.; Harrar, E.S.; Bethel, J.S.; and Baker, W.J. 1962. *Forest products: Their sources, production and utilization.* 2nd ed. Toronto: McGraw-Hill

Schuerch, C. 1964. Wood plasticization. *For. Prod. J.* 14 (9): 377–81

U.S. Forest Products Laboratory. 1974. *Wood handbook: Wood as an engineering material.* Agric. Handbook 72 (rev.). Washington, DC: U.S. Dep. Agric.

13

Pulp and Paper

J.L. KEAYS

Deceased, formerly with Western Forest Products Laboratory, Vancouver

Papermaking is one of the world's oldest industries. Its origins may be traced to the Chinese philosopher Ts'ai Lun, who is credited with the invention of paper in 105 AD. The materials he used were the inner bark of the mulberry tree and, later, young bamboo. The art of manufacturing paper spread slowly; it was not until 650 years later that paper was produced outside of China, first in Samarkand in 751 AD and then in Baghdad in 795 AD.

Papermaking traveled slowly to the West. The Arabs brought the art from Baghdad to Spain, where the Moors manufactured the first paper in Europe in 1050. By the year 1200 paper was widely manufactured in France, Italy, and Holland. The spread of papermaking to other European countries was slow, however, and it did not reach England until about 1500. By 1700 papermaking was widely practiced throughout the countries of the Old World. In the New World the first paper mill was started near Philadelphia in 1690, and by 1810 there were 110 paper mills in the United States and Canada.

For the first seventeen hundred years of paper manufacture, the most difficult and time-consuming operation was the formulation of the sheet from a watery suspension of paper-making material called a pulp slurry. The first paper machine was put into use in France in 1798, and by 1804 the first efficient machine was in operation. Invented by Louis Robert, and subsequently improved by the Fourdrinier brothers, the Fourdrinier paper machine established the basic processing principles on which most paper would be made for the next two hundred years. The first multivat cylinder machine, which permitted the production of multilayered paperboard, was invented in England in 1809. Paper machines increased in size, speed, and sophistication, and production per machine rose from a few tonnes per day to as high as 500 tonnes (550 tons) per day.

The invention of printing in Germany in 1450, and the increase in literacy throughout the Western world thereafter, led to an increase in

Figure 13.1 Aerial view of Canadian International Paper Company's newsprint mill at Gatineau, Quebec. (Photo: Pulp & Paper Canada)

the demand for paper. Until the middle of the nineteenth century most of the world's paper was manufactured from rags, and an increasing amount of straw was used. The supply of rags and straw was not sufficient to keep up with the demand, however. A long search for new papermaking materials and methods ensued, and many different raw materials, such as bamboo, grasses, and reeds, were tried. The search culminated in the successful use of wood for pulp, first in the mechanical process in 1844 and then in the chemical processes a few years later.

Once the limitation on the supply of raw material was removed, there was a rapid increase in the production of pulp and paper throughout the world, which has continued to the present day (Table 13.1).

Today a vast pulp and paper industry exists throughout the world. The total world production of paper and paperboard in 1978 was 160 million tonnes (176 million tons), and that figure increases by approximately 5% per year. Pulp and paper are produced in thousands of mills located in practically every country in the world; 147 of these mills are in Canada. The basic requirements for the development of a pulp and paper industry are plentiful supplies of wood, water, and power, and Canada is wealthy in all three. Since the first paper mill was built in Canada in St Andrews East, Quebec, in 1803, pulp and paper have grown together to become one of Canada's largest industries,

Table 13.1 *World production of paper and paperboard*

Year	World production: Millions of tonnes	Millions of tons
1945	23	25
1950	38	42
1955	55	61
1960	74	81
1965	98	108
1970	128	141
1972	139	153
1973	149	164
1974	152	167
1975	132	145
1976	149	164
1977	154	170
1978	160	176

Source: FAO (1978)

Table 13.2 *Canada's pulp and paper industry*

Year	Production of paper and paperboard: Millions of tonnes	Millions of tons	Value in millions of dollars
1945	3.9	4.3	
1950	6.0	6.6	200
1955	7.1	7.8	1000
1960	7.9	8.7	1200
1965	9.4	10.4	1400
1970	11.1	12.2	1800
1972	12.3	13.5	1900
1973	13.0	14.3	2300
1974	13.2	14.5	3200
1975	10.1	11.1	2900
1976	11.8	13.0	3400
1977	12.1	13.4	4000
1978	13.4	14.7	4700

Source: Statistics Canada (1978a)

supporting a large part of the nation's total economy (Table 13.2).

In Canada today, 70% of all pulp is made from spruce and fir, 10% from pine, 10% from hemlock, and the remainder from a number of other wood species, including several hardwoods. In the east, the preferred pulpwood species are the spruces and balsam fir, but other species are also used, such as jack pine, aspen, birch, and maple. The pulp mills on the coast of British Columbia use mainly western hemlock, Douglas-fir, and western red cedar, along with lesser amounts of the spruces and true firs. The mills in the interior of British Columbia use white spruce and lodgepole pine, as well as small amounts of the true firs and Douglas-fir. In general the use of Canada's available forest resources by the pulp and paper industry matches the demand for pulp and paper in Canada and in foreign markets. The one exception is aspen, which is underused at present.

A brief descriptive overview of the pulp and paper industry is presented in this chapter. More detailed information on the complex and highly developed technology of the industry may be found in the sources cited in the bibliography at the end of the chapter.

PULPING AND BLEACHING PROCESSES

Pulping

The pulp and paper industry uses a variety of raw materials and processes to produce a large number of products. The purpose of all pulping processes is to separate the wood into individual fibers, or parts of fibers, or into small fiber bundles or fragments. These can then be freed of contaminants by washing and screening, made into a uniform slurry in water, and recombined into continuous sheets, which we call paper and paperboard.

Wood can be separated into fibers or fiber fragments for paper manufacture by either mechanical or chemical means. In the mechanical process the fibers are torn from the wood. In chemical pulping most of the lignin holding the individual fibers together is dissolved, freeing the fibers for further processing. Although individual fibers or fiber fragments can be separated from the solid wood by a great many chemical agents and mechanical means, over 90% of the world's pulp, and essentially all the pulp produced in Canada, results from three main pulping methods: the chemical, semichemical, and mechanical methods. All pulping methods use common techniques, even though they differ in detail, and each gives a pulp with unique properties.

Chemical Pulping Methods

Both the kraft and sulfite processes are chemical pulping methods, although the chemicals used in the two processes are quite different. In both subdivided wood in the form of chips, sawdust, or shavings is pulped at high temperature and pressure. The pulp is thoroughly washed to remove cooking liquor and screened to remove contaminants or material that has been only partially pulped. The screened pulp may then be bleached to a higher brightness or treated mechanically (beaten or refined) to improve its papermaking characteristics, or it may be used directly. Finally the pulp is made up into a uniform slurry, free of all contaminants, ready for final processing.

Figure 13.2 Immense pile of debarked pulpwood logs at Grand Falls, Newfoundland. (NFB Photothèque. Photo by Chris Lund)

Pulp in this form may be sent directly to a paper machine to be converted into a product such as kraft linerboard, which is used in corrugated shipping containers, or it may be mixed with one or more other pulps and the mixture fed to a paper machine (newsprint is manufactured in this way). Pulp may also be formed into heavy sheets, which, when folded, are known as wet laps, and shipped to other mills for further processing. Alternatively, pulp may be dried in the form of a thick web, which is shipped in roll form or cut into sheets and shipped as baled market pulp.

The basic principles of the kraft and sulfite processes are described in the following sections.

The Kraft Process

The kraft process is a pulping process in which lignin is dissolved by a solution of sodium hydroxide and sodium sulfide. It is essentially an offshoot of the soda process, which was developed in England in 1851. In that process wood chips were cooked at 170°C (340°F) for 4 hours in a solution of approximately 8% sodium hydroxide. The soda process has now been almost completely replaced by the kraft, or sulfate, process, although it was used widely for many years, particularly for the manufacture of straw pulps.

The kraft process dates back to 1883, when it was found that the addition of sodium sulfide to the soda cooking liquor reduced cooking

Figure 13.3 Washing chemical (kraft) pulp to remove cooking liquor. (Photo: Pulp & Paper Canada)

time and gave a stronger pulp. Development of the kraft process was slow at first, but it accelerated rapidly as modern bleaching methods, highly efficient chemical recovery systems, and large-scale continuous digesters became available. Not only has the kraft process replaced the soda process almost entirely, but it is rapidly displacing the sulfite process as well. Over 80% of all chemical pulps are produced by the kraft process, and the percentage is increasing.

The kraft process has a number of advantages over the other conventional chemical pulping processes. The pulping chemicals can be economically recovered for reuse, almost any wood species can be used, and the pulp is strong and suited to almost all types of paper and paperboard. The kraft process is also suitable for the production of dissolving pulps used in the manufacture of rayon and other products.

Kraft pulp is produced by cooking wood chips in a digester in a cooking liquor containing sodium hydroxide and sodium sulfide in the ratio of approximately 3:1. About 15 kg, or 15 lb, of chemicals are used for every 100 kg, or 100 lb, of chips. Cooking temperature is usually 165–175°C (330–347°F), and the cooking time at maximum temperature is 1–3 hours. When the pulp is to be bleached, the wood is cooked to a yield of approximately 44%; that is, 100 kg, or 100 lb, of oven-dry and bark-free wood gives 44 kg, or 44 lb, of unscreened pulp out of the digester. The yield may be up to 50% or higher if the unbleached pulp is to be used for products such as linerboard.

Regardless of the end use, the pulp is washed free of cooking liquor. Uncooked knots and oversized material are separated and returned to the digester, and the pulp is screened. The screened pulp is essentially free of fiber bundles and other contaminants, such as sand, and is now ready for further processing in a paper mill, bleach plant, or pulp dryer.

A major advantage of the kraft process is its recovery system. The waste liquor, termed black liquor, containing roughly 50% of the wood substance dissolved during the pulping process, is concentrated to over 50% solids by evaporation and fed to a recovery boiler. The concentrated black liquor solids are burned to provide most of the heat required by the pulp mill and to produce a smelt consisting mainly of sodium carbonate. The smelt is dissolved in water to give green liquor, to which is added calcium hydroxide, formed by the burning of limestone. The resulting mixture of sodium hydroxide and sodium sulfide (white liquor) can be used for further pulping. Chemicals used or lost during pulping and pulp processing are made up in the form of sodium sulfate added to the black liquor just before its combustion – hence the name sulfate process. In a modern, efficient kraft mill, as little as 25 kg (50 lb) of sodium sulfate makeup is required per tonne (ton) of pulp produced. A kraft mill that does not bleach the pulp is close to self-sufficiency in heat, and the makeup chemical, sodium sulfate, is low in cost and plentiful in supply.

In the past, kraft mills were characterized by a strong odor of organic sulfur compounds. Developments in process flow and processing equipment over the years have reduced the discharge of sulfur compounds into the atmosphere, and in a modern, well-designed kraft mill odor is minimal. Treatments have also been introduced to recover the portion of fibers and dissolved solids that were formerly discharged into rivers and tidewaters.

All new kraft mills and most of the older mills now have savealls to recover useful fiber and primary sedimentation tanks or clarifiers to remove suspended solids. These solids are recycled to the mill system, burned, or otherwise disposed of. The mill waste liquors from the primary treatment system, which are free of suspended solids but contain a variety of dissolved solids, may be fed to a secondary treatment system. This system consists of one or more biological basins, where the toxicity to fish and the oxygen demand of the organic content of the wastes are reduced to an acceptable level before discharge.

Sulfite Process

The sulfite process was developed in 1866 by an American, Benjamin Tilghman, who found that pulp could be produced by heating chips in a bisulfite–sulfurous acid solution. It became the dominant pulping process from the late 1800s to the 1940s. The sulfite process has a number of advantages. The cooking chemicals are sufficiently inexpensive and plentiful to make a liquor recovery system unnecessary, and the pulp is admirably suited to the manufacture of many paper products as well as dissolving-grade pulp. In unbleached form, sulfite pulp has been found to be very satisfactory as the chemical component in newsprint, and its use has been widespread.

In spite of these advantages, the amount of sulfite pulp manufactured as a percentage of total pulp production has been declining steadily over the past 30 years. Several critical factors have been responsible for this decline: the pulp is not as strong as kraft pulp; a number of wood species, including many of the pines, western red cedar, and Douglas-fir, cannot be pulped by the sulfite process; and chemical recovery systems that are as economical as the standard kraft recovery system have not been developed for the sulfite process. Sulfite pulp in bleached form, however, is brighter than bleached kraft and is less expensive to bleach.

The term sulfite process covers a wide range of pulping systems. In the original sulfite process, a strong solution of sulfur dioxide, formed by burning sulfur or by roasting iron pyrites, was reacted with limestone to give strongly acid cooking liquor made up of calcium bisulfite with a high content of free sulfur dioxide. This liquor was the basis of the conventional sulfite process that was dominant for many years in the chemical pulp field.

In the late 1940s a magnesium-base sulfite process was developed in Canada, at Howard Smith Paper Mills, which permitted the recovery of chemicals and energy by the concentration and burning of the waste liquor in a manner similar to that of the kraft recovery process. Because environmental damage caused by waste-liquor discharges was not recognized as an important problem until some 20 years later, only a few mills converted to this new sulfite process at that time.

In the late 1960s consideration of waste-liquor recovery systems became important for all sulfite mills. The primary reason was not to recover chemicals but to reduce the discharge of polluting organic material. At a sulfite-pulp yield of 55% and with no recovery system, 45% of the wood, which is dissolved in the waste sulfite liquor, is discharged to the stream. Thus, the discharge of organic materials from a mill producing 550 tonnes (600 tons) per day of unbleached sulfite pulp is considerable at approximately 450 tonnes (500 tons) per day. Increasing environmental concern has resulted in the establishment of stringent mill effluent standards in various countries. Because it was difficult for conventional calcium-base sulfite mills to meet these regulations, processing techniques were changed.

By the use of a sodium, magnesium, or ammonium base in place of calcium in the sulfite process, the cooking liquors could be prepared over a range of acidities, and a greater number of wood species could be pulped. Effective chemical recovery systems could also be developed. The development of recovery systems for such soluble-base sulfite mills has slowed, but not stopped, the decline in use of the sulfite process.

In Canada the main use of sulfite pulp is in the manufacture of newsprint, which traditionally contains 75–85% mechanical pulp and 15–25% chemical pulp. Unbleached sulfite pulp at appreciably higher than normal yields can also be used in newsprint manufacture. Beginning in the 1950s there was an increase in the number of newsprint mills using unbleached sulfite pulp in yields of 50–75%. Today all the newsprint manufactured in western Canada contains semibleached kraft pulp, and a number of eastern mills are converting to the use of kraft pulp, rather than sulfite pulp, as the chemical component in newsprint manufacture.

Semichemical Pulping Methods

The term semichemical pulp covers a wide range of pulping chemicals and products. In general, semichemical pulping involves a partial removal of lignin from the chips so that the pulp yield is high – from 75 to 85%. Semichemical pulps can be prepared from a variety of pulping liquors containing bisulfites, sulfites, carbonates, and hydroxides. Whereas calcium-base sulfite cooking liquor is highly acidic and kraft cooking liquor is highly alkaline, the semichemical pulping liquors are generally close to neutral – hence the term neutral sulfite semichemical (NSSC), the most commonly used semichemical process.

This NSSC process was developed in the 1920s at the Forest Products Laboratory in Madison, Wisconsin. It was found that when hardwood chips were pulped in a solution of sodium sulfite (prepared from sulfur dioxide and sodium hydroxide or carbonate), the wood was softened and could then be converted to a pulp suitable for the manufacture of a high-quality corrugating medium by the mechanical action of beating or refining. Widespread use of the process was delayed, however, until the development of modern, high-precision refiners.

In the NSSC process, hardwood chips are pulped for 2–4 hours at 160–190°C (320–375°F) in a solution containing 12% sodium sulfite and 3% sodium carbonate. After the cook the pulp is washed and the partially pulped or defibrated chips are further broken down into fiber bundles and individual fibers by the mechanical action of a refiner. Most of today's refiners consist of rotating plates or discs. The pulp slurry is pumped between these discs and flows radially from the inner to the outer part of the discs. The refiner-plate surfaces are grooved in order to break down the softened chips into progressively smaller fiber bundles and, ultimately, into individual fibers. Most chemical pulps such as kraft pulp for linerboard and sulfite pulp for fine papers require similar mechanical refining in order to obtain pulps suitable for the papermaking process.

The NSSC process not only has a high yield but produces a sheet that is particularly resistant to bending. Consequently most NSSC pulp is converted to a corrugating medium, a use for which its characteristic low brightness is not detrimental. In its final processing step, the corrugating medium is fluted, or given a uniform wavy shape, by passing the sheet through steam and through grooved rolls. This sheet is then used as the middle laminate between two linerboard sheets in the manufacture of container-board for packaging.

A number of other semichemical processes are used, but they are restricted to specialty products that are produced in small numbers.

Mechanical Pulping Methods

The first grinder that made possible the conversion of wood to a usable pulp was developed in Germany in 1844. In the traditional groundwood or mechanical pulp process, debarked and washed bolts of wood are pressed against rapidly revolving pulpstones. The protruding particles, or grits, in the pulpstone press into the wood and tear off fiber fragments, individual fibers, and small debris or wood flour to give a dilute water suspension of mechanical pulp or pulp slurry. Originally pulpstones were manufactured from natural sandstone, but they are now manufactured artificially, the grits being composed of carborundum or alundum. Considerable heat is generated during the grinding process, and the pulp slurry from the grinders is quite hot. This slurry is passed through successively finer knotters or screens to remove large chunks of wood, knots, small pieces of unprocessed wood, large fiber bundles, and small fiber bundles called shives. In modern practice, the pulp is finally passed through banks of centrifugal cleaners to remove contaminants, such as sand and other heavy particles.

The stone groundwood process produces pulp of a fairly high brightness, low strength, and high yield (90% or higher), but it requires a large amount of power. The high yield occurs because the pulp is formed by mechanical means, which preserves most of the original wood substances. Consequently the cost per tonne (ton) of production is considerably cheaper than that of the chemical pulping processes.

Most of the mechanical pulp manufactured is used in making newsprint. Other uses include the manufacture of absorbent papers, wallpapers, tissues, and inexpensive writing and printing papers. In general, mechanical pulp is used when low-cost pulp with low strength requirements is sufficient but good absorbency, good opacity, high bulk, and a good printing surface are also required.

In the 1950s and 1960s a new process, the refiner mechanical process, was developed for the conversion of wood to a mechanical-type pulp. In this process wood in the form of chips, sawdust, or shavings is passed between two flat discs, one or both of which may be power driven. The refiner plates have a grooved pattern designed to give the desired abrading action, and the wood is sequentially reduced from chips to matchsticks, to large fiber bundles, to smaller fiber bundles, and finally to individual fibers, fiber fragments, and fine

Figure 13.4 Pulpwood grinding room in newsprint mill – MacMillan Rothesay Ltd., Saint John, New Brunswick. (Photo: Pulp & Paper Canada)

material. A pulp is produced that is similar to, but stronger than, stone groundwood pulp.

The refiner mechanical pulping process has several advantages. Wood in practically any form – chips, sawdust, shavings, or chip fines – can be converted to mechanical pulp; the pulp is stronger than stone groundwood pulp; and, finally, the wood process flow is simpler than the conventional stone groundwood process. Refiner mechanical pulp is used for essentially the same products as stone groundwood: in newsprint, in limited quantities in tissues, in a variety of writing and printing papers, and in many low-grade papers.

The main reasons for the strong interest in refiner mechanical pulp are the potential advantages of its use in the manufacture of newsprint. Because of its higher strength, it can be substituted for some of the chemical pulp in newsprint, and in time it may eliminate chemical pulp altogether. Traditionally newsprint has been manufactured from 75–85% mechanical pulp and 15–25% sulfite or kraft pulp. Elimination of the chemical pulp would reduce the manufacturing costs of newsprint, give a higher yield of product from a given forest area, and effect a substantial reduction in pollution of the environment.

The amount of refiner mechanical pulp used throughout the world has slowly increased. A number of mills in Canada pulp sawdust by this process, and an increasing number of mills are now converting logs to chips to mechanical pulp through refiners. Most new refiner plants use pressurized refiners to produce thermomechanical pulps. These pulps are being used in an increasing number of paper and paperboard products as well as in newsprint.

Miscellaneous Pulping Methods

Over the years a vast number of other pulping methods have been proposed, tested in laboratories, or used on a small scale. Wood can be converted to pulp by digestion with acids (such as nitric acid), by organic solvents, by concentrated aqueous solutions of certain salts (such as sodium xylene sulfonate), by bases (such as concentrated ammonium hydroxide or lime), by oxidizing agents (such as chlorine or nitrogen dioxide), and by reducing agents (as in hydrogenation). In spite of the vast amount of research carried out in the development of new pulping methods, few have developed beyond pilot plant operation or use in small mills.

The reasons for the unsuccessful development of these methods include poor pulp characteristics, such as strength and brightness, serious pollution problems, low pulp yields, excessively high costs, and corrosion problems. Only about 5% of the world's pulp is produced by these methods.

Bleaching

In chemical pulping and bleaching the principal objective is to remove as much of the lignin as possible so that the relatively undamaged cellulose fibers can develop maximum strength when formed into a pulp or paper sheet. Most of the lignin is removed in chemical pulping during the severe treatment of wood chips by chemicals under conditions of high temperature and pressure. The bleaching process is much milder and removes the remaining lignin with a minimum loss of valuable carbohydrate material. In theory it would be preferable to remove most of the original wood lignin by bleaching rather than by pulping, but high costs and the increased discharge of water pollutants render this practice impractical.

There are two main reasons for bleaching pulps: to remove contaminants and to improve brightness. Contaminants must be removed from dissolving-grade pulps because they would otherwise interfere with further chemical processing of these pulps. Many of the other pulps in unbleached form have a brightness that is unacceptably low for the production of white- or colored-paper products. Unbleached kraft pulp, for example, is dark brown because it contains approximately 5%

Figure 13.5 Bleach washer line at Northwood Pulp and Timber Ltd., Prince George, British Columbia. The pulp passes through several bleaching stages until the desired brightness is reached. (Photo: Pulp & Paper Canada)

lignin and 95% carbohydrates; it is suitable for use only in a limited number of products. When this pulp is bleached, practically all the residual lignin is removed. The fiber that is left is made up essentially of carbohydrates – cellulose and hemicelluloses – which are white.

The main bleaching chemicals in commercial use are chlorine, hypochlorites, chlorine dioxide, and caustic soda. Because of the explosive nature of chlorine dioxide and the problems arising from its large-scale production, handling, and transportation, it was not fully developed as a commercial bleaching agent until just before the Second World War. Today chlorine dioxide is used in almost every pulp mill producing bleached kraft and sulfite pulps, largely because of Canadian research and development that has taken place over the past 40 years.

Kraft and sulfite unbleached pulps differ significantly in brightness. Kraft pulps are quite dark (like the typical brown grocery bag), whereas sulfite pulps are as bright as newsprint. For this reason, unbleached kraft pulps require a more severe bleaching process than do sulfite pulps to reach the same brightness.

A typical and widely used bleaching sequence for unbleached kraft pulps is the CEDED sequence, where C stands for chlorination, E stands for caustic extraction, and D stands for chlorine dioxide treatment. Most of the residual lignin in the pulp is made soluble and removed in the first two stages; the last three stages brighten the pulp to the desired level. Sulfite pulps require fewer stages of bleaching because they contain less lignin and have a much higher unbleached brightness. A typical bleaching sequence is CED. A less commonly used bleaching chemical is hydrogen peroxide, which is used as a final stage when a bleached pulp with a high and particularly stable brightness is required.

Since bleaching chemicals and processing are expensive, bleaching is normally restricted to the so-called bleachable grades of pulps. These grades are produced at a low lignin content (approximately 5%) so that the cost of removing the residual lignin is not too high. The shrinkage in bleaching – that is, the loss in pulp yield – is usually around 8%; thus, an unbleached kraft pulp at 44% yield is reduced to a high-brightness pulp at approximately 40% yield after a full bleaching sequence.

Brightening

The main action of bleaching chemicals is to dissolve lignin, and this results in a lower pulp yield. The brightness of high-yield pulps, such as mechanical or semichemical pulps, can be improved by treatment with various chemicals or brightening agents that bring about structural changes in the lignin without removing it. In this way the high pulp yield is preserved. Agents in commercial use include zinc or sodium hydrosulfite and sodium peroxide.

PULP CONVERSION

Pulp may be considered a partially processed raw material midway between the tree and the end product used by the consumer. This product may be paper, a container, a place mat, tissue, or any one of the tens of thousands of products into which pulp is converted. Pulps produced for conversion fall into three broad categories: paper and paperboard, fiberboard, and dissolving grades.

Approximately 85% of the world's pulp is converted to paper and paperboard. The terms paper and paperboard cover an extremely broad range of felted, or matted, fiber sheets formed by draining the water from a suspension of fiber through a fine screen. Compared with paperboard, paper is thinner, has a lower weight per unit area, and is finer and more flexible. The term paper comes from the word *papyrus*, which refers to an ancient writing material made by the early Egyptians by pasting together thin sections of a Nile reed.

Some 10% of the world's production of pulp is in the fiberboard category. In a modification of the mechanical pulping process, wood chips are presteamed, with or without the addition of small amounts of chemicals, and then refined under pressure to give a coarse type of fiber that is low in brightness and strength. Traditionally this type of fiber has been used in the manufacture of coarse papers and boards, such as asphalt papers, roofing felts, and insulation board. In the original process developed during the early 1920s, wood chips were heated by steam to high temperature and then suddenly released to atmospheric pressure to provide a coarse fiber suitable for the manufacture of such papers and boards. Through recent modifications of this process, a finer, stronger, and brighter type of pulp, which gives a superior fiberboard, is obtained.

Dissolving-grade pulps constitute approximately 5% of all pulps produced. In the manufacture of these grades, wood chips are pulped by either the kraft or the sulfite process; the resultant pulp is then highly bleached and chemically refined by treatment with a solution of caustic soda to give a bleached, dissolving-grade pulp at 33–35% yield. These pulps form the basis for the manufacture of rayon and cellulose-derived plastics and other products. Additional information on dissolving-grade pulp is given in Chapter 5.

PAPER AND PAPERBOARD MANUFACTURE

Machines

In the long history of the paper and paperboard industry, two basic types of papermaking machines have been developed: the Fourdrinier machine for the manufacture of paper and the cylinder machine for the manufacture of paperboard.

Fourdrinier Paper Machine

The basic principle of the Fourdrinier paper machine was developed in France in 1799. In this machine a dilute pulp slurry of approximately 0.5 kg (1 lb) of fiber in 100 kg (200 lb) of water is fed into a reservoir called a headbox. From the headbox the slurry passes through a long slit, called a slice, onto a moving, continuous screen of wire mesh known as the wire. As the slurry progresses down the wire, some of the water drains away, assisted by the partial vacuum created by the rollers, or table rolls, that support the wire. More water is removed from the slurry by various suction boxes toward the end of the wire.

At the end of the Fourdrinier wire section the sheet of wet fiber

Figure 13.6 Wet end of high-speed paper machine in newsprint mill at MacMillan Rothesay Ltd., Saint John, New Brunswick. Headbox and Fourdrinier wire section are in the foreground; press rolls and drying section are on the right. (NFB Photothèque. Photo by B. Brooks)

passes onto another belt, known as a press felt. The sheet proceeds through one or more presses, which squeeze out more water, and then goes into the drying section. Here the sheet travels over a series of hot dryer rolls; when it emerges, the remaining water has been removed by evaporation. From the drying section the sheet passes through a vertical arrangement of heavy rolls called a calender stack, where a combination of high temperature and high pressure gives a finish or smoothness to the paper. After it leaves the calender stack the paper is wound into large rolls, which are later cut into smaller rolls for shipment.

The Fourdrinier paper machine has a number of disadvantages. Only a limited amount of water can be removed from the wet fiber sheet during the short time when it is moving down the wire, and the machine gives a sheet with different properties on the two sides. In efforts to develop a more uniform sheet and to overcome some of the other weaknesses of this machine, new types of machines called formers have been developed in recent years. In these machines the dilute pulp slurry passes between travelling parallel twin wires, permitting water to be removed from both sides of the fiber web. From the sheet former the web passes through press rolls and into a dryer section. The purpose of these formers is to produce a more uniform sheet at higher speed and at lower cost. The Pulp and Paper Research Institute of Canada made a substantial contribution to this technology by developing the Papriformer, which is now used in several mills.

Figure 13.7 Dry end of high-speed paper machine showing calender stack and winder.

An extremely wide range of products is manufactured on paper machines. These products include newsprint, which is used directly on newsprint presses; linerboard and corrugating medium used in the manufacture of corrugated shipping containers; and many writing and printing papers for which further processing consists mainly of cutting the rolls into sheets, followed by packaging.

Although any Fourdrinier machine can produce a wide range of products, in general paper machines tend to fall into a small number of classifications. These classifications include newsprint machines that produce newsprint only; machines that produce linerboard and corrugating medium for container board; fine paper machines that produce a wide range of writing and printing papers; and tissue machines that produce napkins, towelling, and a variety of tissue papers.

Cylinder Machines

Many paperboard grades are made on cylinder machines. On these machines the Fourdrinier wire section is replaced by several wire-covered cylinders, each partially submerged in a vat containing a dilute suspension of fibers. As each cylinder rotates, it picks up a layer of pulp fibers to form a sheet, which is transferred to a felt pressed against the cylinder by a couch roll. The sheets removed from the cylinders are combined into a single sheet that is passed through press rolls, a dryer section, and a calender stack, similar to those of the Fourdrinier machine. The layers making up the sheet may be from the same fiber stock or may be from different types of pulp.

The advantages of the cylinder machine are that the paperboard may be made to any thickness and that inexpensive materials may be used in the inner plies.

Table 13.3 *The main categories of paper*

Major category of paper and paperboard	Approximate percentage of total paper and paperboard production
Newsprint	20
Communications papers	20
Personal papers	10
Construction papers	10
Industrial papers	5
Packaging	35

Materials

As a rough average, for every 100 kg, or 100 lb, of paper or paperboard manufactured, 80 kg, or 80 lb, of new pulp is used, and the remaining 20 kg, or 20 lb, is made up of recycled fiber, fillers, pigments, additives, and coatings. Some papers, such as newsprint, contain only mechanical and chemical pulp and perhaps extremely minute amounts of dye. At the other extreme, coated boxboards may contain virgin fiber, several types of recycled waste paper fiber, chemicals to impart water resistance, pigments, and clay coatings. Heavily pigmented coated papers may contain up to 30% fillers and coating. Literally hundreds of dyes, fillers, additives, coatings, chemicals, glues, sizes, and so on may be added in the manufacture of the thousands of different kinds of paper products.

Products

Pulp is used mainly for the manufacture of a wide variety of cultural, communications, construction, industrial, and packaging papers. The main categories of paper products and the percentage of each produced are shown in Table 13.3.

Each of the main categories may be subdivided into a number of products; the number of subdivisions varies greatly. Newsprint has only three main subdivisions – rotogravure, letterpress, and offset – depending upon the type of printing press to be used. Communications, construction, and personal papers are each made up of a dozen or so main product groupings. Packaging and industrial papers, by contrast, may be subdivided into hundreds of thousands of products.

Following is a brief description of each of these categories:

Newsprint Canada is one of the world's largest manufacturers of newsprint, producing 8.8 million tonnes (9.71 million tons) annually, or 37% of the total world production.

Communications papers These include writing, printing and book, and magazine papers. Large amounts of Canada's bleached grades of hardwood and softwood pulps are exported for conversion in other paper mills into a variety of communications papers.

Personal papers These include napkins, tissue, and towelling.

Construction papers These include insulation, insulating papers, soundproofing papers, ceiling tiles, and a range of building paperboards.

Packaging papers These include an almost infinite variety of products: boxes, boards, cardboards, and so on. They constitute the largest single use of pulp.

FUTURE TRENDS IN PULP AND PAPER

Raw Materials

Traditionally pulp has been manufactured from the merchantable stem of mature trees. The preferred species have been primarily the softwoods: spruces, pines, true firs, hemlocks, Douglas-fir, and cedar. Although some hardwood species have been used, Canada's hardwood resources, particularly aspen, have been underused. There will be an increasing trend toward more complete use of Canada's

Figure 13.8 Loading newsprint on a freighter for shipment to Australia. (NFB Photothèque. Photo by Chris Lund)

hardwood species for pulp manufacture as the world reserves of unused fiber become scarcer.

Although in the past pulpwood has been made almost exclusively from the merchantable stems of trees, in the future increased use will be made of other tree components – the tops and branches. Similarly, although only mature trees of the preferred pulpwood species were used previously, more use will be made of small-diameter trees of any species for pulp manufacture. In general the shift will be from the selective use of the stems of the preferred wood species to the use of all trees and all parts of the tree.

Processing

Established trends show that all parts of wood procurement will become more mechanized. Moves will be made to increase chipping of small-diameter trees, treetops, and limbs or branches in the woods rather than at the pulp mills.

Rapid changes in the pulping processes are unlikely, but production of refiner mechanical pulp will increase. More kraft pulp will be manufactured, accounting for an increasing percentage of the total pulp produced. Pulp production will benefit from modifications in processing, especially in bleaching, such as gas-phase bleaching, and increased use will be made of an oxygen bleaching stage. The amount of sulfite pulp produced will continue to decline, and the neutral sulfite process will probably account for about the same percentage of the total production as it does now. More use will be made of the semichemical processes, particularly for the hardwood resource that is not used at present.

Pulp and Paper Manufacture

A major trend in paper manufacture will be an increase in the amount of paper made on modern formers instead of on the older, conventional cylinder and Fourdrinier machines. Further improvements can be expected in all aspects of pulp processing and paper conversion.

Pulp and Paper Products

The end products, whether manufactured in Canada or elsewhere, can be expected to change slowly. Newsprint will probably capture a decreased percentage of total production, although the demand will remain high. The increased percentage of total paper and paperboard used in communications and personal papers will offset a decrease in construction papers. Industrial and packaging papers will remain relatively unchanged.

As has happened in the past, some paper products will rise in popularity while others will disappear altogether. Paper products expected to increase in use are copying papers, computer papers, and disposable paper products. Finally, new paper products will be developed that are now unknown.

In broad context, the pulp and paper industry in Canada and the world can be expected to remain an important and integral part of the economy for some time to come.

BIBLIOGRAPHY

Casey, J.P., ed. 1960. *Pulp and paper – Chemistry and chemical technology*. 2nd ed. 3 vols. New York: Interscience

FAO. 1978. *1978 yearbook of forest products*. Rome

Libby, C.E., ed. 1962. *Pulp and paper science and technology*. 2 vols. New York: McGraw-Hill

Macdonald, R.G., and Franklin, J.N. eds. 1969. *Pulp and paper manufacture*. 2nd ed. 3 vols. New York: McGraw-Hill

Rydholm, S.A. 1965. *Pulping processes*. New York: Interscience

Statistics Canada. 1978. Cat. 36-204 Annual. *Pulp and paper mills. 1978*. Ottawa

14

Residues

JAMES DOBIE
Council of Forest Industries of British Columbia, Vancouver

Residues are generated in all stages of forest use from the harvesting of trees to the final manufacturing processes. Traditionally, in forest harvesting in Canada, those portions of the tree not considered merchantable have been left in the woods to decompose. Normally only the stem of the tree is considered worthy of harvest, and the proportion of the stem harvested depends on the standards applied.

Residues like bark, sawdust, pulp chips or solid residue, shavings, plywood trims, and sander dust, which are produced in manufacturing lumber, plywood, and shingles and shakes, can provide raw material for other products. These products include pulp and paper, panel products, furniture, and miscellaneous wood products, which in turn create their own residues.

The economical use of residues is beneficial to the economy because more value is obtained from each unit of raw material. As the cost of harvesting roundwood increases, more demands are being made on wood residues to help meet the increasing output of such products as pulp and paper and particleboard. Furthermore, rising energy costs have created considerable interest in the potential of wood residues for power generation and heating. These demands have resulted in higher dollar returns for residues, which now make an important contribution to the product volume of many wood enterprises.

FOREST RESIDUES

Merchantable Components

A tree may be divided into merchantable and unmerchantable components. Merchantable means that which is of commercial quality. Today in Canada the merchantable component is normally regarded as the stem of the tree (Figure 14.2) from stump height to a top diameter that varies according to the degree of utilization practiced. The degree of

Figure 14.1 Residues provide raw material for other products and industries.

utilization in turn depends on the economics involved at the time of cutting or on the government regulations in force where timber on government land is concerned.

In most of Canada trees are logged to a close utilization standard, which implies recovery of the stem from a 30.4 cm (1 ft) stump height or lower to a 10 cm (4 in.) diameter top or less. In areas such as the west coast, where the topography is rough and breakage of trees in felling is considerable, utilization standards are of necessity less rigorous, and more waste is generated.

Unmerchantable Components

Portions of the tree normally considered unmerchantable include the top of the stem, the crown and branches, and the stump-root system. The proportion of the total tree represented by these residues depends on many factors including tree size, tree form, branch and crown characteristics, and type of root system. Estimates (Keays 1971) for Canadian pulpwood species are 50% of the oven-dry and bark-free stem for stump-root systems, 5–15% for branches, and 6–15% for unmerchantable tops of small trees utilized to top diameters of 10–15 cm (4–6 in.).

Studies (Young 1968) of eight commercial species in Maine show that the merchantable stem for all species contains about 65% of the total fiber of the tree; 25% is contained in the stump and roots, and the remaining 10% is found in the upper stem and branches.

The foregoing estimates of residues show that 31 to 54% of the merchantable stem is residue. Although crude, the estimates give a rough indication of the proportion of the tree normally considered unmerchantable but which could be used as raw material for other products.

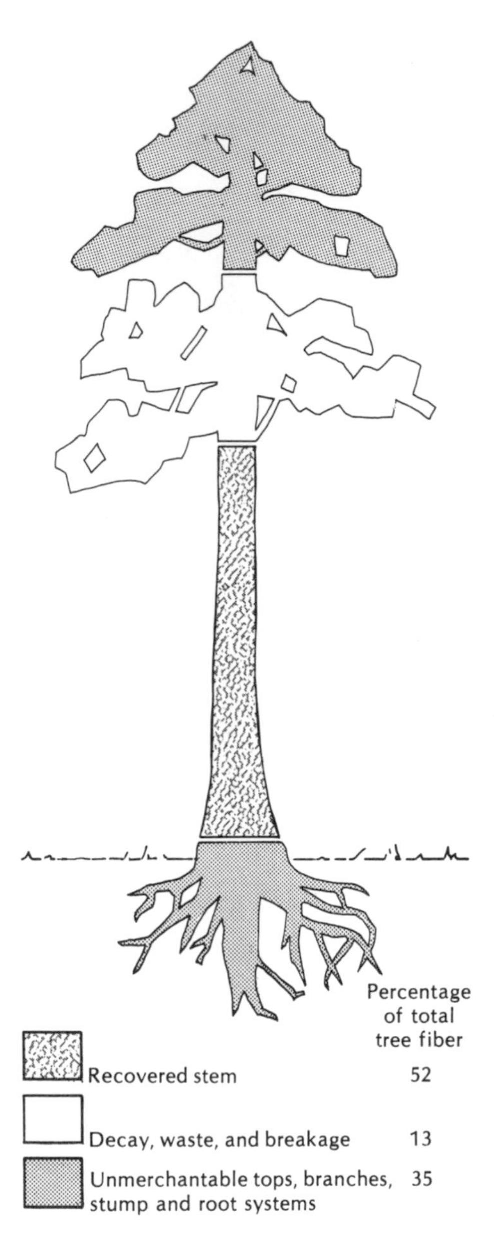

Figure 14.2 Disposition of complete tree biomass in recovered stem and residues.

Logging Residue

In addition to the unmerchantable portions of the tree, considerable volumes of wood fiber are left behind after logging because of waste and breakage in harvesting and the nonrecovery of material with a high proportion of decay or of undesirable species.

In a study conducted in the western United States, logging residues from sound, live trees were estimated at 14% of the reported 1969 log harvest (Howard 1971). A further volume equivalent to 15% of the log harvest was left in the form of dead trees and cull trees, which are uneconomical to use because of defects.

These percentages probably apply as well to coastal British Columbia, where topographical and stand conditions are similar, but they are probably excessive for Canada as a whole, where a total of 20% of the log harvest is probably more accurate. With increasing pressure from governments and environmentalists to increase forest utilization, it is likely that forest residues will be reduced considerably in the future.

Because of the variability anticipated in the volume of residues and the crudity of available estimates, readers are cautioned that local estimates of residues should be made where important strategies rest on the findings.

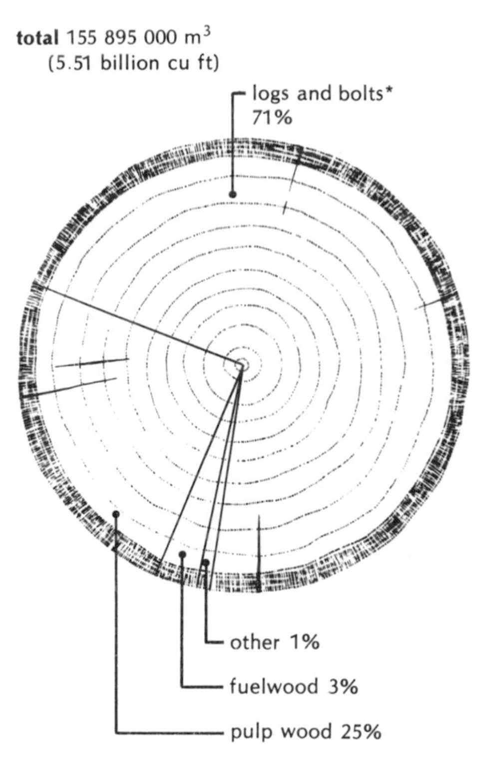

Figure 14.3 Designation of roundwood harvested in Canada, 1978. (Source: Statistics Canada Cat. 25-202 Annual 1978a)

Table 14.1 *Canadian primary forest production by provinces, 1978*

Province	Percentage of production
Newfoundland	1.47
Prince Edward Island	0.10
Nova Scotia	2.67
New Brunswick	5.46
Quebec	21.86
Ontario	12.95
Manitoba	1.16
Saskatchewan	1.84
Alberta	4.18
Yukon and Northwest Territories	0.10
British Columbia	48.21

Total production: 155.895 million m^3 (5.51 billion cu ft)

Source: Statistics Canada, Cat. 25-202 Annual, *Canadian forestry statistics*, 1978

MANUFACTURING RESIDUES

The estimated volume of roundwood produced in Canada in 1978 (Statistics Canada 1978a) was 155.895 million m^3 (5.51 billion cu ft), 71% of which was logs and bolts and 25% pulpwood (Figure 14.3). Distribution of production by province is shown for the year 1978 in Table 14.1. About 29% of the volume of logs, bolts, and pulpwood was actually used by the pulp industry. The remaining 71% was used in the manufacture of lumber, shingles and shakes, veneer and plywood, board products, and other miscellaneous wood products. In 1978 the manufacture of lumber consumed roughly 56% of the volume of roundwood used for the foregoing items (Figure 14.4).

Lumber

The major residues produced in sawmills are bark, sawdust, pulp chips or solid-wood residue, and planer shavings.

Bark

In most instances bark is removed from the log before sawing. Debarking the log first prolongs saw-tooth life by providing a clean surface for sawing and allows bark-free pulp chips to be produced from the sawn slabs. In areas where there is no market for pulp chips, logs may be sawn without first removing the bark. The bark and slabs are then burned, but the amount of residues disposed of in this way is insignificant.

The volume of bark produced at sawmills where debarking takes place depends on the bark characteristics of the species being processed. These characteristics vary within and between species (Hale 1955; Smith and Kozak 1967). Bark volume as a proportion of wood volume is estimated at about 30% for Douglas-fir and 11% for lodgepole pine (Dobie and Wright 1975). Not all the bark on logs removed from the forest arrives at the processing facilities. Losses are experienced in log handling and in transit, and these losses vary with the method of harvesting and transportation. Studies of water-transported logs in British Columbia show that bark coverage for individual logs varies from 0 to 100% and averages 30–86% for rafts of logs (Dobie and Wright 1975).

Sawdust

The volume of sawdust produced in sawmills depends on the lumber product, sawing methods, and saw types. Conventional large-log mills convert a greater percentage of the log volume to sawdust than modern mills of the small-log chipper-headrig type do (Dobie and Wright 1975). The sawdust output of large-log mills is roughly twice that of chipper headrigs, or approximately 6–12% of wood volume, because more saw cuts are made. In addition, the percentage of log volumes converted to sawdust in a given mill normally increases as the dimensions of the product decrease. For example, a mill producing 17 mm (1 in. nominal) lumber creates more sawdust than the same mill producing 38 mm (2 in. nominal) lumber.

Chippable Residue

The majority of medium- and large-size sawmills in Canada manufacture pulp chips from solid-wood residues because of the financial benefits of doing so where a chip market exists. The volumes of pulp chips produced depend largely on lumber recovery practices. If an effort is made to recover the maximum amount of lumber from a log, then the proportion of the log chipped will be low. Chipper headrigs normally convert 40–50% of the log volume to pulp chips, while band-saw operations convert only 30–35% to pulp chips.

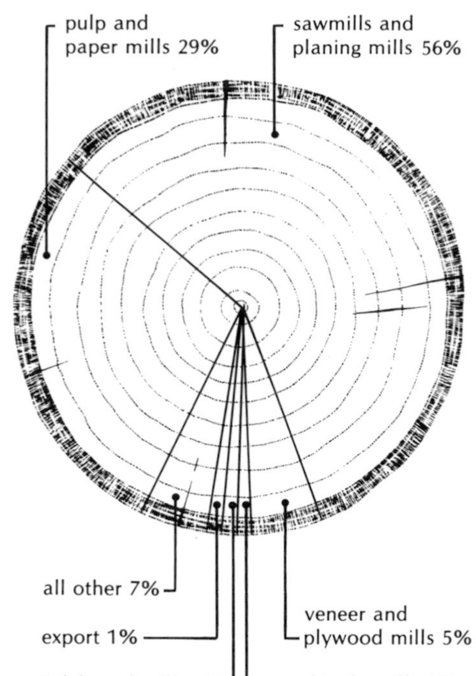

Figure 14.4 Use of logs, bolts, and pulpwood in Canada, 1978. (Source: Statistics Canada Cat. 25-201 Annual 1978b, 35-204 Annual 1978c, 35-206 Annual 1978d, 36-204 Annual 1978e, and 65-202 Annual 1978f)

Planer shavings

Surfacing, planing, or dressing lumber produces residue in the form of wood shavings. Different volumes of planer shavings are produced, depending on the relation between the size of rough, dried lumber and the final dressed size. The volume depends on sawing accuracy and on the surface smoothness created by the saws or chipping heads. Poor sawing control and rough lumber surfaces require generous rough-lumber dimensions in order to meet dressed standards for marketing. Lumber cut to excessive thickness in the sawmill produces a high proportion of planer shavings in the planer mill. As a result, some mills convert as much as 30–35% of rough-lumber volumes to shavings; others convert as little as 12%.

Total Sawmill Residues

Fluctuations in residue yields may be expected throughout the industry. Again, local studies of yields should be made if accurate estimates are important.

An indication of average product and residue yields from band mills cutting large logs and from chipping headrigs cutting small logs is provided in Figure 14.6.

Veneer and Plywood

The softwood veneer and plywood industry in Canada is concentrated in British Columbia; a minor proportion of the output comes from Alberta. Hardwood veneer and plywood are produced almost exclusively in Quebec and Ontario.

Residues produced in the manufacture of veneer and plywood may be classified as green-end residues and dry-end residues. Green-end residues include bark, log trim, sawdust, core, green veneer clippings, and roundup and spur trim derived from shaping and cutting veneer bolts so that they can be peeled to produce veneer. Dry-end residues include veneer breakage, jointer trims, reject veneer, reclipping losses, loss at the glue spreader, panel trim, sawdust, and sander dust.

Bark

More than 50% of the softwood veneer logs are Douglas-fir; the remainder consist largely of spruce and pine and a smaller number of hemlock and true firs. Major species for hardwood veneer and plywood manufacture are birch and aspen. The percentage of bark residue in veneer and plywood manufacture is probably slightly higher than in lumber manufacture because a greater proportion of Douglas-

Figure 14.5 Debarking large veneer logs.

Figure 14.6 Product and residues as a percentage of the total volume of wood and bark for two sawmill types.

fir logs is used and Douglas-fir has a higher percentage of bark than most other species (Smith and Kozak 1967).

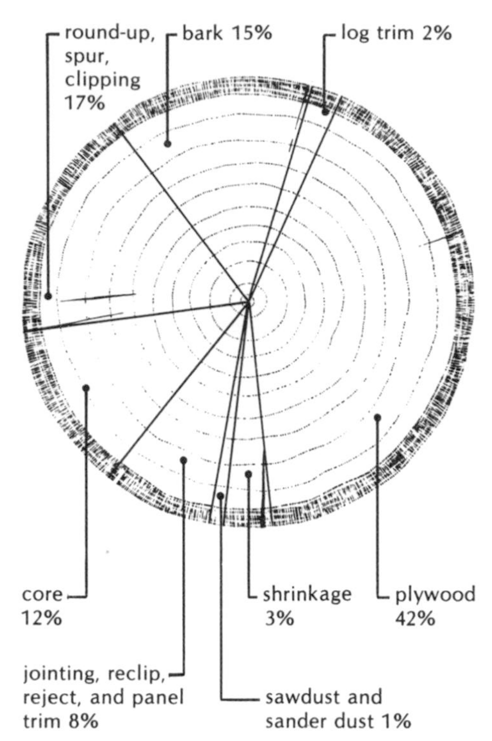

Figure 14.7 Product and residues as a percentage of the total volume of wood and bark for plywood manufacture.

Log Trim, Roundup, Spur Trim, and Green Clippings
The volume of green-end chippable residue available depends on the warpage or curvature in long log lengths sent to the plant, the deviation of the log from the round, log taper, and quality. Thus, the volume of these residues may be extremely variable.

Log trims occur when improper lengths of long logs are sent to the plant for manufacture into 1.4 or 2.6 m (4.5 or 8.5 ft) veneer bolts. For example, a 11 m (36 ft) log produces four 2.6 m (8.5 ft) veneer bolts and leaves a 60 cm (2 ft) chunk, or 5.5% of the log input, as residue.

Roundup and spur-trim volumes depend on the shape of the log and the accuracy of veneer-bolt manufacture. Out-of-round and highly tapered bolts require more roundup than cylindrical bolts do. Similarly, volumes removed by spur-trim knives depend on how accurately the log is sawn into the veneer bolts.

Green-veneer clipping volumes are greater for logs with large knots and other defects than for high-quality logs.

Cores
Many sawmills now peel veneer bolts down to a core diameter of around 13 cm (5 in.). Therefore, core as a percentage of bolt volume varies with initial bolt size. Because of differences in value, there is good reason to manufacture lumber rather than pulp chips from cores, and many plywood plants now have integrated sawing units for this purpose.

Dry Veneer Residues
Jointing is necessary to prepare random-width veneer strips for edge-gluing into half- or full-width sheets of veneer. The cutting head of the jointer removes approximately 1.6–9.5 mm ($\frac{1}{16}$–$\frac{3}{8}$ in.) of veneer from the pieces to be edge-glued. Other losses at the jointer include veneer breakage and mismanufacture. After jointing there are reclipping losses, followed by losses at the glue-spreader-panel layup operation. The latter include veneer discards and core-layer overhangs in excess of the untrimmed panel size. Finally, the panels are trimmed to size, resulting in sawdust and panel-trim residues; if the panels are sanded, sander-dust residue also develops.

Total Veneer and Plywood Residues
Estimates of product yields and residues in veneer and plywood manufacture are given in Figure 14.7, based on studies in the United States and Canada (Hunt and Woodfin 1970; Dobie and Hancock 1972; Woodfin 1973).

Shingles and Shakes
In Canada this industry is concentrated in British Columbia, which accounts for around 91% of total Canadian shipments. Shingles make up about 52% of this output; shakes, 48%. Western red cedar is used for most of the production of both shingles and shakes. Residues from

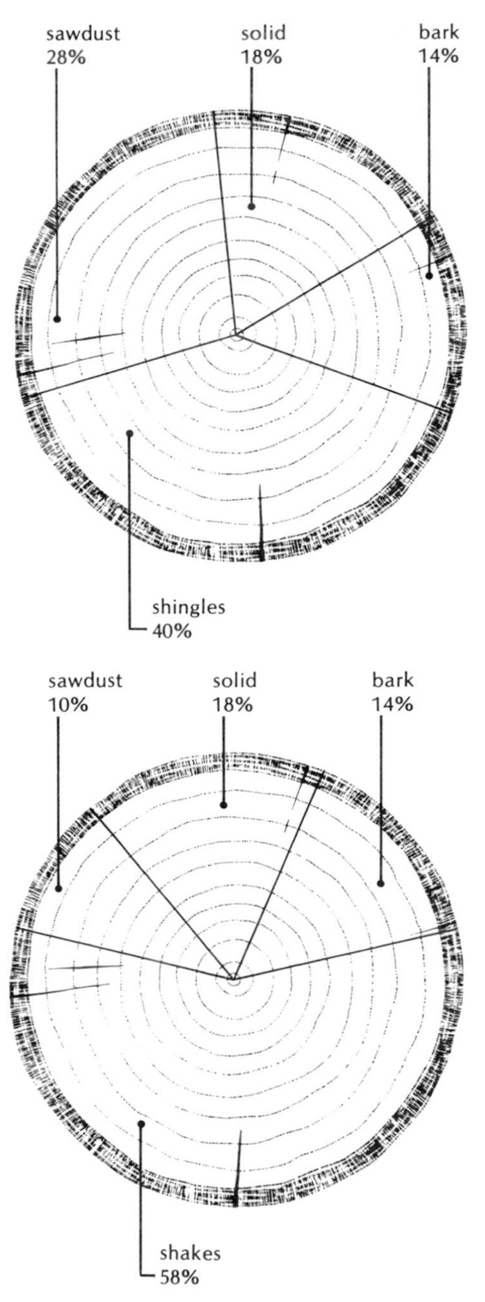

Figure 14.8 Product and residue yields from shingle and shake manufacture.

their manufacture include bark, sawdust, and solid or chippable residue.

Bark

The bark content of western red cedar is estimated at 14% of the total volume of wood and bark (Dobie and Wright 1975). However, about 50% of cedar bark is lost in transit, so the amount available at the plant would be around 8% of the total volume.

Sawdust

The manufacturing process for shingles differs from that for shakes in that shingles are sawn and shakes are split, and only a portion of the shakes are resawn for certain applications. Thus, there is proportionately more sawdust created in shingle manufacture than in shake manufacture.

Chippable Residues

These residues are created at the log cut-off saw and at the shingle saws. Portions of shingle logs not suitable for shingle manufacture are removed at the cutoff saw, where the blocks are manufactured to length. When the log is split into wedge-shaped sections and subsequently processed for maximum recovery of the shingles or shakes, a pie-shaped portion known as a spalt remains. Trims from the edges of shingles and shakes are known as splints.

Total Shingle and Shake Residues

Little information exists on residue yields from shingle and shake manufacture. On the basis of one study (McBride 1959), the residue yield would be about 60% for shingles and 42% for shakes (Figure 14.8).

Pulp and Paper

In pulp and paper manufacture, where the fiber input is obtained from roundwood in the wood room of the pulp mill, bark is the only solid residue available in appreciable quantities.

Some modern pulp mills do not have wood rooms and obtain their supply of raw fiber from other manufacturing residues. In 1978, of a total of 81 million m^3 (2.87 billion cu ft) of wood used by the pulp and paper industry in Canada, 54% was from round pulpwood and the remainder from purchased wood residue, of which 91% was pulp chips.

Wood-pulp production in Canada in 1978 amounted to 20.153 million tonnes (22.214 million tons) (Statistics Canada 1978d), of which 40% was groundwood or mechanical pulp, 44% was sulfate, 12% was sulfite, and the remaining 4% was other minor pulp products. Thermomechanical pulp produced in Canada is likely to occupy a role of increasing importance in Canadian pulp production because of its high yields of over 90%. Groundwood or mechanical-pulp yield is about 95% of the wood input; sulfate and sulfite, about 45%.

In the sulfate process and some sulfite processes, the black liquor is recovered and used as fuel within the system. In the remaining sulfite processes, which are gradually being phased out because of pollution-control regulations, the waste-liquor residue is discharged into waterways because the chemicals used make it unsuitable for recovery. An insignificant amount of fiber, as far as calculable residues are concerned, is lost in the pulping process and can be safely ignored for the purpose of this chapter.

Product and residue yields are illustrated in Figure 14.9.

Secondary Wood Industries

Secondary wood-manufacturing industries using roundwood, lumber, veneer, and plywood include those industries that manufacture products such as boxes and pallets; household furniture; sashes, doors, and other millwork; coffins and caskets; and parts of boats. Estimated wood input to these industries is about 7.7 million m^3 (275 million cu ft) per year. Residues include bark, sawdust, shavings, sander dust, and miscellaneous solid pieces, but there are no published data on the proportion of residue to wood input for many of these secondary industries.

Studies of furniture-component yield (Neilson, Bousquet, and Pnevmaticos 1970; Applefield 1971) show that only about 40–50% of the total volume of lumber purchased is actually used in furniture manufacture. About 80% of the residues are solid wood, 10% shavings, and 10% sawdust and sander dust.

If furniture-component yields are indicative of utilization in secondary wood industries in Canada, then 50–60% of the total wood input in these industries becomes residue. In 1975, based on the above studies, this residue amounted to 2.5–3.1 million m^3 (90–108 million cu ft) of solid-wood equivalent. If bark for the proportion of roundwood used is included, these estimates increase by about 8%.

SUMMARY OF RESIDUE VOLUMES

Estimates of product and residue yield in Canada's forest utilization for a typical year (1975) are illustrated in Figure 14.10. The proportions of each product or residue are given on the directional arrows from each circle. For example, 71% of the logs and bolts were used as sawlogs, 16% as pulpwood, 7% as veneer logs, 1% as shingle logs, 1% for particleboard products, 1% for miscellaneous uses, and 3% for export. Likewise 37% of sawlog input was recovered as lumber, 31% as chippable residue, 14% as bark, 9% as shavings, 8% as sawdust and 1% as shrinkage.

USES OF RESIDUES

The uses made of residues from forest-products manufacturing depend to a great extent on the concentration and integration of different facets of the industry. Where there is a concentration of diversified manufacturing complexes, markets can be developed for residues as fuel for power generation or fiber for pulp mill and fiberboard input as well as many other miscellaneous uses. Population centers are usually good markets for wood residues for fuel, soil conditioning, and horticultural

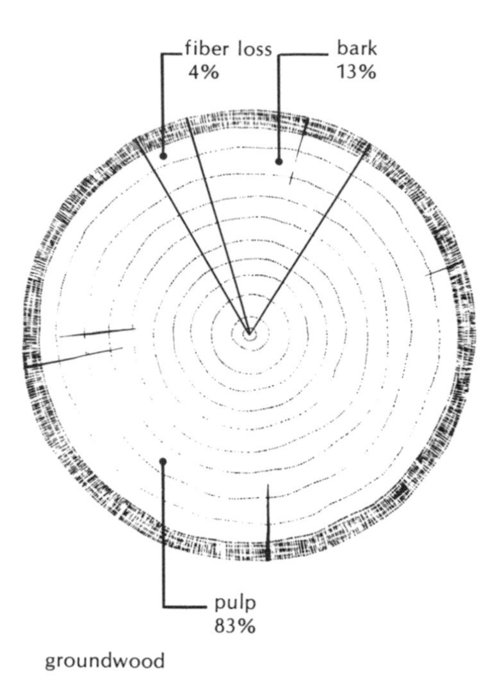

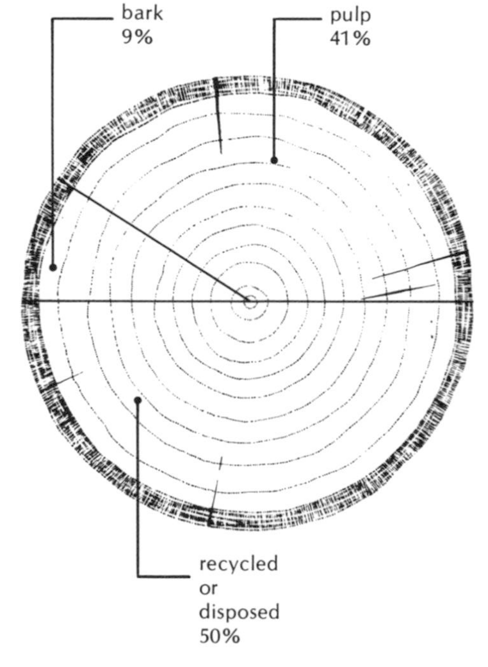

Figure 14.9 Product and residue yields from pulp manufacture.

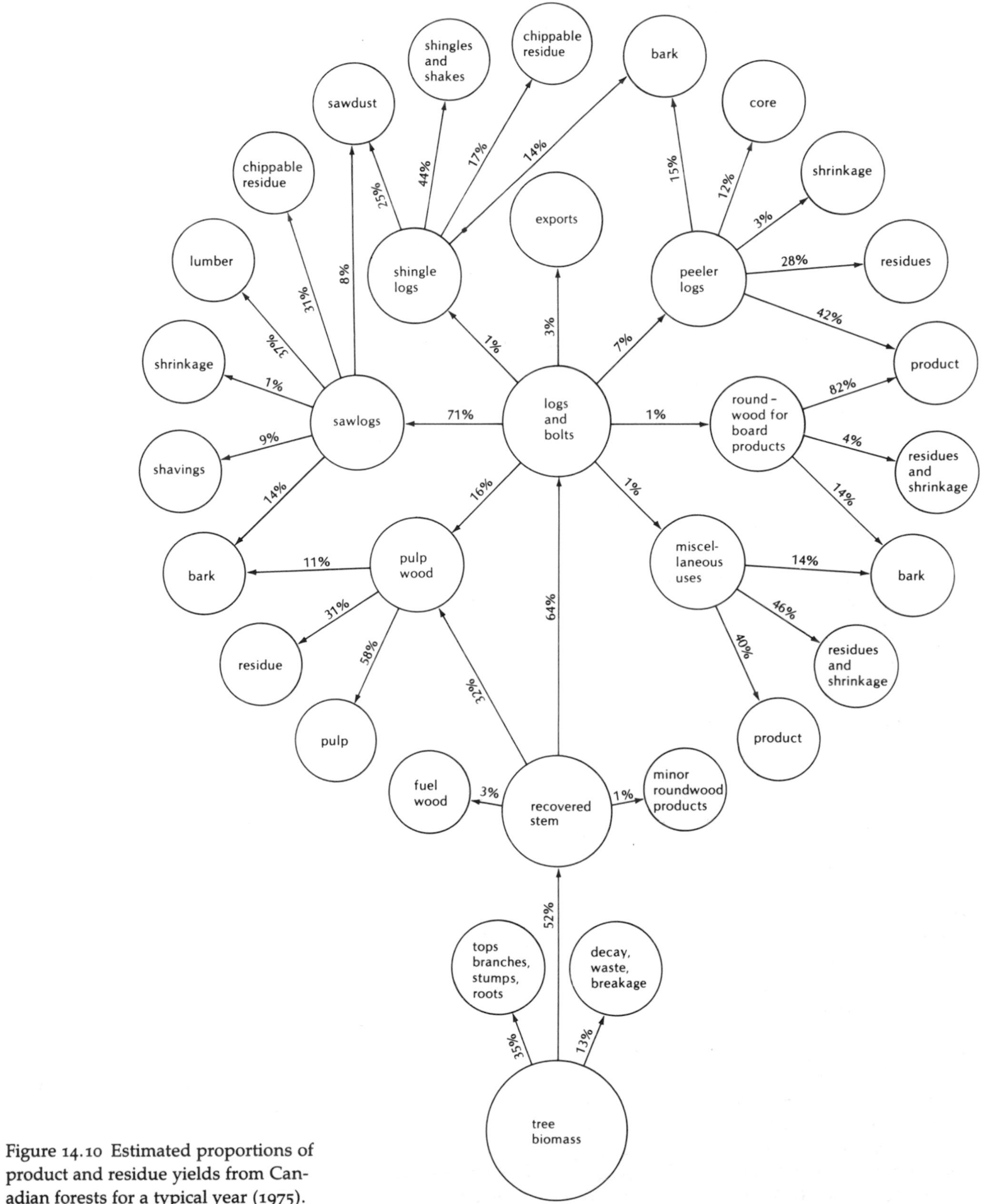

Figure 14.10 Estimated proportions of product and residue yields from Canadian forests for a typical year (1975).

purposes. Concentrations of agricultural and animal-breeding enterprises offer markets for wood residues as mulch and animal bedding.

Where these markets do not exist, however, residue disposal creates a problem, particularly because of environmental considerations. Where residue burning is permitted, teepee burners are still common in the sawmill and plywood industry in Canada, even though this method of disposal represents a cost to the industry, and the energy potential could better be used for power generation or heating (Corder 1974; Hammond et al. 1974; Keeley 1974).

Where waste burning is not permitted, landfill is a popular method of disposal of excessive residues (Evans 1973). This practice also introduces a cost to the industry, however, and it creates an environmental nuisance because chemicals are leached from the wood into waterways. Further, the discharge of pulp-mill wastes into waterways is now being restricted by most government pollution-control agencies. Adherence to the strict environmental standards requires process modifications and extra costs to eliminate such discharges.

Thus, the use of residues is advantageous because it eliminates disposal costs and often results in a net positive return, even though it may be small. Many uses have been found for forest-product residues, and research is continuing to develop new products from them. The literature on the use of residues is voluminous, and space does not permit extensive treatment of the subject in this book; emphasis is given to only the major uses of each residue. For readers interested in pursuing aspects of wood-residue use in depth, a useful starting point would be the Institute of Paper Chemistry's bibliographic series, which includes over 1400 abstracts.

Bark

Probably the largest use of bark is as fuel to generate power and steam. For this purpose it is usually mixed with other mill residues such as sawdust, planer shavings, and wood chunks, which are pulverized into a mixture commonly referred to as hogged-fuel. If hogged-fuel is in excess of requirements for power or steam generation in a mill, it may be marketed, burned, or disposed of as land fill.

The market for bark is expanding in horticulture and agriculture for mulching, composting, soil amendment, animal bedding, and decorative purposes. Unprocessed bark used in this form is worth the cost of disposal up to a few dollars per tonne (ton) net at the mill. Processed bark for use as soil conditioners and mulches can yield profits of 25–30% on invested capital (Sarles 1973).

Bark may also be used for other purposes: absorbing agents for oil spills; additives to well-drilling muds; charcoal; chemicals; extractives and medicines; compressed fireplace logs; extenders and fillers for plastics; groundcover for picnic sites, playgrounds, and bridle paths; loose-fill insulation material; particleboard, hardboard, and insulation board; a plant growing medium; and tanning materials (Aaron 1966; Van Vliet 1971; Maloney 1973).

Sawdust and Shavings

Like bark, most of these residues are used as fuel for power or process-steam generation. Heat values for various Canadian species are estimated at 18.6–23.2 MJ/kg (8000–10 000 BTU/lb), dry basis (Dobie and Wright 1975).

Quantities of sawdust and shavings are also burned in the traditional teepee burners of the wood-processing industry. Where burning is not allowed, alternative disposal methods are burying or dumping.

There is a growing use of sawdust in pulp manufacture. Some companies use a specially designed digester for sawdust, and mix sawdust pulp with pulp from regular furnish in the proportion of 10–20%. Another growing use for sawdust and shavings is in board products such as hardboards and particleboards. These products are replacing plywood in some of its traditional uses such as for furniture core stock, floor underlayment, door cores, wall panelling, counter tops, kitchen cabinets, and shelving, as well as in other uses (see Chapter 10).

Residues from some hardwood species may be used as a diet supplement for animals (Kitts et al. 1969; Heaney and Bender 1970; Satter, Baker, and Millett 1970; Shelford, Kitts, and Krishnamurti 1970; Steinberg 1973). Sawdust and shavings are also used in animal bedding, poultry litter, domestic fuel, fuel logs, briquettes, mulch and soil conditioner, packing, thermal insulation, meat smoking, sweeping compounds, porous brick and tile, and molded articles.

Pulp Chips and Chippable Residue

The manufacture of pulp chips from wood residues that previously would have been discarded has benefited both the wood manufacturing and the pulp industries. Besides providing additional revenue for the lumber and plywood industry, pulp chips provide another source of fiber for pulp mills.

Figure 14.11 Pulp chips from wood residues form a substantial part of the pulp-industry fiber input. (Photo: British Columbia Forest Service)

Most of the pulp chips produced by the sawmill and plywood industries in Canada are consumed domestically; the remainder is sold to the United States. In 1978 domestic pulp mills received about 38 million m^3 (132 million cu ft) of residual chips and other chippable wood residues (Statistics Canada 1978e). Total pulpwood and wood-residue consumption in Canada was 81 million m^3 (2.88 billion cu ft), so pulp chips and residues from other industries represented about 46% of pulp-industry wood-fiber input in 1978.

Veneer Log Cores

Cores remaining after veneer peeling are used for the manufacture of lumber or pulp chips. Because lumber has a higher value than pulp chips, cores should be large enough to yield merchantable lumber. When market prices favor lumber over veneer, larger cores should be left in order to increase lumber production at the expense of the production of veneer (Dobie and Neilson 1973).

Veneer and Plywood Trims

Much of the volume of veneer trims is suitable for pulp-chip manufacture or for use in particleboard and fiberboard products.

Plywood panel trims are unsuitable for pulping because they contain cured glue, but they are still usable as particleboard and hardboard furnish. In many cases these trims are used as fuel for power generation or are burned as waste.

Sander Dust

Only about 15% of Canadian plywood is sanded, and most of the resulting sander dust is used as fuel for power generation or is burned as waste. In some instances it is mixed with other wood residues in fuel logs.

In areas of eastern Canada, sander dust from several plants is sold to asphalt-roofing manufacturers for use as filler. It can also be ground to a wood flour and used as an extender in plastics, molded shapes, and other applications.

Pulping Residues

In chemical pulping somewhat more than half the original wood is contained as dissolved components in the spent liquor. In the kraft process the spent liquor may be collected and concentrated to permit the solids to be burned as fuel and the chemicals to be recovered for further use. With the sulfite process recovery of the calcium-base spent liquor is difficult and expensive. A viable recovery system comparable in economy and practicality to the recovery system of spent kraft liquor does not exist, and a common disposal method has been to discharge the spent sulfite liquor into waterways. Because this method is now prohibited under environmental regulations, considerable effort has been expended to find uses for the residue (Forss 1968).

Two of the most successful large-scale uses for spent sulfite liquor have been as a road binder and as a binder for animal-feed pellets. Minor applications for the liquor are in linoleum pastes, foundry core binders, soil stabilization, emulsion, Portland cement, and ceramic mixes (Pearl 1969). Because the lignosulfonates obtainable from spent sulfite liquor have good dispersant properties, they are used as additives in oil-well-drilling muds and as dispersing agents for carbon black in rubber applications. They are also used in vat dyestuffs in the textile industry, in pesticide and agricultural sprays, in ore flotation, and in special applications in the cement and ceramic industry.

Ethyl alcohol is being produced commercially at two pulp mills, in Ontario and Quebec, from the dissolved hexose sugars in pulp liquors. The sugar contents have also been used commercially in the manufacture of torula yeast, a protein concentrate and good source of vitamin B. Other wood chemicals and derivatives obtainable as byproducts of the pulp industry are described in Chapter 5.

TRENDS IN THE USE OF RESIDUES

About 50% of the biomass in a stand of timber is extracted from the woods for use. At present the remainder is not used, because it is uneconomical to do so. If the shortages of wood fiber that are forecast do occur, increased use will be made of some of the fiber components now left in the woods.

Of the volume of timber consumed by wood-using industries such as lumber, plywood, and pulp manufacturing, probably less than 50%, on the average, is finally converted into products. The remainder is residue of one form or another.

In areas where there is a concentration of diversified forest-products industries, the use of residues is fairly complete because of demands for fiber for fuel and for pulp and particleboard products. In other areas of Canada, however, an estimated 50% of the residues available is still being dumped or burned as waste.

There are indications that the use of wood residues will increase in several different ways. Because of the escalating costs of natural gas and oil, the most important use of wood residues will be as fuel for power generation and process steam. It is doubtful that these cost increases for natural gas and oil will ever be reversed, so major expenditures may now be considered for the right kind of equipment to make full use of the energy potential of wood residues.

The long-range outlook is that residue-using activities will expand proportionately more than residue-supplying activities. Increased demand creates increased prices and provides an incentive to look for new sources of residues and to use material that was not used before.

If the economics of whole-tree chipping of logging residues is satisfactory, less residue will be left in the woods. If the demand for all forest products increases at the rate generally forecast, increased prices will make the use of forest residues economically feasible for the forest-product industries in Canada. The result should be better use of forest residues.

REFERENCES

Aaron, J.R. 1966. *The utilization of bark.* For. Comm. Res. Dev. Pap. No. 32. London

Applefield, M. 1971. Wood and the North Carolina furniture industry. *Woodwkg. Furniture Dig.* 73 (12): 28–30

Corder, S.E. 1974. Wood-bark residue is source of plant energy. *For. Ind.* 10 (2): 72–3

Dobie, J., and Hancock, W.V. 1972. Veneer yields from B.C. interior Douglas-fir and white spruce. *Can. For. Ind.* 92 (7): 32–3

Dobie, J., and Nielson, R.W. 1973. These equations tell when to cut larger veneer cores for more lumber. *Can. For. Ind.* 93 (2): 50–1

Dobie, J., and Wright, D.M. 1975. *Conversion factors for the forest products industry in western Canada.* West. For. Prod. Lab. Inf. Rep. VP-X-97. Vancouver

Evans, R.S. 1973. *Hogged wood and bark in British Columbia landfills.* West. For. Prod. Lab. Inf. Rep. VP-X-118. Vancouver

Forss, K. 1968. *Spent sulphite liquor – An industrial raw material.* Finnish Pulp Pap. Res. Inst. Rep. No. 417. Helsinki

Hale, J.D. 1955. Thickness and density of bark; trends and variation for 6 pulpwood species. *Pulp Pap. Mag. Can.* 56 (12): 113–17

Hammond, V.L.; Mudge, L.K.; Allen, C.H.; and Schiefelbeing, G.F. 1974. Energy from forest residuals by gasification of wood wastes. *Pulp Pap.* 48 (2): 54–7

Heaney, D.P., and Bender, F. 1970. The feeding value of steamed aspen for sheep. *For. Prod. J.* 20 (9): 98–102

Howard, J.O. 1971. Forest products residues – Their volume, use and value. *For. Ind.* 14 (20): 22–7

Hunt, D. L., and Woodfin, R. O., Jr. 1970. *Estimate of dry veneer volume losses in Douglas-fir plywood manufacture.* U.S. Dep. of Agric., Pac. NW For. Range Expt. St. Note PNW-134. Portland, Oreg.

Keays, J. L. 1971. *Complete tree utilization; an analysis of the literature.* West. For. Prod. Lab. Inf. Rep. VP-X-69, 70, 71, 77, 79. Vancouver

Keeley, M. 1974. Energy in perpetual motion – Or how to burn your hog and make money. *B.C. Lumberman.* 58 (4): 34–6

Kitts, W.D.; Krishnamurti, C.R.; Shelford, J.A.; and Huffman, J.G. 1969. Use of wood and woody by-products as a source of energy in beef cattle rations. *Ad. Chem. Ser.* 95 (17): 279–97

Maloney, T.M. 1973. Bark boards from four west coast softwood species. *For. Prod. J.* 23 (8): 30–8

McBride, C.F. 1959. Utilizing residues from western red cedar mills. *For. Prod. J.* 9 (9): 313–16

Neilson, R.W.; Bosquet, D.W.; and Pnevmaticos, S.M. 1970. Sawing pattern effect on the yield of furniture components from high quality hard maple logs. *For. Prod. J.* 20 (9): 92–8

Pearl, I.W. 1969. Utilization of by-products of the pulp and paper industry. *Tappi* 52 (7): 1253–60

Sarles, R.L. 1973. Using and marketing bark residues. *For. Prod. J.* 23 (8): 10–14

Satter, L.D.; Baker, A.J.; and Millett, M.A. 1970. Aspen sawdust as a partial roughage substitute in a high concentrate dairy ration. *J. Dairy Sci.* 53 (10): 1455–60

Shelford, J.A.; Kitts, W.D.; and Krishnamurti, C.R. 1970. Utilization of alder sawdust (*alnus rubra*) by ruminants. *Can. J. Anim. Sci.* 50: 208–9

Smith, J.H.G., and Kozak, A. 1967. *Thickness and percentage of bark of commercial trees of British Columbia.* Univ. BC Fac. For., Vancouver

Statistics Canada. 1978a. Cat. 25-202 Annual. *Canadian forestry statistics.* Ottawa

– 1978b. Cat. 25-201 Annual. *Logging.* Ottawa

– 1978c. Cat. 35-204. Annual. *Sawmills and planing mills and shingle mills.* Ottawa

– 1978d. Cat. 35-206 Annual. *Veneer and plywood mills.* Ottawa

– 1978e. Cat. 36-204 Annual. *Pulp and paper mills.* Ottawa

– 1978f. Cat. 65-202 Annual. *Exports.* Ottawa

Steinberg, J.M. 1973. Let them eat wood. *B.C. Lumberman* 57 (10): 76–8

Van Vliet, A.C., ed. 1971. *Converting bark into opportunities.* Conf. Proc. Nov. 1971, Dep. For. Corvallis: Oreg. State Univ.

Woodfin, R.O., Jr. 1973. Wood losses in plywood production – Four species. *For. Prod. J.* 23 (9): 98–106

Young, H.E. 1968. Quantum increases in fiber production. *Proc. For. Eng. Cong.* pp. 102–5. St. Joseph, Mich.: Am. Soc. Agr. Eng.

15

F.A. TAYELOR
Western Forest Products Laboratory, Vancouver
V.N.P. MATHUR
Canadian Forestry Service, Ottawa

Codes and Standards

THE SIGNIFICANCE OF CODES AND STANDARDS

Standards are documents that set forth for the manufacturer, marketer, and consumer the required quality, physical characteristics, and strength properties of a material or product. In Canada standards are produced under the auspices of a number of standards-writing organizations. Voluntary committees that have a balance of interest between producers on the one hand and consumers and general-interest groups on the other hand do the actual preparation of the standards.

Codes are legal documents – frequently bylaws – that specify how products manufactured to certain standards are to be used to provide a specified end result. For example, building codes are promulgated to specify how forest products, which are produced to a number of standards, are to be used to provide housing and other buildings that will have a specified quality and level of performance.

Codes and standards provide the basis for the orderly flow of products of specified quality from the producer to the user. Without voluntary acceptance of codes and standards, our affluent standard of living would be impossible. Voluntary standards enter our society in such forms as lumber grades, dimensions of nuts and bolts, sizes of food packages, tire grades, and the designs for electrical-appliance outlets that permit one to plug in a television set or teakettle anywhere in the country.

An individual organization or company uses internal standards to ensure that its products are unique, to maintain its competitive position, and to satisfy the requirements of its quality-control program. The same organization or company also uses external standards to ensure that its products are interchangeable with similar products made by other companies and that they meet the requirements of codes

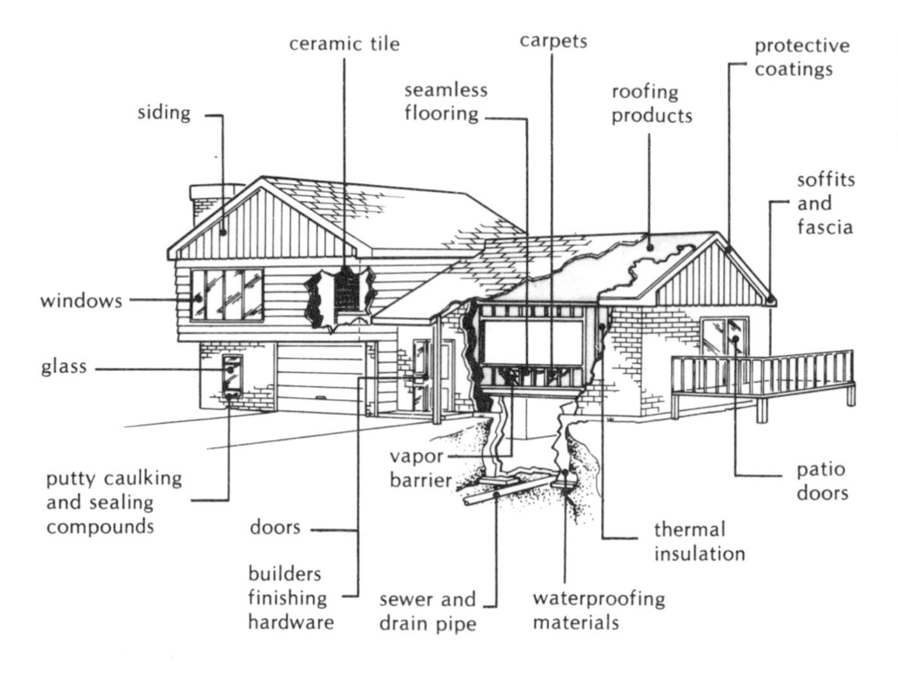

Figure 15.1 Typical areas of building construction for which standards have been developed. (Source: Canadian General Standards Board)

of practice and of trade promotion associations.

Within an industry, standards ensure uniformity of product from a number of producers. In the lumber industry, the most common standard is the lumber-grading rule, which describes in detail the whole range of lumber products. The grading rule may be used by the machine operator in the plant to produce a product of the desired quality, and by the grader or quality-control inspector to assign the appropriate grade to a product. It may be used by the engineer or architect to specify the grade and type of product required for a specific purpose, by the builder or consumer to ensure that the material he or she receives is of the quality required, and by the code official to ensure that the material used meets the requirements of the applicable code.

Nationally, standards promote the orderly conduct of business in manufactured products. Internationally, they facilitate the production of goods for foreign markets and help those goods to gain acceptance by foreign consumers. Without national and international standards for manufacturing, the interchangeability of products between different markets and from different sources would be impossible.

Codes and standards protect society by ensuring that the goods and services purchased are safe and of good quality. They contribute to the flow of commerce by serving as guidelines for the production of goods, ensuring uniformity of products both nationally and internationally. In the forest-products industry, standards provide the basis for industrial quality control for a wide range of products.

DEVELOPMENT AND USE OF CODES AND STANDARDS

In practice codes and standards are so interrelated that it is often difficult for the general public to understand the significant differences between them or their development and application.

Codes

Codes are systematic statements of a body of law, and those governing the use of forest products are given statutory force. In Canada building codes are the responsibility of the provinces, which delegate this responsibility to their municipalities. To promote uniform regulations across Canada, the National Research Council has established, through the Associate Committee on the National Building Code, a model building bylaw entitled The National Building Code of Canada. This code has no legal significance unless it is adopted by a jurisdictional authority, but the trend has been for provincial governments to adopt it as the provincial code. The use of this code is also mandatory for all construction financed through the Canada Mortgage and Housing Corporation. Over 90% of the population lives in areas that have adopted the National Building Code in whole or in part. Nevertheless, some large cities and municipalities still operate under their own special local codes or under one of the four model codes used in the United States.

The Associate Committee on the National Building Code has 30 members, who are chosen for their broad, general outlook on building construction. Since this is the policy-making body, members include architects, engineers, building officials, material suppliers, and others from relevant disciplines, selected from across the country in order to give regional representation.

The National Building Code is developed on the consensus principle and is not prepared by building officials, as the model codes in the United States are. The code is actually prepared by several standing committees, dealing with residential standards, construction safety, farm buildings, plumbing, structural design, fire safety, and other matters.

Code enforcement is in the hands of local building officials. For housing built under the National Housing Act, construction is also regulated by the Canada Mortgage and Housing Corporation, which administers this act. The groups that produce building codes have a moral and professional responsibility to assure public safety and health; yet they must take an objective attitude since their decisions are based on facts or on the most reliable and experienced opinions available. Building officials who administer the code are faced with a similar set of moral and professional responsibilities. In addition, they have the legal responsibility for enforcing the building bylaws in the area of their jurisdiction.

Building codes are not new. The earliest known written code was created by the Babylonian king Hammurabi in 2000 BC (DeGrace 1960). Public regulation of the design and construction of buildings is now a generally accepted part of the overall building process. From the outset the purpose of building regulations has been to protect the public. Because complete freedom in building can lead to community chaos, the government, through building and planning legislation, places limits on this freedom in the general interest of the community. Building codes or regulations are purely technical documents covering design standards, materials, and their applications, and should be restricted to those technological matters relating to structural safety, fire protection, and health aspects of the building. Matters of planning, such as zoning, density, and building heights, are usually kept out of building codes except when they have a bearing on safety or health.

There are two basic types of building codes: specification codes and performance codes. The first codes, and those in effect until recently, were nearly all specification codes, which state what materials may be used and how. Modern practice favors performance codes, which give the designer and the builder freedom to use any material or design technique they wish, provided the completed structure meets the required standards of structural performance, fire prevention, and health.

The challenge of devising suitable regulatory devices that permit these goals to be attained is met by the use of 'escape clauses' in most current codes. In the National Building Code of Canada, for instance, a clause specifically empowers the authority having jurisdiction to accept new techniques and materials where test, experience, or accepted engineering analysis has established their suitability. The fundamental principle is that no form of construction should be excluded as long as it satisfies the intent of the code. The difficulty of application lies in the lack of criteria against which the performance of proposed innovations may be assessed and in the slow development of evaluation techniques for measuring criteria.

One technique used in drafting regulations in performance or functional form that provides practical guidance for the designer, builder, or code authority is the use of 'deemed-to-satisfy' clauses. These clauses may take the form of brief specifications or reference to other documents, such as codes of practice. Thus, traditional materials and methods cease to be mandatory, as in the older specification codes, but are recognized as examples of compliance with the required standards of performance. By comparison with these examples, new materials of equal or better performance may then be used.

Wood is one of those materials that can gain from performance codes, since arbitrary restrictions are sometimes placed on the use of wood in buildings because of concern that it is combustible. Although it is true that wood burns, it actually provides an extra margin of safety during a fire because it maintains its structural integrity much longer than noncombustible building materials that are subject to sudden structural failure or collapse from heat.

Standards

Standards have been in widespread use in North America less than 200 years. In 1798 the United States government placed an order for 10 000 muskets to be produced within two years. At that time, when guns were handcrafted and top production of the government armory was about 300 muskets a year, such a target was unheard of.

The contract was won by an extraordinary man, Eli Whitney, who was already famous for his invention of the cotton gin. To meet the contract's delivery date, he designed a factory where the guns would be manufactured on a 'new principle,' that of standardized, interchangeable parts. Production fell short of the target, peaking in 1807 at 2000 units – six times the production of the government's skilled craftsmen – but this number was achieved by unskilled workmen who

Figure 15.2 Codes and standards ensure that quality and performance are built into modern residential construction. (NFB Photothèque. Photo by Ted Grant)

had to assemble 50 interchangeable parts, mass-produced by machine in standard shapes and sizes. Whitney was the first to use standardized parts in mass production, and his new principle revolutionized manufacturing.

Standards – prescribed ways of making, using, and doing things – had evolved casually over the centuries. They were in existence long before Whitney's time, but he introduced the concept of organized standardization. By the end of the nineteenth century, precise standards governed the design, manufacture, and use of innumerable goods, materials, and services. The definitions of size and shape, grade, safety, performance, measurement, and terminology permitted the technological explosion of the nineteenth century to proceed in an orderly manner. In effect, standards became the language of commerce.

For most of that century, each manufacturing plant developed its own standards for its products and processes. Toward the end of the century standardization entered its second stage, when group standards developed by professional societies or trade associations were adopted by entire industries. With the realization that different industries had common problems came the third, or institutional, stage of standards development. Organizations were created for the sole purpose of developing standards to provide common solutions for industry as a whole. Because these standards were sound they became essentially national standards and the organizations became national institutions. Finally, because of the importance of standards in world trade, standardization has proceeded to the fourth and most difficult stage, that of international cooperation.

Historically, standards have been developed by engineers and technicians, with little or no input from others. Today nontechnical groups are becoming involved as never before. The reasons for the change are essentially trade and consumer demand for standards of quality and performance in addition to safety standards, which previously were the main consideration in consumer goods.

To succeed in the export market, a product not only must conform to the buyer's national standards, or to international standards, but also must have inherent quality and give good performance. Developing nations tend to favor international standards, and it is essential that Canada, which is dependent on export trade for its well-being, establish credibility in export markets through judicious use and application of standards.

Standards development is a complicated undertaking. To achieve a standard acceptable throughout the industry, every party to the decision must have a high level of technical competence. Standards developed *in camera* may be issued quickly, but standards developed by consensus represent informed agreement. Such standards represent the decision of a body of technical people who not only have agreed to the standard but also have acquired sufficient knowledge, understanding, and skill to put the new standard into practice.

STANDARDS ORGANIZATIONS

The National Standards System

The Standards Council of Canada was established by an act of Parliament on October 7, 1970. The council was assigned the task of developing a national standards system to provide a medium through which Canadian standards-writing organizations could recognize, establish, and improve the Canadian standards program. Before the establishment of the Standards Council and the National Standards System, there were no national standards in Canada.

The National Standards System comprises five accredited standards-writing organizations: the Canadian Standards Association (CSA), the Underwriters' Laboratories of Canada (ULC), the Canadian Gas Association (CGA), the Canadian General Standards Board (CGSB), and the Bureau de normalisation du Québec (BNQ). These organizations are responsible for the development and preparation of standards in their designated subject areas under the overall direction of the Standards Council of Canada. The council provides coordination and support for the national program for standardization and also manages Canada's participation in international standards work as the member body for Canada in the International Organization for Standardization (ISO). Involvement in this organization is important for Canada as one of the world's major trading nations.

The Standards Council of Canada is a national, nongovernment agency, consisting of not more than 57 members and headed by a president and vice-president. Membership is broadly representative of all levels of government, primary and secondary industries, distributive and service industries, trade associations, labor unions, provincial associations, consumer associations, and the academic community. Members are appointed for a term not exceeding three years. The

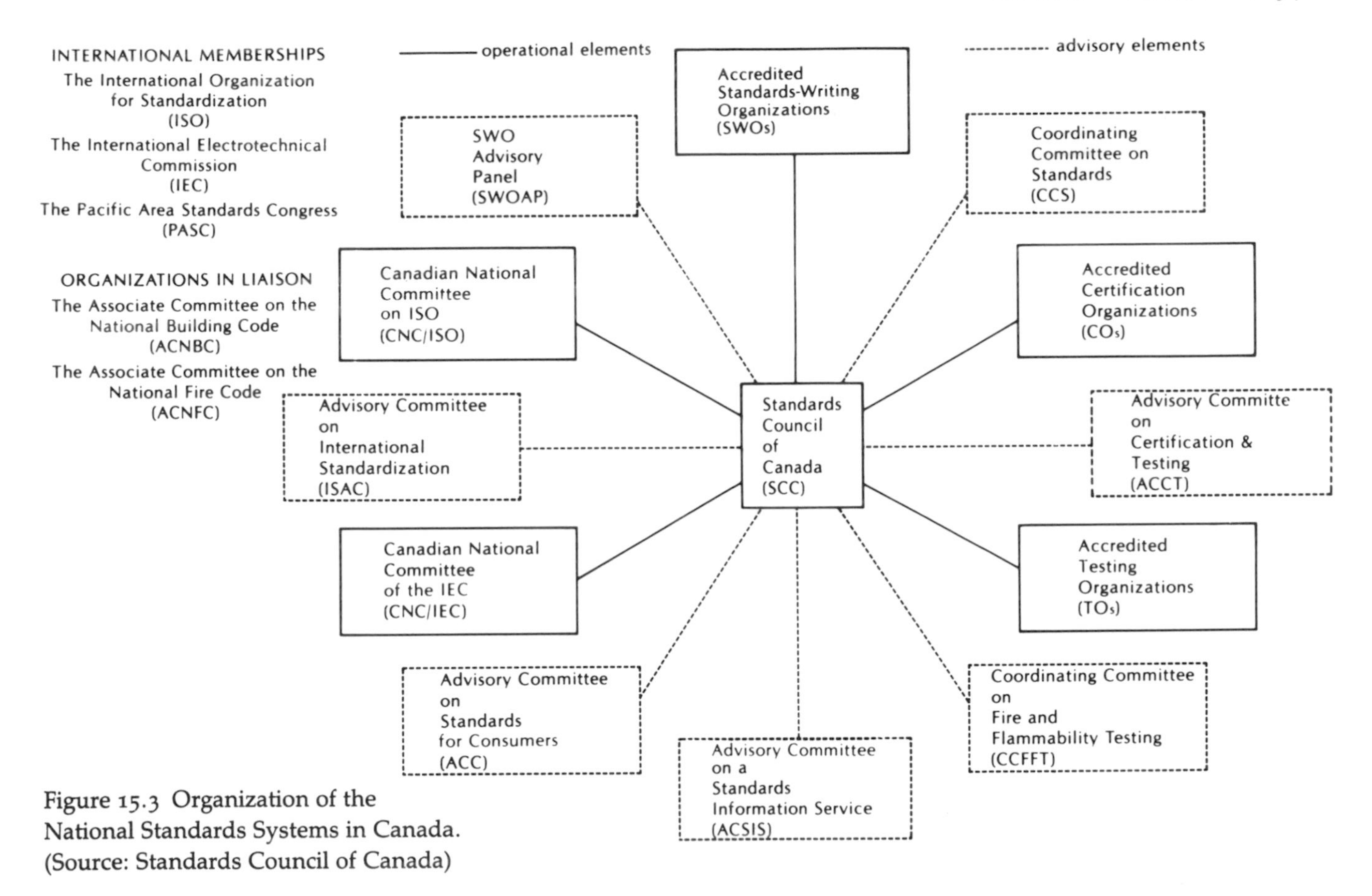

Figure 15.3 Organization of the National Standards Systems in Canada. (Source: Standards Council of Canada)

administrative function is carried out by a permanent staff in Ottawa, headed by an executive director who is appointed by the Governor in Council.

Standards developed by the Canadian Standards Association, the Canadian General Standards Board, and the Underwriters' Laboratories of Canada, within the framework of the National Standards System, and by the American Society for Testing and Materials are used by the lumber and forest-products industries in Canada. In addition, Canadian members on the ISO technical committees assist in the coordination and unification of the National Standards of Canada and the ISO standards. The structure and function of these organizations are briefly outlined in the following sections.

Canadian Standards Association (CSA)

The Canadian Standards Association (Figure 15.4) was chartered in 1919 by a group of Canadian industrialists as a self-governing national association concerned with the development and application of standards and related services. The hundreds of committees that produce standards under CSA auspices are a balanced national representation of more than 2500 companies and organizations that are affiliated with the association as sustaining members. These include governments, utilities, professionals, manufacturers, educational bodies, and trade associations.

All standards issued by the CSA reflect a national consensus of

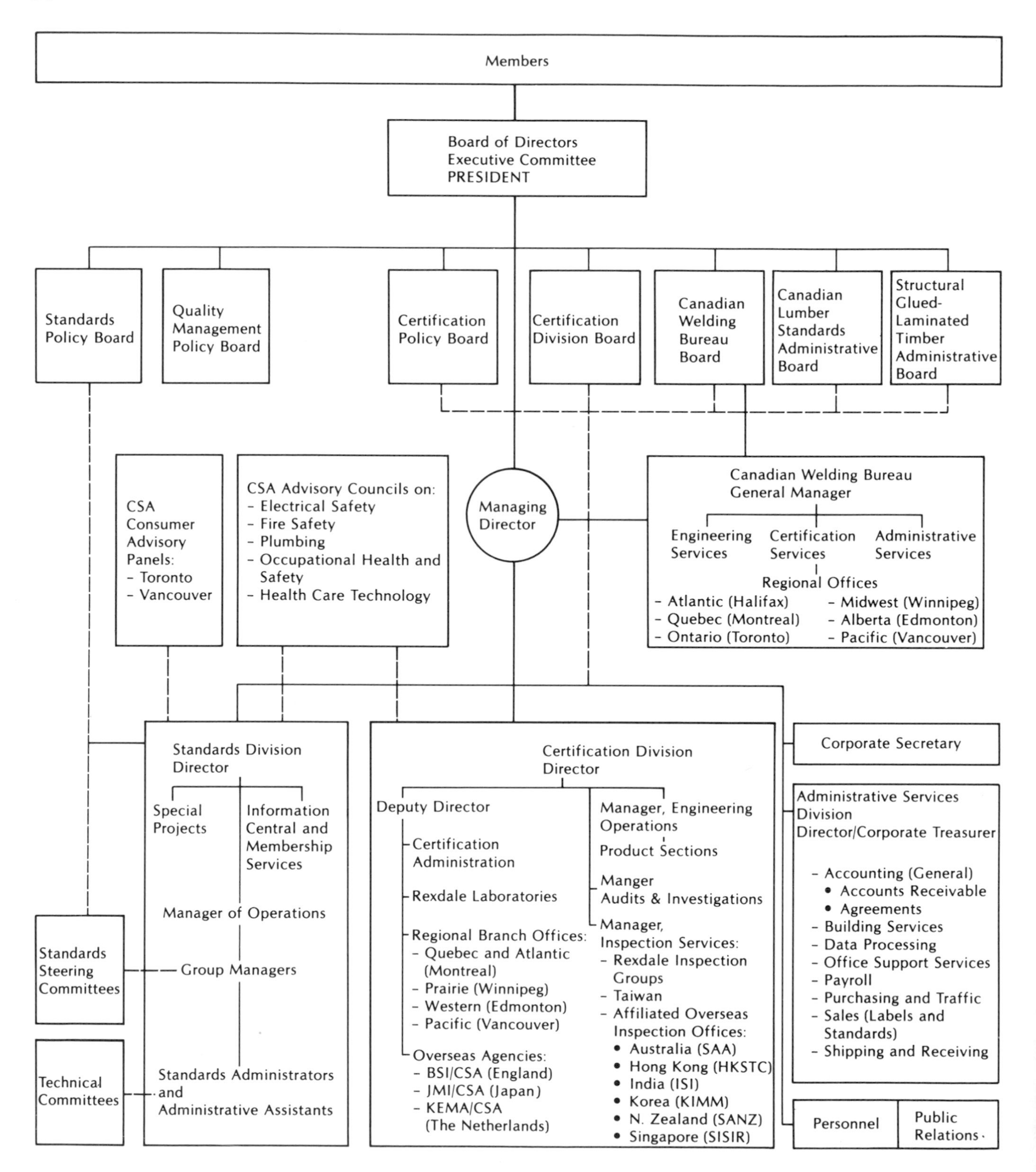

Figure 15.4 Organization of the Canadian Standards Association. (Source: Canadian Standards Association)

manufacturers; scientific, technical, and professional organizations; government agencies; and consumers. The standards are developed in accordance with the CSA's published policies and procedures and are widely used by industry and commerce. They are frequently used by municipal and provincial governments and by the federal government as well.

Where circumstances are appropriate for a CSA standard to be designated as a national standard of Canada, the standard is written to conform to the criteria and procedures set forth by the Standards Council of Canada.

In addition, the CSA provides comprehensive certification, testing, inspection, and related information services for its members and clients. Most consumers are probably familiar with the CSA logo or symbol that appears as a seal of approval on goods and equipment manufactured in accordance with CSA standards. The association maintains six testing stations across Canada, in addition to agencies in Britain, Europe, Japan, Southeast Asia, India, Australia, and New Zealand.

Canadian General Standards Board (CGSB)

The Canadian General Standards Board has evolved from the Canadian Government Purchasing Standards Committee, which was formed in 1934 to prepare federal government purchasing standards outside the engineering field, which was already covered by the then Canadian Engineering Standards Association.

Since that time the scope of activities of the CGSB has continued to expand, and it is now responsible for the development of national standards of Canada in numerous subject areas. These cover many areas of interest to the wood-products industry, including wood fiberboard (hardboard), packaging and packaging materials, furniture, and thermal insulation.

Underwriters' Laboratories of Canada (ULC)

The Underwriters' Laboratories of Canada is a nonprofit organization sponsored by the Canadian Underwriters' Association. The ULC develops and publishes standards, classifications, and specifications for products that have a bearing on fire or accident hazards or crime prevention.

Several of the ULC standards cover products and requirements for fire protection for products of interest to the wood-products industry. These standards are referenced in government regulations and the National Building Code and support the Canada Mortgage and Housing Corporation's lending activities. The ULC also provides certification and testing services covering fire-retardant treatments and pressure-treating facilities for wood products.

International Organization for Standardization (ISO)

The International Organization for Standardization was formed in 1947, with the Canadian Standards Association, as one of the charter members, representing Canada. The Standards Council of Canada (SCC) assumed the responsibility of the Canadian Standards Association on April 1, 1972. The object of ISO is to promote the development of standards throughout the world in order to facilitate international exchange of goods and services and to develop cooperation in the spheres of intellectual, scientific, technological, and economic activity. Canadian representation on ISO technical committees is drawn from the five accredited standards-writing organizations of the National Standards System, which work together within the guidelines developed by SCC for the integration of international and national standards.

American Society for Testing and Materials (ASTM)
The ASTM Committee D-71, which deals with standards relating to wood and test methods for wood, has provided the basis for the use of wood as an engineering material in Canada and in our major export market, the United States. Many Canadians representing government and wood-industry technical organizations are members, hold office, and play an active role in formulating ASTM standards.

ASTM is a nonprofit corporation with headquarters in Philadelphia. Essentially, ASTM attempts to provide a technical language of communication within any given field between the producer and consumer, as well as between the expert in a particular technological discipline and the layman. To this end, ASTM publishes standardized techniques, practices, definitions, specifications, criteria, and procedures, all resulting from the consensus of its standards-promulgating committees.

STANDARDS RELATED TO WOOD

The following is a partial list of typical standards for wood and wood products used in Canada. These standards and others concerning wood and wood products are continually being evaluated by the various committees of the standards-writing organizations and the ASTM. The standards are promptly amended whenever necessary to recognize current practice and to incorporate the latest technology.

CSA Standard 086-1976 Code for the Engineering Design of Wood This standard provides design criteria for structurally graded lumber, lumber that is not structurally graded, glued-laminated timber, plywood, piling, pole construction, and major fastenings. It is intended for use in the design or appraisal of structures or structural elements made from wood or wood products.

CSA Standard 0121-M Douglas-fir Plywood This standard covers minimum requirements for construction, sizes, regular grades, specialty panels, manufacturing tolerances, and glue bond for Douglas-fir plywood intended for structural, construction, and industrial application.

CSA Standard 0122-1977 Structural Glued-Laminated Timber This standard governs the manufacture of structural glued-laminated timber and may be applied in its entirety or in part for products other than structural glued-laminated timber.

CSA Standard 0141-M Softwood Lumber This standard covers the principal trade classifications and sizes of softwood lumber for yard, structural, and shop use. It provides a common basis of understanding for the classification, measurement, grading, and grade marking of rough and dressed sizes of various items of lumber, including finish, boards, dimension stock, and timbers.

CSA Standard 0188.1-M Mat-Formed Particleboard This standard applies to mat-formed wood particleboard; it specifies the requirements for various physical and mechanical properties and includes the test methods to be used for measuring these properties.

ASTM Designation D-245-70 Standard Methods for Establishing Structural Grades and Allowable Properties for Visually Graded Lumber These methods cover the basic principles for visually grading structural lumber and for establishing related unit stresses and stiffness values for design.

In addition to the standards listed above, other wood-related standards cover wood windows, wood doors, structural requirements for mobile homes, highway bridges, scaffolding, bleacher seats, fire-test methods, lumber pallets, air-pollution control, adhesives, panel products, and many other items.

Acceptance (Certification) of Wood Materials

The first examination of a product for quality and for adherence to specified standards is at the manufacturing plant. For example, in a lumber mill the various products are graded for class and quality by mill graders according to the NLGA 1979 Standard Grading Rules for Canadian Lumber, which are published by the National Lumber Grades Authority (NLGA).

The NLGA was incorporated under the Department of Consumer and Corporate Affairs, January 11, 1971, 'to establish, issue, publish, amend and interpret grading rules, for all species of lumber manufactured in Canada.' Membership in the NLGA may include the 12 regional rules-writing or lumber-inspection agencies in Canada and 'a corporation society or agency in Canada representing manufacturers of lumber.'

The NLGA grading rules are revised periodically by the issuing of supplements. Their purpose is to maintain a uniform standard of quality among mills manufacturing the same or similar woods. Lumber manufactured to these rules may be regarded as Canadian Standard Lumber meeting the provisions of CSA Standard 0141-M, Softwood Lumber. The rules set forth sizes, characteristics of all grades, measurement procedures, grading and reinspection procedures, species and commercial species groupings, and moisture content provisions, and show facsimile grade stamps of CSA-approved grade-marking agencies (Figure 15.5).

Although lumber is graded according to the rules of the National Lumber Grades Authority, overall supervision of grading, and control and certification of agencies and associations grading lumber, are the responsibility of the Canadian Lumber Standards (CLS) Division of the CSA. The CLS Division consists of:

1. The CLS Administrative Board (CLSAB), comprising representatives of industry and of large wood users such as Canada Mortgage and Housing Corporation, the National House Builders Association, and the federal government. The CLSAB is responsible to the CSA board of directors for the policy and control of the grade marking of Canadian lumber. Its office is in Vancouver and includes a field staff for inspection and control of the grade-marking process. In carrying out

A.F.P.A.® 00
S—P—F
S-DRY STAND

Albert Forest Products Association
11710 Kingsway Avenue
Edmonton, Alberta T5G 0X5

CFPA® 00
S-P-F S-DRY
CONST

Central Forest Products Association
14-G 1975 Croydon Avenue
Winnipeg, Manitoba R3P 0R1

CLA
S-P-F
100
No. 2
S-GRN.

Canadian Lumbermen's Association
27 Goulbourn Avenue
Ottawa, Ontario K1N 8C7

ILMA® S-DRY 1
00 S—P—F

Interior Lumber Manufacturers Association
295-333 Martin Street
Penticton, B.C. V2A 5K8

0 0
No 1
S-DRY
D FIR (N)
NLGA RULE

MacDonald Inspection
125 East 4th Avenue
Vancouver, B.C. V5T 1G4

CLMA® 1 1
S-GRN 1
D FIR (N)

Cariboo Lumber Manufacturers Association
301-197 2nd Avenue North
Williams Lake, B.C. V2G 1Z5

10
CONST S-P-F
S-GRN

N.W.T. Grade Stamping Agency
P.O. Box 2157
Yellowknife, N.W.T. X0E 1H0

COFI®
100
S-P-F
S-GRN
No 1

Council of Forest Industries
of British Columbia
1500-1055 West Hastings Street
Vancouver, B.C. V6E 2H1

Northern Interior Lumber Sector
803-299 Victoria Street
Prince George, B.C. V2L 2J5

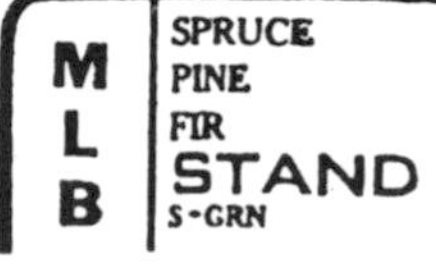

MILL 11 — 466

Maritime Lumber Bureau
P.O. Box 459
Amherst, Nova Scotia B4H 4A1

PLIB®
NLGA RULE
No 1
S-GRN
00
HEM-FIR-N

Pacific Lumber Inspection Bureau
Ste. 1130, 1411 Fourth Avenue Building
Seattle, Washington 98101
B.C. Division: 1460 - 1055 West Hastings St.
Vancouver, B.C. V6E 2G8

SEL. STR.
SISA® ÉPINETTE PIN - SAPIN | ALIB® SPRUCE PINE - FIR
NOM ET/OU N° DU MOULIN MILL'S NAME AND/OR NUMBER
CLASS R1-1108 GRDR
R VERT S GRN

Service d'inspection des sciages
de l'Atlantique
Atlantic Lumber Inspection Bureau
A Branch of Quebec Lumber
Manufacturers Association
Suite 200, 3555 Boulevard Hamel ouest
Québec, P.Q. G2E 2G6

O.L.M.A.® 01-1
CONST. S-DRY
SPRUCE - PINE - FIR

Ontario Lumber Manufacturers
Association
Suite 414, 159 Bay Street
Toronto, Ontario M5J 1J7

L'association des manufacturiers de bois de
sciage du Québec
Quebec Lumber Manufacturers Association
Suite 200, 3555 Boulevard Hamel ouest
Québec, P.Q. G2E 2G6

Figure 15.5 Typical grade-mark facsimiles. (Sources: National Lumber Grades Authority; National Research Council of Canada)

this function the CLSAB delegates certain administrative responsibilities to the CLS Industry Committee, which acts as its adviser on grade-marking problems.

2. The CLS Industry Committee, comprising representatives of the lumber-manufacturing and distributing associations, which is responsible to the CLSAB for:
a) ensuring that board policy and requirements are carried out;
b) administration of the grade-checking service;
c) investigating complaints such as violation of policy, poor grading standards, and misuse of stamps;
d) recommending to the board necessary policy changes;
e) recommending to the board additions to or deletions from the list of associations or agencies authorized to grade-mark lumber in Canada;
f) raising and administering required funds; and
g) preparation of information reports for the board.

The mill graders form the first link in the quality-control chain. Their grade evaluation is periodically checked by one of the 12 independent inspection agencies in Canada. Because the products of the mills are used to manufacture other products, further certification and inspection activities are carried out to ensure quality and safety for the consumer. The Canadian Standards Association provides certification service for manufacturers who, under license from CSA, wish to use the appropriate CSA marks on certain products (e.g., glued-laminated beams) to indicate conformity with a CSA standard.

CSA certification of a number of products is provided voluntarily in the interest of maintaining agreed-upon standards of quality, performance, interchangeability, and safety. CSA certification may also form the basis for acceptance by inspection authorities.

Thus, standards, quality inspection, and certification programs provide the guidelines for codes to be used and for regulatory authorities to use in inspection programs to ensure that housing, structures, and other wood components are of adequate quality.

Every citizen's life is influenced each day by many codes and standards. The most obvious, of course, are those affecting housing, foodstuffs, and appliances. Virtually every item of commerce is regulated by one or more standards, including weight, color, size, and quality. Without these codes and standards, the consumer would have no assurance of product quality, the safety of hazardous materials, or the durability of construction for the design life of a product. There would be no effective regulation of import and export trade, no standard parts, and no interchangeability of components. Without these codes and standards, our complex, technological society would come to a halt.

REFERENCE

DeGrace, R.F. 1960. Facts about building codes that every supplier should know. *For Prod. J.* 10 (10): 538

ADDITIONAL SOURCES

Standards Council of Canada
350 Sparks Street
Ottawa, Ontario K1R 7S8

Canadian Standards Association
178 Rexdale Boulevard
Rexdale, Ontario M9W 1R3

Canadian General Standards Board
Department of Supply and Services
Ottawa, Ontario K1A 0S5

Underwriters' Laboratories of Canada
7 Crouse Road
Scarborough, Ontario M1R 3A9

Standards Council of Canada
International Standardization Branch
Meadowvale Corporate Centre
2000 Argentia Road, Suite 2-401
Mississauga, Ontario L5N 1P7

American Society for Testing and Materials
1916 Race Street
Philadelphia, Pennsylvania 19103

16

R.W. KENNEDY
University of British Columbia,
Vancouver

The Future of Wood

PREDICTIONS FOR FUTURE USES OF WOOD

Wood is our major renewable resource, and its products are biodegradable. This advantage is becoming increasingly important in a society concerned more and more about the environment. In Canada and the United States we process annually a greater tonnage of wood than of steel, cement, plastics, aluminum, and all other metals combined (Brown 1970).

Will wood products be able to maintain their leading position as we progress into the twenty-first century? Predictions for the future are based on assumptions regarding the socioeconomic structure of the nation and the world. Let us look ahead to perceive what conditions might be like.

Population growth in Canada, the United States, the European Economic Community, and developing and underdeveloped countries will have an important bearing on the future economic growth of Canada. Demographic projections in developed Western countries point to a possible stabilization of population early in the twenty-first century, with a lower percentage of the population under 40. Nevertheless, the population of North America is expected to rise to at least 300 million by the end of the century. Despite reductions in birth rates in the more industrialized societies, world population may double by the year 2000.

Increased population implies that more mouths must be fed from our limited and dwindling supply of land available for crops and livestock; therefore, more food must be produced from the land and other sources. Since the 1930s it has been technically possible to propagate yeast protein on a growth medium consisting of sulfite-pulping waste liquor or of wood hydrolyzed to low-molecular-weight sugars. This protein food can be eaten by both humans and animals. Livestock can also be fed on a diet of steamed-wood residues or on wood molasses.

Figure 16.1 The forest is one of Canada's major resources.

Thus, it is conceivable that wood will be deemed as an indirect source of food to sustain the anticipated increase in world population.

Escalating population and human desires on the one hand conflict with limited resources and a polluted environment on the other. If the problems of increasing urbanization are resolved and we succeed in conserving our resources, pollution may be reduced as remedial measures are applied. But the burgeoning population, affluence, increased leisure time, and urbanization will continue to promote economic growth, increased literacy, and the desire for better housing, all of which will require wood products in the form of paper and building materials. Continued economic strength in the forest-products industry thus seems assured.

Shortages and high manufacturing costs of other raw materials should also help to keep wood in its preeminent position. The outlook for an expanded plastics industry is constrained by its heavy reliance on petrochemicals, while other nonrenewable materials, competitive with wood, require substantially higher costs in extraction and manufacture. Wood also enjoys a certain competitive advantage over other building materials. In general wood is situated closer to markets than other materials, such as iron, are, and power requirements to convert wood from its raw state to a finished state, regardless of form, are appreciably less. For example, steel for exterior-wall framing in residential construction requires over three times the processing energy needed for an equivalent installation using lumber. For aluminum or concrete blocks, the energy requirements are more than eight times that of lumber (U.S. Forest Service 1973). In spite of developments in nuclear power and geothermal power, these facts should work to the advantage of wood consumption as energy costs rise.

The annual world timber harvest for industrial purposes is about 1.4 billion m^3 (48.9 billion cu ft) (FAO 1978). It has been estimated that by the year 2000 the world demand for industrial wood will be 2.5 billion m^3 (88.3 billion cu ft), or, if fuel needs are taken into account, a total world demand of 4 billion m^3 (141.3 billion cu ft) (Keays 1974). Even to approach this latter figure, world industrial wood production must increase by an additional 1.1 billion m^3 (39.4 billion cu ft) over today's level. Assuming the land is kept stocked, the wood-growing capacity of the world's forests is estimated to be in excess of 4.5 billion m^3 (158.9 billion cu ft) per year. Extrapolating from these figures, which are admittedly very rough approximations, it will not be many years beyond 2000 before the world's total annual growth of wood will only balance the demand. Our recent search for additional markets for wood products will shift to a search for additional wood resources to meet market demands. Demands on forests for parks, watersheds, and wildlife or ecological reserves may intensify anticipated shortfalls in timber availability. Alternatively, intensive silvicultural techniques and improved forest management, leading to decreased rotation periods, may expand the allowable cut to meet this increased demand for wood.

Today annual wood production in Canada for all purposes is approximately 156 million m^3 (5.5 billion cu ft) (FAO 1978). This volume is expected to rise to about 235 million m^3 (8.3 billion cu ft) by 1990 (Jegr,

Figure 16.2 Chronology of some important developments in the forest-products industry since 1900. (Photo: Council of Forest Industries of British Columbia)

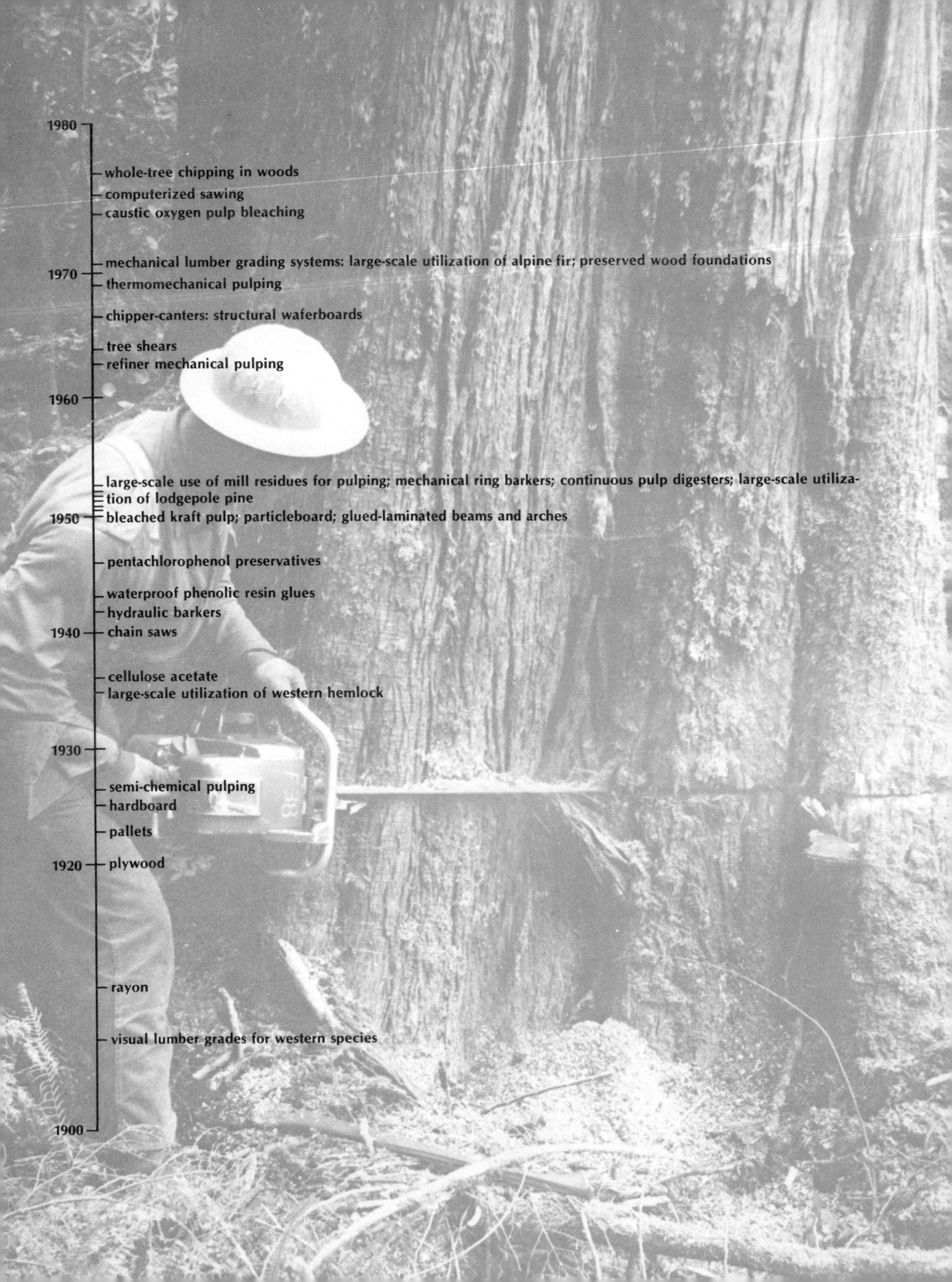

1980

- whole-tree chipping in woods
- computerized sawing
- caustic oxygen pulp bleaching
- mechanical lumber grading systems: large-scale utilization of alpine fir; preserved wood foundations

1970

- thermomechanical pulping
- chipper-canters: structural waferboards
- tree shears
- refiner mechanical pulping

1960

- large-scale use of mill residues for pulping; mechanical ring barkers; continuous pulp digesters; large-scale utilization of lodgepole pine
- bleached kraft pulp; particleboard; glued-laminated beams and arches

1950

- pentachlorophenol preservatives
- waterproof phenolic resin glues
- hydraulic barkers

1940

- chain saws
- cellulose acetate
- large-scale utilization of western hemlock

1930

- semi-chemical pulping
- hardboard
- pallets

1920

- plywood
- rayon
- visual lumber grades for western species

1900

Miller, and Thompson 1976) and to 257 million m³ (9.1 billion cu ft) by the year 2000 (Manning 1973). Counting available timber on *all* inventoried forest land, an annual harvest of 300 million m³ (10.6 billion cu ft) could be taken without depleting the resource base, but an allowable annual cut of 226 million m³ (8.0 billion cu ft) is perhaps more realistic (Reed and Assoc. Ltd. 1973). In other words, the allowable cut on a sustained yield basis may be in balance with, or even less than, requirements by the year 2000. Nevertheless, Canada is one of the few highly developed countries that has reasonably adequate resources to meet projected demands; thus, our forest-based economy should continue to expand.

Undoubtedly there will be severe local shortages as accessible larger timber in the more southerly latitudes of Canada is given preference in harvesting. Different patterns of species utilization are already evident. For example, between 1950 and 1976 the amount of Douglas-fir harvested in British Columbia remained relatively constant but dropped from 40% of the total cut in 1950 to only 13% in 1976 (British Columbia Forest Service 1977). In the interim, other species, particularly the spruces (white and Engelmann), lodgepole pine, and the true firs (amabilis and alpine) gained ascendancy. The absolute amount of Douglas-fir available for harvesting will diminish until somewhat after the turn of the century; then it may reassert itself as the predominant coastal species when second-growth and plantation timber reach maturity.

The poplar species are at present greatly underused throughout Canada. No more than 5% of the allowable annual cut was estimated to have been harvested in 1968 (Maini and Cayford 1968), and the percentage has not changed appreciably since then. This species group has potential for much more extensive use in the future, primarily for pulpwood, structural waferboard, and framing lumber.

FROM 1900 TO THE PRESENT

Early in this century, the recovery of solid-wood products from trees cut in the forest was approximately 30%. Now a much greater proportion of the stem of a tree is used. This development is the culmination of many advances in products and processing technology made over the past 75 years. Some of the most significant events are outlined in Figure 16.2.

Consolidation of sawmills into fewer but larger units, many of which became integral parts of complex forest-products enterprises, has been a leading factor in more efficient wood use. More complete use of the stem of a tree has been made possible by the development of chain saws and tree shears, which permit tree harvesting to stump heights only a few centimeters from ground level. Because of greater demands for pulpwood and new provincial forest-management policies, the top diameter of the merchantable stem has been changed from 20 to 30 cm (8 to 12 in.) to 8 or 10 cm (3 or 4 in.). The widespread introduction of barkers in sawmills in the 1950s and 1960s allowed bark-free slabs, trims, and edgings to be chipped and directed to pulp mills. Now even the intermediate production of slabs and edgings in a sawmill can be eliminated through the use of chipper-canters, which reduce barked logs directly to pulp chips and cants or to dimension lumber.

More recently sawdust and shavings have been pulped and the resulting product mixed with pulp obtained from whole chips. Pulp is also manufactured from sawdust and shavings by mechanical refining techniques and blended with chemical pulp for use in newsprint. Sawdust and shavings for pulping have been exported, especially from British Columbia mills to Japan, but competing domestic demand for use of these residues for newsprint furnish and for fuel may reduce this activity.

The development of waterproof phenol- and resorcinol-formaldehyde resin adhesives led to the accelerated growth of the softwood-plywood industry after the Second World War. This development permitted the supply of panels for use in structural and exterior applications. These adhesives also spurred the development of glued-laminated beams, arches and columns, and, most recently, particleboard and structural waferboard suitable for exterior use. Particleboard plants are now second only to pulp mills in using residues from sawmills and plywood plants.

PRODUCT TRENDS AND PROJECTIONS

A glance at past projections shows just how far they are from the mark. For example, in 1952, a sophisticated effort was made to estimate the demand for forest products in the United States by 1975 (Stanford Research Institute 1954). If actual 1975 production figures are used for comparison, it can be seen that plywood production was underestimated by more than 200% and pulpwood consumption by nearly 50%. Particleboard was not mentioned at all in the 1952 forecast, despite subsequent experience showing an annual growth of 25% for this product.

In spite of the inaccuracies to be expected when estimating future trends, it is important, nevertheless, to make these estimates for short- and long-range production planning. Figure 16.3 shows the Canadian production and consumption of major forest-product commodities based on such economic projections (Manning and Grinnell 1971; U.S. Forest Service 1973). The trends suggest that Canadian production of all commodities will increase by the year 2000 as follows: lumber, 50%; plywood, 110%; pulp and paper, 70%; particleboard, 150%; hardboard, 140%.

Lumber is the only product whose per capita consumption is expected to decrease (Figure 16.3), an extension of a long-time trend begun in 1909. Increased total lumber production is largely a reflection of expected population trends coupled with higher exports to the United States as its available timber supplies shrink.

Panel products (plywood, particleboard, and hardboard) promise to grow at an accelerated rate. These products are highly desired because of their general utility, performance, and lower labor costs in frame construction. Particleboard in the form of structural waferboard has the potential to capture a significant portion of the lumber and plywood sheathing and flooring markets. Particleboard is now available in forms other than the traditional thin sheets. Particles have been reconstituted between veneer sheets for studs and joists, but end-finger-jointing and

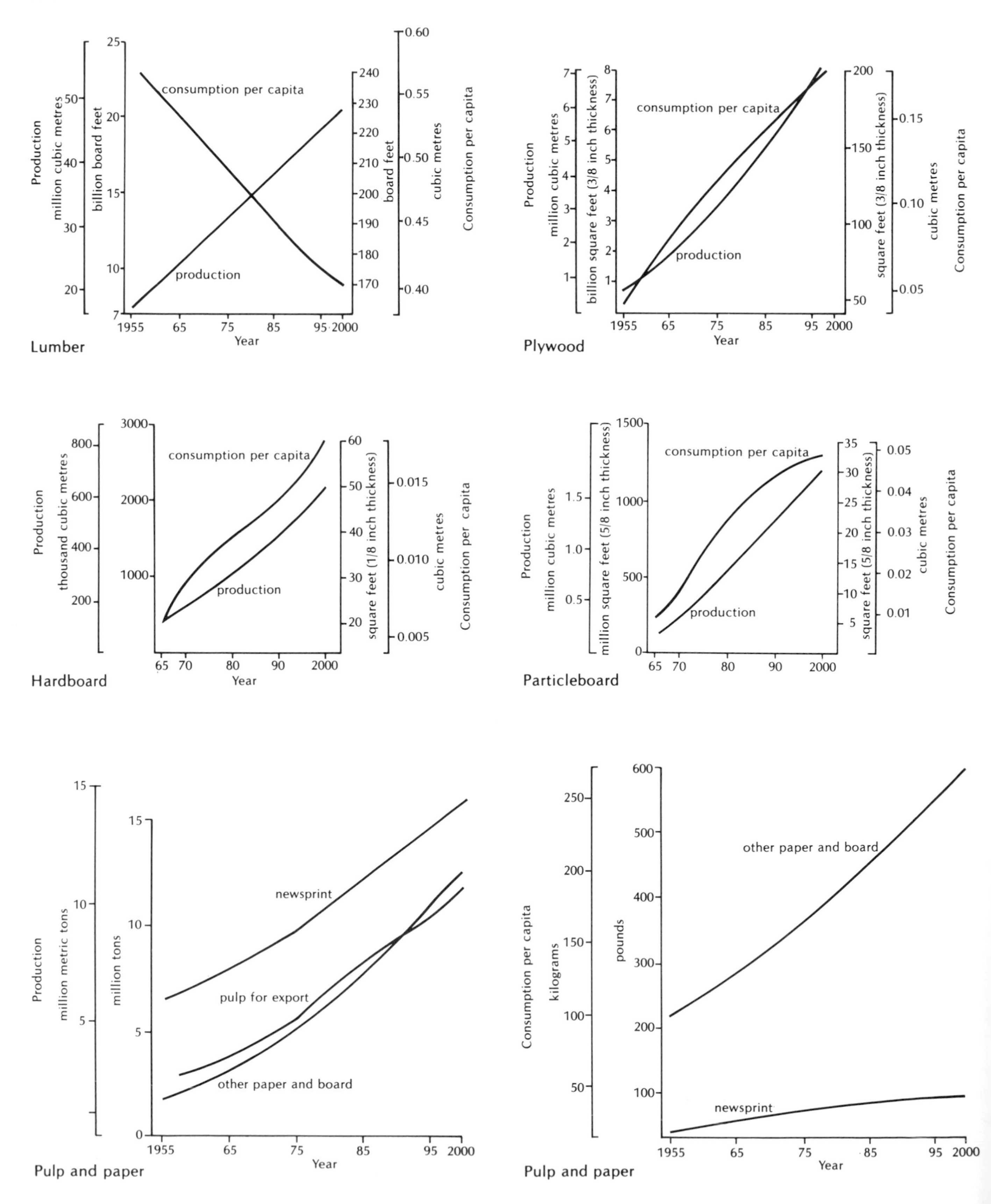
consumption per capita
production
Production
million cubic metres
billion board feet
board feet
cubic metres
Consumption per capita
Year
Lumber
consumption per capita
production
Production
million cubic metres
billion square feet (3/8 inch thickness)
square feet (3/8 inch thickness)
cubic metres
Consumption per capita
Year
Plywood
consumption per capita
production
Production
thousand cubic metres
square feet (1/8 inch thickness)
cubic metres
Consumption per capita
Year
Hardboard
consumption per capita
production
Production
million cubic metres
million square feet (5/8 inch thickness)
square feet (5/8 inch thickness)
cubic metres
Consumption per capita
Year
Particleboard
newsprint
pulp for export
other paper and board
Production
million metric tons
million tons
Year
Pulp and paper
other paper and board
newsprint
Consumption per capita
kilograms
pounds
Year
Pulp and paper

Figure 16.3 Product trends and projections for major forest-product commodity items in Canada. (Sources: Lumber, plywood, pulp, and paper from Manning and Grinnell 1971. Particleboard and hardboard from Statistics Canada Cat. 36-003 and 36-001, December issues, 1966–76. Because the 1970 figures for per capita consumption of these board products were similar in the United States and Canada, projections for Canada are based on those made in USDA 1973, *The Outlook for Timber in the United States*, Forest Resource Rep. No. 20, Washington, DC. All Canadian production is assumed to be consumed internally, with no imports or exports of these board products.)

edge-gluing of small, solid-wood pieces might be regarded, initially, as the more likely alternatives to sawn studs and joists. The availability and cost of resins and the standards of wood use demanded by forest-resource managers will also play a role in determining the future of particleboard and similar reconstituted material. The growth of reconstituted wood products at the expense of lumber and other primary products may be tempered by the energy required to disassemble wood and reassemble pieces into a saleable commodity. Nevertheless, they will be produced at an accelerating rate; only the magnitude of the increase is unclear.

The forest-products industry has an excellent opportunity to restrain escalating power costs by using mill residues for energy. Although a start has been made in firing dry kilns directly from mill residues, improvements in efficiency and engineering are required. Large-scale gasification of residues could provide the quality of gas required for pulp-mill lime kilns and internal combustion and gas turbine engines. In addition, cogeneration of steam for conversion to electric power is possible in specific regions where there is a large concentration of primary wood-conversion plants.

THE CHALLENGES AHEAD

Although significant improvements in standards of tree use have been made, much remains to be done. Many sawmills and plywood plants continue low-yield, inefficient practices, producing a nonuniform product. Fully 60% of the weight of the logs reaching many sawmills is converted into byproducts having an average value considerably less than the cost of logging and delivery to the mills.

Productivity in many sawmills is still strongly conditioned to high daily throughput of logs without adequate consideration of the amount or quality of lumber recovered. Better saw guides, log scanners, improved bucking practices, computerized setworks, and statistical quality-control techniques are all examples of tools that can be used to improve lumber recovery. Techniques for increasing the rate of log processing are required to make other tools, such as high-strain band saws, more attractive for log breakdown.

Chemical pulp mills are likewise inefficient in their use of raw material. The yield of softwood pulp is commonly in the range of 45% of the dry weight of the wood entering the digester. Process improvements to increase and retain higher proportions of the carbohydrate fraction without sacrificing strength are desirable incremental goals. Mild chemical treatments need to be developed that, when followed by mechanical-disc refining, will produce strong, ultra-high-yield pulp with less energy input than is now required in conventional thermomechanical pulping. The potential for reducing pollution might be an added incentive for such development if higher yields could be realized with smaller amounts of environmentally benign chemicals. Research directed to making stronger pulps might also extend wood-fiber resources, thereby reducing the basis weight now required for adequate performance of papers. By attaining higher yields of better-performing products from a constant timber supply, all of these areas for development should be looked on as potentials for greater profits and an expanded industry.

Figure 16.4 The use of bark, chips, and other wood residues will intensify as more use is made of Canada's forest resource. (Photo on left: NFB Photothèque. Photo by G. Lunney)

Because logs of smaller diameter and lower quality are arriving at the sawmill, more reliance on glued wood products seems inevitable. The potential of laminated veneer lumber as an alternative to structural lumber grades has already been demonstrated. Assembled as parallel laminations of thick veneer, this product results in a much higher yield from a sawlog than traditional sawn products do. It should find an initial market in the specialty field for millwork, ladder stock, and roof-truss chords, and later it should enter into more general use.

The potential use of logging and processing residues in composition panels and pulp manufacture should intensify the trend toward whole-tree logging and full-forest utilization. Stump-root systems, tops and branches, small stems, and cull trees that have many defects may contain as much as 50% of the world's fiber supply, but some handling and processing problems must be solved before they can be used economically on a massive scale. A start has been made with the introduction of whole-tree chippers that reduce tops and branches to chips, but the bark content of the chips and the extraction and cleaning of stump-root systems before chipping remain obstacles to complete use of the tree. Public dissatisfaction with slash burning is stimulating greater use of forest residues for panel and pulp manufacture. Increasing use of forest residues and whole-tree logging will generate increasing amounts of foliage for processing. Animal-fodder supplements, glue extenders, pharmaceuticals, and essential oils can be produced from this foliage. Much of the chemical and technical knowledge for manufacturing these products from foliage already exists, but it has not been used before because of the prohibitive costs of collection.

Bark presents an enormous opportunity for new forest products. In British Columbia alone, nearly 1.8 million tonnes (2 million tons) of dry bark are available annually. Bark includes rich reserves of organic chemicals, some of which have potential use in manufacturing adhesives. Strong, dimensionally stable panels have been developed from bark by capitalizing on the thermal properties of its extractives, which soften under high temperatures and thus can replace the waterproof adhesives normally added to traditional wood particleboard. Preliminary research has shown that high-quality waterproof adhesives similar to those derived from petroleum may be made from bark. With these adhesives, the forest-products industry could achieve an increased degree of independence from outside suppliers and reduce the burden of high costs of products made from petroleum and its derivatives.

With energy costs increasing, there is renewed interest in bark as an alternative to the fossil fuels used by wood-converting plants. This use may compete with the other purposes for which bark can be used, although chemicals could be extracted from the bark before the solid residue is burnt. As an added incentive, the efficient use of bark should reduce current problems caused by its pollution of land, air, and water when it is burned to dispose of it or when it is dumped as waste. Research is underway to acquire basic information about how the chemical and physical properties of bark are affected by species, tree age, log-transport systems (water or land), and the method and duration of log storage.

To extend our wood-fiber resources, panel products of reconstituted wood and bark can be developed and engineered to meet particular specifications and to improve the consistency of products. Orientation and adhesion of fibers, particles, or flakes, surface coatings, ease of machining, and market and building code restrictions are examples of the production variables being investigated and the problems to be overcome.

Improved engineering design with wood and wood-based materials is needed to keep wood competitive with other materials and to effect a measure of forest conservation through more efficient use of trees. Combining the various forms of wood, or combining wood with such materials as aluminum, steel, styrofoam, and fiberglass, may help change the emphasis in forest-products manufacturing from primary commodities to secondary components. Complete, prefabricated floor-wall and roof-ceiling sections are practicable, particularly with the diversity of panel products now available and anticipated in the near future. Such developments are coming into increasing use as the building industry moves away from the 'stick-built' custom dwelling to industrialized building concepts. More information about the behavior of full-sized wood floor, wall, and roof systems is required to permit the most efficient use of lumber in housing and engineered timber structures.

Recycling fiber is another way to stretch our existing timber resources. Present rates of recovery are only about half those during the Second World War. Recycling might become more attractive economically as the density of population increases in large urban centers. In order for recycling to be successful, however, the public must be educated to collect and segregate paper from other trash, and in industry, paper must be separated from other residues and additives, such as inks, resins, waxes, and gums, that are added to paper and board during their manufacture. Waste paper used for items such as fluffy, bulk insulation requires no pretreatment but does require chemical treatment after pulping to impart resistance to fire and biological deterioration.

Figure 16.5 Demand for panel products of reconstituted wood fiber is increasing. Aspen waferboard is widely used for sheathing walls and roofs in modern housing.

Alternatively, some radical changes might be made in pulp or printing additives that would not interfere with or detract from repulping and reuse. The use of light-sensitive or soluble inks to ease deinking problems are examples of such a development. Finally, paper, along with other municipal wastes, can be used for fuel. No special prehandling is required, and fuel may be a more economical use for the bulk of our discarded fiber than recycling it for further use.

MEETING THE CHALLENGES

In an increasingly complex, mechanized, and synthetic world, wood is one of the few reminders of our pioneer inheritance and dependence on nature. Frank Lloyd Wright expressed these sentiments by noting that 'wood is universally beautiful to man. It is the most humanly intimate of materials.' The almost elemental need for man to work with wood leads one to be optimistic about the continued demand for traditional wood products. In addition, wood can replace products made from nonrenewable resources – materials that require great amounts of energy to be processed into nonbiodegradable products.

As the public becomes aware that wood is a renewable resource, traditional and new wood products should be accepted more readily, both for their own merits and for their contribution to conservation. Greater product recovery per unit volume of wood is a way of extending our use of this renewable resource. Major increases in product yield from the same timber base can be achieved by the simple expedient of applying research results to practice.

For example, as a conservative estimate, Canadian lumber production could be increased by 12½% and plywood by 7% merely by implementing certain efficiencies in processing (Western Forest Products Laboratory 1978). These incremental improvements in yield are possible *now* and thus far outstrip the return on investment that might eventually be realized from increasing timber volume by intensive forest-management practices. Similar economies may be achieved through more effective engineering use of wood structures and preserved timber. Put another way, forest conservation can be achieved by these measures equally as well as by fire, insect, and disease control. It is the responsibility of industry to implement these developments, however, and it is up to research organizations to continue to improve products and processes. In this manner, Canadian industry may continue to expand without violating sustained-yield principles, keeping in mind the critical break-even point that is forecast around the turn of the century.

The rate at which new developments will occur depends on funds available for research and development and the aggresssiveness of industry in seizing opportunities. Considering the critical importance of wood for shelter and paper production alone, research effort seems to be important. With the additional realization that exported forest products currently earn about 18% of Canadian foreign exchange (U.S. Forest Service 1973), a strong research and development program seems mandatory.

The research and development effort now underway on behalf of wood and forest products is not impressive. Among the 20 leading manufacturing industries in Canada, the lumber, wood-products, and furniture sectors have the lowest research and development expenditures per thousand dollars in sales (Cordell 1971). Paper and allied products are higher on the ladder, but they are still below the Canadian average. The average expenditure for all forest-products sectors in Canada is less than one-tenth the expenditure of the electrical-products sector and only one-third that of petroleum and coal products. More research funds and effort are required if Canada is to make the most efficient and economical use of its significant forest resource over a broad range of products for national and international consumption.

Research efforts in forest products encompass industrial, government, and university sectors in a ratio of 81:13:6 (Science Council of Canada 1970). The proportionate effort in these sectors may not be out of balance. Government research is required particularly for those segments of the wood-converting industry that are fragmented and scattered, such as sawmilling. Government involvement is also required to protect the consumer in an era of more sophisticated products to ensure that quality, stability, and endurance specifications are met. As managers of the forest resource, provincial governments are continually concerned with optimum resource allocation and the development of efficient converting techniques for various classes of timber that will sustain and extend local manufacturing centers. University research, directed at improving our basic understanding of the properties of wood as a chemical, physical, and biological material, provides an expanding fund of knowledge on which product and processing developments can be based and later put into effect by industry.

Although the balance of research expenditures within the forest-products sector may be reasonable, a greater total effort in research and development is required. Whether the challenges of more complete forest utilization and high-quality products are met depends to a large extent on an increased commitment in Canada to forest-products research.

REFERENCES

British Columbia Forest Service 1977. Annual Report. Victoria

Brown, H. 1970. Human materials production as a process in the biosphere. *Sci. Am.* 223 (3): 194–208

Cordell, A.J. 1971. *The multinational firm, foreign direct investment and Canadian science policy*. Spec. Stud. No. 22, Sci. Counc. Can. Inf. Can., Ottawa

FAO. 1978. *1978 yearbook of forest products*. Rome

Jegr, K.M.; Miller, E.R.; and Thompson, K.M. 1976. *Paprisim I. A dynamic model of the Canadian pulp and paper industry*. Pulp and Pap. Repts. PPR 154. Pulp and Pap. Res. Inst., Montreal

Keays, J.L. 1974. Full-tree and complete-tree utilization for pulp and paper. *For. Prod. J.* 24 (11): 13–16

Maini, M.S., and Cayford, J.H., eds. 1968. *Growth and utilization of poplars in Canada*. Dep. For. Rural Dev., For. Branch Publ. No. 1205. Ottawa

Manning, G.H. 1973. Canada's role in future North American forest products markets. *For. Prod. J.* 23 (9): 50–4

Manning, G.R., and Grinnell, H.R. 1971. *Forest resources and utilization in Canada to the year 2000*. Dep. Environ., Can. For. Serv. Publ. No. 1304. Inf. Can., Ottawa

Reed, F.L.C., and Assoc. Ltd. 1973. *Canada's reserve timber supply*. Prepared for Dep. Ind. Trade and Comm., Ottawa

Science Council of Canada. 1970. *Seeing the forest and the trees*. Rep. No. 8. Ottawa: Queen's Printer

Stanford Research Institute. 1954. *America's demand for wood*. Rep. to Weyerhaeuser Timber Co., Tacoma, Wash.

U.S. Forest Service. 1973. *The outlook for timber in the United States*. USDA For. Resour. Rep. No. 20. Washington, DC

Western Forest Products Laboratory. 1978. *The other way – A prescription for increased timber supply*. West. For. Prod. Lab. Inf. Rep. VSP-102. Vancouver

Glossary

Words set in italic type are defined elsewhere in the Glossary. Definitions of additional terms can be located through the Index. This glossary was derived from various sources and represents generally accepted usage of terms in the forest products industry.

air dried Dried by exposure to air in a yard or shed, without artificial heat.

allomone Transpecific *pheromone*. A chemical odor stimulating behavior response in a wide variety of insects.

allowable cut The volume of wood that may be harvested from a specified area, under management, for a given period of time.

allowable unit stress The value of a strength property normally published for design use. Allowable unit stresses are identified with grade descriptions and standards, reflect the anisotropic structure of wood, and anticipate certain end uses.

angiosperms Various orders of hardwoods that have true flowers and seeds enclosed in a fruit.

anisotropic Not isotropic; that is, not having the same properties in all directions.

annual layers (rings) The layers of wood grown by a tree during a single growing season; in the temperate zone, annual layers of many species are readily distinguished because of differences in the cells formed during the early and late parts of the season.

apical meristems Tissue at the tips of all twigs and roots of a tree that repeatedly subdivides to form new cells. Responsible for the lengthening of branches and roots and for extending the height of a tree.

assembly time The minimum and maximum time allowed after glue spreading before pressure must be applied to form a satisfactory wood-glue bond.

balloon frame system A framing system, principally used for two-story housing, in which the studs run the full height of the building from the foundation wall to the top plate supporting the roof. Floor joists of the upper story rest on a sill nailed to the studs.

bandsaw A band of steel with teeth on one edge (single cutting) or both edges (double cutting), running on a set of large wheels. Used for longitudinal cutting of wood. See also *twin and quad bandsaws*.
barker A machine for removing bark either mechanically or by high-pressure water jets.
bark pockets Small patches of bark that have become partially or wholly enclosed by the growth of a tree.
beam A structural member that supports a load applied transversely to it. See also *timbers, rectangular*.
bent wood Curved wood formed by steaming or boiling, or by special finishing, and then bending to a form.
bird's-eye figure Figure produced on flat-sawn or rotary-cut surfaces by small, conical depressions of the fibers, which form numerous rounded areas of the grain remotely resembling small eyes. Generally limited to hard maples.
black liquor The dark, alkaline waste liquor from the manufacture of pulp by the kraft (sulfate) process or the soda process. Usually concentrated and burned in a furnace to recover heat and chemicals.
bleaching, pulp The process of removing residual lignin from pulp to improve the brightness and strength.
bleed-through The exudation of colored wood extractives or of coating materials through a paint film.
blooming The formation of crystals on the surface of treated wood by exudation and evaporation of the solvent in preservative solutions.
board Lumber that is less than 38 mm (2 in.) thick and wider than 38 mm (2 in.).
bolt A short section of wood, as cut for shingles, shakes, rough dimension stock, stakes, pallet and crating material, and rotary-cut veneer.
bound moisture See *bound water*.
bound water Water contained within the cell walls of wood and held by hygroscopic forces.
bow The distortion of lumber along the face of a piece from end to end, measured at the point of greatest deviation from a straight line.
brash wood Wood with low resistance to shock and with a tendency to sudden and complete breakage across the grain without splintering.
brightening A process to improve the color of mechanical pulps.
brown rot A condition caused by fungi that decompose the cellulose and associated carbohydrates in wood rather than the lignin. The result is a brown, friable residue. Sometimes called 'brown cubical rot' because of the formation of cracks caused by shrinkage.
brown stain See *stain*.
bucking Cross-cutting felled trees into logs or bolts.
bull edger. A combination circular gang resaw and edger used to break down small cants as well as for edging.
burl figure Swirled figure produced by cutting through burls, which are hard, woody outgrowths on trees.
butt joint An end joint formed by abutting the squared ends of two pieces of wood.
calorific value The potential heat-production value of a wood source. Depends on the cellulose-lignin ratio, the percentage of extractives, and the moisture content.

cambium A thin layer of tissue between the bark and wood that repeatedly subdivides to form new wood and bark cells.

cant A log that has been slabbed on one or more sides by the headrig for subsequent breakdown into lumber by other machines. See also *flitch*.

canting A sawing method that is a combination of two other sawing methods, sawing around and live sawing. The most common sawing pattern used for softwoods. See also *sawing around* and *live sawing*.

capillary forces The forces of liquid adhesion and cohesion combined with surface tension by which a liquid moves through a cellular structure. Also called 'capillary action.'

case hardening A condition of stress and set in dry lumber characterized by compressive stress in the outer layers and tensile stress in the center or core.

cell A general term for the structural units of plant tissue, including wood fibers, vessel members, and other elements of diverse structure and function.

cellulose The carbohydrate that is the principal constituent of wood and forms the framework of the wood cells.

checks Lengthwise separations of wood that usually extend across the annual layers and commonly result from stresses set up in wood during drying.

chipper canter A headrig machine that reduces barked logs directly to chips and cants without producing sawdust.

chlorosis A condition in green foliage marked by yellowing or blanching resulting from mineral deficiency.

circular saw A circular metal plate with teeth on the circumference that rotates on a drive shaft.

cladding See *siding*.

collapse The flattening of single cells or rows of cells in the heartwood during the drying or pressure treatment of wood. The wood surface is often characterized by a caved-in or corrugated appearance.

composites Built-up, bonded products consisting wholly of natural wood, or in combination with metals, plastics, etc.

compression failure Deformation of wood fibers resulting from excessive compression along the grain either in direct end compression (as sustained by columns) or in bending (as on the upper side of a beam under load). It may develop in standing trees as a result of bending by wind or snow or internal longitudinal stresses developed in growth; it may also result from stresses imposed after the tree is cut. In surfaced lumber, compression failures may appear as fine wrinkles across the face of the piece.

compression wood Abnormal wood formed on the lower side of branches and inclined stems of softwood trees. Compression wood is identified by its relatively wide annual layers and dark reddish color. Compared with normal wood, it shrinks excessively lengthwise. See also *reaction wood*.

conditioning The use of humidity in a dry kiln to produce a uniform distribution (equalization) of moisture in timber and to reduce drying stresses.

conifer See *softwoods*.

cooperage Containers, such as barrels and kegs, consisting of two round head pieces and a body composed of staves held together with hoops.

core board A solid or discontinuous middle layer or ply used in panel-type glued structures (such as furniture panels and solid- or hollow-core doors). In furniture manufacture, also known as 'furniture board' or 'industrial board.'

core gap Spaces in the cross-bands of plywood in which the veneers do not butt tightly together.

creep The increase in deflection of a beam under load after the passage of time.

crook The distortion of lumber from a straight line along the edges from end to end of a piece, measured at the point of greatest deviation from a straight line.

cross-band In plywood, a layer of veneer whose grain direction is at right angles to that of the face plies; also, to place layers of wood with their grains at right angles in order to minimize shrinking and swelling.

cross-cutting Sawing wood across the grain to expose an end called a cross-section or transverse section. See also *transverse*.

cross grain Wood in which the fibers are not aligned parallel to the axis of the piece. See also *diagonal grain* and *spiral grain*.

crotch figure Figure produced by the grain when the junction of two or more branches, or the stem and a branch, is cut in a suitable direction.

cubical rot See *brown rot*.

cupping Distortion of a board whereby the faces become concave or convex across the grain or width. This condition usually occurs in drying.

cure The change in properties of an adhesive by chemical reaction which results in the development of maximum strength of the adhesive. Generally accomplished by the action of heat or a catalyst, with or without pressure.

curly grain Wavelike undulations in the orientation of wood cells that cause light to be reflected at different angles from the surface and result in a pleasing effect of alternating light and dark bands.

decay The decomposition of wood substance by fungi. The destruction is readily recognized because the wood has become punky, soft and spongy, stringy, ring-shaked, pitted, or crumbly. Decided discoloration or bleaching of the rotted wood is often apparent.

deciduous See *hardwoods*.

decking Lumber used for pallets, roofs, and walls – usually tongued and grooved.

degrade A reduction in the quality of wood due to defects that result from seasoning.

delamination The separation of layers in a laminate through failure within the adhesive or at the bond between the adhesive and the lamination.

density As usually applied to wood of normal cellular form, density is the mass of wood substance enclosed within the boundary surfaces of a wood-plus-voids complex having unit volume. It is variously expressed as kilograms per cubic metre or pounds per cubic foot at a specified moisture content.

depression, wet-bulb The difference between the dry-bulb and wet-bulb temperature.

diagonal grain Wood in which the annual layers are at an angle with the axis of a piece as a result of sawing at an angle to the fiber direction. A form of cross grain.

diffuse-porous wood Wood from certain hardwood species whose pores are nearly uniform in size and distributed evenly through the annual layer (e.g., birch and maple). Annual layers are sometimes difficult to identify.

diffusion Spontaneous movement of heat, dissolved material, moisture, or gas through a body or space. Movement is from high points to low points of temperature, concentration, or partial pressure.

dimensional stabilization Special treatment of wood to reduce swelling and shrinking caused by changes in its moisture content that accompany changes in relative humidity.

dimension lumber Lumber with a thickness of 38 mm (2 in.) up to, but not including, 114 mm (5 in.) and a width of 38 mm (2 in.) or more.

dimpled grain A distinctive figure produced on flat-sawn or rotary-cut surfaces of certain softwoods, notably lodgepole pine, by small, conical depressions of the fibers.

dogs Steel, teethlike projections usually attached to the knee of a headrig carriage to hold the log firmly in position on the carriage headblock.

dressed lumber Lumber surfaced on one or more sides by a planer.

dry-bulb temperature The temperature of air as indicated by a standard thermometer.

drying (seasoning) Removing moisture from green wood to improve its serviceability and utility. See also *air dried* and *kiln dried*.

dryout Glue line failure caused by exceeding the maximum *assembly time* (q.v.).

dry rot A condition caused by the attack of a specific fungus, *Merulius lacrymans* or *Poria incrassata*, in which the fungus is capable of transferring water to 'dry' wood, resulting in brown rot. Sometimes erroneously applied to all decay.

durability A general term for permanence or resistance to deterioration. Frequently used to refer to the degree of resistance of a species of wood to attack by wood-destroying fungi under conditions that favor such attack.

earlywood The portion of the annual layer that is formed during the early part of the growing season. It is usually less dense and weaker mechanically than latewood.

edge-glued Where two pieces of wood are joined edge to edge by gluing.

edge grain Lumber in which the annual layers form an angle of 45–90° with the wide surface of the piece. Also referred to as 'quarter sawn' or 'vertical grain.'

edger A machine used to produce two parallel sides (wide face) by removing the rounded edges of a board (wane). Lumber is edged to specified width (softwoods) and to random width (hardwoods).

empty-cell process Any process for impregnating wood with preservatives in which air in the wood is maintained at or above atmospheric pressure before injection of the preservatives under pressure. After the pressure is released, a vacuum is drawn to drive out a portion of the preservatives from the wood cell cavities.

encased knot See *knots*.

end joint The place where two pieces are joined end to end, commonly by scarf-jointing or finger-jointing.

equilibrium moisture content The moisture content at which wood neither gains nor loses moisture when surrounded by air at a given relative humidity and temperature.

essential oils Pleasant smelling oils prepared by passing steam through foliage and finely divided twigs of several conifer species (mainly cedars, Douglas-fir, and western hemlock).

extractives Substances in wood, not an integral part of the cellular structure, that can be removed by solution in hot or cold water, ether, benzene, or other solvents that do not react chemically with wood components.

face checks Small longitudinal splits or separations visible on the surface of wood.

fiberboard A broad, generic term inclusive of sheet materials of widely varying densities manufactured of refined or partially refined wood (or other vegetable) fibers. Bonding agents and other materials may be added to increase strength or resistance to moisture, fire, or decay.

fiber saturation point The stage in the drying or wetting of wood at which the cell walls are saturated and the cell cavities are free from water. It is usually taken as approximately 25–30% moisture content, based on oven-dry weight.

fiber, wood Long, thin, cylindrical wood cells, tapered and closed at both ends. Also a general term of convenience for any long, narrow cellular tissue.

fiddleback figure Figure produced by a type of fine wavy grain wood. Wood with such figure is traditionally used for the backs of violins.

figure Any characteristic pattern produced in a wood surface by annual growth rings, rays, knots, deviations from regular grain such as interlocked and wavy grain, and irregular coloration.

filler Any substance used to fill the holes and irregularities in planed or sanded surfaces so as to decrease the porosity of the surface for finish coatings.

fine grain A nontechnical term variously used to describe wood with narrow, inconspicuous annual layers or with relatively small or uniform cell diameters.

finger joint An end joint made up of several meshing fingers of wood bonded together with adhesive. Fingers may be sloped or cut parallel to either the face or the edge of the piece.

finish Wood products such as doors, stairs, and other fine work required to complete a building, especially the interior.

finishes Coatings of paint, varnish, lacquer, wax, and so on applied to wood surfaces to protect and enhance their durability or appearance.

finite element method A method used in wood engineering for detailed stress analysis and precise calculations of deflections.

fire endurance A measure of the time during which a material or assembly continues to withstand fire or to give protection from fire under specified conditions of test and performance.

flat grain The figure produced when lumber is sawn approximately tangent to the annual layers. Lumber is considered flat-grained when the annual layers make an angle of less than 45° with the surface of the piece. (Also referred to as 'flat sawn' or 'plain sawn.')

flat sawn Another term for *flat grain*.

flitch A portion of a log sawn on two or more sides, frequently with wane on one or both edges, and intended for further conversion into lumber. See also *cant*.

framing Dimension lumber used for the structural members of a building, such as studs, joists, and rafters. Light and Structural Light Framing are grades of lumber 38–89 mm (2–4 in.) thick and 38–89 mm (2–4 in.) wide.

free water Moisture that is contained in cell cavities and intercellular spaces and is held by capillary forces only.

full-cell process Any process for impregnating wood with preservatives or chemicals in which a vacuum is drawn to remove air from the wood before admitting the preservative. This process favors heavy absorption and retention of preservative in the treated portions.

fungi A lower form of chlorophyll-less nonvascular plant life. Wood-inhabiting fungi use constituents of wood as food and also require moisture, oxygen, and suitable temperatures in order to develop.

grain In its restrictive meaning, grain designates the direction of alignment of wood elements that determines a plane of cleavage. This term is also used in a variety of ways to describe the size, arrangement, appearance, or other qualities of wood fibers. (See also *cross grain, curly grain, diagonal grain, edge grain, fine grain, flat grain, interlocked grain, open grain, spiral grain, straight grain, texture, figure.*)

green Used in referring to freshly sawn or undried wood. Wood that has become completely wet after immersion in water is not considered green but may be said to be in the 'green condition.'

growth-ring figure See *figure.*

gum A comprehensive term for nonvolatile, viscous plant exudates which either dissolve or swell in contact with water. Many substances referred to as gums, such as pine and spruce gum, are actually oleoresins.

gymnosperm A term signifying plants bearing exposed seeds, usually borne in cones. See also *softwoods.*

gypsumboard A panel material formed of gypsum plaster faced on both sides by a sheet of structural paper.

hardboard A generic term for a panel manufactured primarily from interfelted lignocellulosic fibers (usually wood), consolidated under heat and pressure in a hot press to a density of 497 kg/m^3 (31 lb/cu ft) or greater.

hardwoods Generally one of the botanical groups of trees that have broad leaves in contrast to the conifers or softwoods. The wood produced by these trees contains pores. The term has no reference to the actual hardness of the wood.

headrig The first machine in a sawmill to start the breakdown of logs into lumber products.

headsaw The principal saw in a sawmill used for the breakdown of logs by cutting parallel to the grain.

heart check A radial *shake* originating from the heart or central portion of a log. Also called 'heart shake' and 'rift crack.'

heart shake See *heart check.*

heartwood The inner core of a woody stem, where the cells no longer participate in the life processes of the tree. Usually contains extractive materials that give it a darker color and greater decay resistance than the outer enveloping layer (sapwood).

hemicellulose Noncellulosic polysaccharides of the cell wall that are easily decomposed by dilute acid, yielding several different simple sugars.

holocellulose The total carbohydrate fraction of wood – that is, cellulose plus hemicellulose.

honeycombing A term used to describe advanced white rot; also checks, often not visible on the surface, that occur in the interior of a piece of wood, usually along the wood rays during seasoning.

hydrogenation Treatment of wood with hydrogen and suitable catalysts at high temperature and pressure to produce a gas or oils.

hydrolysis Conversion of the polysaccharides in wood or other cellulosic materials into sugars by treatment (hydrolysis) of wood with acids.

hyphae Threadlike strands of fungi.

intergrown knot See *knots.*

interlocked grain A cross grain condition in which the direction of slope of the fibers alternates periodically between left-hand and right-hand spiral arrangements.

isotropic Having identical properties in all directions.

joist One of a series of parallel beams used to support floor and ceiling loads and supported in turn by larger beams, girders, or bearing walls.

juvenile wood The innermost layers of wood adjacent to the pith, formed during the juvenile years of the tree's growth. Certain features, such as cell structure and size, differ from those typical of mature wood.

kerf The narrow slot cut by a saw as it advances through wood, or the thickness of wood removed as sawdust by a saw.

kiln A chamber having controlled air flow, temperature, and relative humidity used for drying lumber, veneer, and other wood products.

– **compartment kiln** A dry kiln in which the total charge of lumber is dried as a single unit.

– **progressive kiln** A dry kiln in which the total charge of lumber is dried as several units, such as kiln truck loads, that are moved progressively through the kiln. The temperature is lower and the relative humidity is higher at the end where the lumber enters the kiln than at the discharge end.

kiln dried Wood dried in a kiln to not more than 19% moisture content.

kiln schedule A prescribed series of dry-bulb and wet-bulb temperatures and air velocities used in drying a kiln charge of lumber or other wood products.

knots Those portions of a branch or limb that have been surrounded by subsequent growth of the stem. The shape of a knot as it appears on a cut surface depends on the plane of the cut relative to the long axis of the knot.

– **encased knot** A knot whose annual layers are not intergrown (i.e., not continuous) with those of the surrounding wood.

– **intergrown knot** A knot whose annual layers are intergrown (i.e., continuous) with those of the surrounding wood.

– **loose knot** A knot that is not held firmly in place or position and that cannot be relied upon to remain in place.

– **pin knot** A knot of not more than 13 mm (½ in.) diameter.

– **spike knot** A knot sawn approximately parallel to its long axis so that the exposed section is definitely elongated.

kraft (sulfate) process A chemical pulping process in which lignin is dissolved by a solution of sodium hydroxide and sodium sulfide.

kraft pulp A chemical wood pulp obtained by cooking wood chips at high temperature in a solution of sodium hydroxide and sodium sulfide.

laminated wood An assembly made by bonding layers of veneer or lumber with an adhesive so that the grain of all laminations is essentially parallel.

latewood The portion of the annual layer that is formed during the latter part of the growing season after the earlywood formation has ceased.

lignin The thin, cementing layer between wood cells, located principally in the secondary wall and the middle lamella. Lignin is the second most abundant constituent of wood. Chemically it is an irregular polymer of substituted propylphenol groups, and thus no simple chemical formula can be written for it.

limit states design A structural design procedure for proportioning to a structure a measured degree of safety against the occurrence of undesirable conditions or limit states in which the structure ceases to fulfill the function(s) for which it is intended. Those exceeding the load capacity, fracture, and so on are called 'ultimate limit states.' Those which restrict the use or affect the appearance, such as minor distress, vibration, cracking, and deformation, are called 'serviceability limit states.'

linerboard A paperboard used as a facing material in corrugated and solid fiber shipping containers. Linerboard is usually classified according to furnish, as for example, kraft linerboard.

live sawing Sawing through and through without turning the log or by turning it only once – that is, sawing with a bandmill headrig or with a circular headrig.

longitudinal Generally, parallel to the direction of the wood fibers.

longitudinal shear strength The capacity of a body to resist longitudinal shearing stresses.

lumber The product of saw and planing mills that is not further manufactured beyond sawing, resawing, passing lengthwise through a standard planing machine, crosscutting to length, and matching.

lumen The cavity within a wood cell.

margo That portion of the membrane of bordered pits that supports the torus; that is, the membrane exclusive of the torus.

matched lumber Lumber that is edge-dressed and shaped to form a tongued and grooved or similar joint when pieces are laid side by side.

mechanical pulping The production of fibers and fiber bundles by grinding wood with pulpstones or by mechanical refiners as opposed to chemical methods.

meniscus The curved upper surface of a liquid in a tube or container. Wood cells serve as containers for liquid water. The surface of the water in the cell is concave, owing to the effect of surface tension.

microfibril A threadlike component of the cell wall structure composed of chain molecules of cellulose extending through regions of parallel order known as crystallites and through regions of disorder known as amorphous regions. Microfibrils are the smallest natural units of cell wall structure that can be distinguished with an electron microscope.

middle lamella The lignin-rich layer that cements adjoining cells together. This layer is dissolved in the chemical pulping processes which separate wood into pulp fibers.

mildew Surface growths of fungi, usually dark gray in color, growing on the exterior wood of buildings.

millwork Planed and patterned lumber for finish work in building, including items such as sashes, doors, cornices, panel work, and other items of interior or exterior trim. Does not include flooring, ceilings, or siding.

mineral stain See *stain*.

modulus of elasticity A measure of the stiffness of wood.

modulus of rigidity A measure of the torsional stiffness of wood.

modulus of rupture A measure of the maximum strength of wood.

moisture content The amount of water contained in wood, usually expressed as a percentage of the weight of the oven-dry wood.

mold Superficial, usually colored growth of fungi on damp wood; also referred to as mildew.

mycelium The mass of interwoven filamentous hyphae that form the vegetative portion of fungi.

naval stores Oils, resins, tars, and pitch extracted from pine and fir trees. Historically, the term was derived to describe those products when they were used in the construction of wooden sailing vessels.

oleoresin A solution of resin in an essential oil that occurs in or exudes from many plants, especially softwoods.

open grain Common classification for woods with large pores, such as oak, ash, and walnut.

overlay A thin layer of paper, plastic, film, metal foil, or other material bonded to one or both faces of panel products, or to lumber, to provide a protective or decorative face or a base for painting.

pallet A horizontal platform device used as a base for assembling, storing, handling, and transporting materials and products as a unit load.

paper Generally, a matted or felted sheet of vegetable fiber, formed on a screen from a water suspension, used for writing and printing as well as for wrapping and many other purposes. Paper is one of two broad subdivisions of the general term, paper; the other is paperboard.

paperboard A general term describing sheets made of fibrous material 0.012 in. or more in thickness. Compared with paper, paperboard is heavier per unit area, thicker, and more rigid. Paperboard is the term used to describe any single variety, or group of varieties, of board materials used in the production of boxes, folding cartons, and solid fiber and corrugated shipping containers.

parenchyma Short cells having simple pits and functioning primarily in the metabolism and storage of plant food materials. They remain alive longer than the tracheids, fibers, and vessel segments, sometimes for many years. Two kinds of parenchyma cells are recognized: those in vertical strands, known more specifically as axial parenchyma, and those in horizontal series in the rays, known as ray parenchyma.

particleboard A generic term for a panel manufactured from lignocellulosic materials – commonly wood – essentially in the form of particles (as distinct from fibers). These materials are bonded together with synthetic resin or other suitable binder, under heat and pressure, by a process wherein the interparticle bonds are created wholly by the added binder.

pheromones Chemical odors conveying messages to individual insects or to insects of the same species. Pheromones regulate the social life and reproduction of insects.

phloem Inner bark tissue, characterized by the presence of sieve tubes and serving for the transport of foodstuffs.

photosynthesis The formation of carbohydrates in green plants from carbon dioxide and water of the air by the action of light on chlorophyll.

pile A long, heavy timber, round or square cut, that is driven deep into the ground to provide a secure foundation for structures built on soft, wet, or submerged sites.

pin knot See *knots*.

pit A discontinuity in the secondary cell wall normally found in adjacent pairs of cells forming a pathway for liquid movement between neighboring cells. The two halves of a pit pair are normally separated by a membrane consisting of the middle lamella and adjacent primary cell walls. Sometimes the central portion of the membrane is thickened to form a torus.

pit aspiration The displacement of the torus of a bordered pit pair against one of the pit borders closing the pit aperture.

pit pair Two complementary pits of adjacent cells.

pitch pocket An opening extending parallel to the annual layers that contains, or has contained, either solid or liquid pitch.

pitch streaks A local accumulation of resin in the form of a streak, occurring in certain softwoods.

pith The small core of soft primary tissue occurring near the center of a tree stem, branch, and sometimes, root.

pith flecks Flecks on planed lumber caused by insects boring in the cambium layer, producing wound tissue with brownish contents.

plain sawn See *flat grain*.

planer-matcher A surfacing machine for lumber. Profiler heads can be inserted to create side-matching pieces of wood – that is, tongue and groove or shiplap. See also *matched lumber*.

plank A piece of square-cut timber, generally more than 25 mm (1 in.) thick and 140 mm (6 in.) wide or greater, and of any length.

platform frame system A framing system in which floor joists of the upper stories rest on the top plates of the story below (or on the foundation wall), and bearing walls and partitions rest on the subfloor of each story.

plywood A composite panel or board made up of cross-banded layers of plies, bonded with an adhesive, of veneer only, or veneer in combination with a core of lumber, or of particleboard. Generally the grain of one or more plies is roughly at right angles to that of the other plies, and almost always an odd number of plies are used.

pocket rot Advanced decay that appears in the form of a hole or pocket.

pores See *vessels*.

post and beam framing Construction in which posts and beams support the loads; the partition walls are not load-bearing. The roof is usually decking.

preservative Any substance that, for a reasonable length of time, is effective in preventing the development and action of wood-rotting fungi, borers of various kinds, and harmful insects that deteriorate wood.

pressure process Any process of treating wood in a closed container whereby the preservative or fire retardant is forced into the wood under pressures greater than 1 atmosphere (101 kPa). Pressure is generally preceded or followed by vacuum, as in the vacuum-pressure and empty-cell processes; or the applications of pressure and vacuum may alternate, as in the full-cell and alternating processes.

pyrolysis Chemical decomposition of wood by the action of heat; that is, burning of wood.

quarter sawn See *edge grain* .

radial Coincident with a radius from the axis of the tree or log to the circumference. A radial section is a lengthwise section in a plane that passes through the center line of the tree stem.

rafter One of a series of parallel structural members of a roof designed to support roof loads. The rafters of a flat roof are sometimes called roof joists.

raised grain A roughened condition of the surface of dressed lumber in which the hard latewood is raised above the softer earlywood but not torn loose from it.

rays, wood Ribbonlike strands of tissue extending radially within a tree and varying in height from a few cells in some species to several centimetres in oak. The rays serve primarily to store food and transport it horizontally through the tree.

reaction wood Wood with abnormal structure and properties formed in parts of leaning or crooked stems and in branches. In hardwoods it is called 'tension wood'; in softwoods, 'compression wood.'

relative density Formerly called specific gravity. As applied to wood, the ratio of the oven-dry weight of a sample to the weight of a volume of water equal to the volume of the sample at a specified moisture content (green, air-dry, or oven-dry).

relative humidity Ratio of the amount of water vapor present in the air to the amount that the air would hold at saturation at the same temperature. It is usually considered on the basis of the weight of the vapor but, for accuracy, should be considered on the basis of vapor pressures.

relaxation Reduction of stress with time on a wood member maintained under constant deflection.

resaw A sawing machine used to break down cants into lumber, for recovering lumber from slabs, and for upgrading lumber by ripping off defective portions.

resin A comprehensive term for secretions of certain trees, or of insects feeding on them, that are oxidation or polymerization products of the terpenes, consisting of mixtures of aromatic acids and esters insoluble in water but soluble in ether, alcohol, and other organic solvents. These secretions often exude from wounds and are obtained commercially by tapping or by extraction with solvents. The term is also applied to synthetic organic products related to the natural resins.

resin ducts Intercellular canals or passages that contain and transmit resinous materials. They may extend vertically parallel to the axis of the tree or at right angles to the axis and parallel to the rays.

resistivity The resistance of a cubic centimetre of material, such as wood, to the direct-current flow of electricity between opposite faces.

ribbon stripe A form of figure produced on the surface of wood because of the presence of interlocked grain.

rift crack. See *heart check*.

rigid frame A rib type of construction, formed from lumber joined at the crown and haunches by plywood gussets, that is easily fabricated and erected on site.

ring-porous Used in referring to a group of hardwoods in which the annual growth layers consist of a more or less continuous zone of large earlywood pores that changes relatively abruptly to a denser latewood zone having smaller pores and an abundance of fibrous tissue (e.g., oak and ash).

ring shake A separation along the grain that occurs most commonly between adjoining annual layers. See also *shake*.

ripping Cutting lengthwise, parallel to the grain.

rotary-cut veneer Veneer cut in a lathe that rotates a log or bolt against a knife set in such a manner as to peel off a continuous thin sheet.

rough lumber Lumber that has been sawn, edged, and trimmed but not dressed (planed).

sap Fluid contents of the living wood cells.

sap stain See *stain*.

sapwood The wood located near the outside of the tree stem containing the tissues actively involved in the transport of sap. It is generally lighter in color than heartwood and has lower natural resistance to decay.

sash A frame structure, normally glazed (e.g., a window), that is hung or fixed in a frame set in an opening.

sash gang saw A sash or frame holding a battery of parallel saws that move up and down on the end of a connecting rod attached to a heavy crank-shaft.

sawing around Breaking down a log by turning it on the carriage of a headsaw to obtain the best yield of lumber from the clear outer portion of the log.

sawn veneer Veneer produced by sawing.

scarf joint An end joint formed by joining with glue the ends of two pieces that have been tapered or beveled, usually to a feather edge, to form a slope of the same length and inclination in both pieces.

scrag saw Two or more pairs of saws, one pair to a drive shaft, or two or more pairs of saws, each saw on an individual drive shaft, all sawing different lines. Saws may be fixed or adjustable to different settings.

sealers Undercoating materials for sealing a surface in preparation for painting, varnishing, or application of final finish.

seasoning. See *drying*.

semichemical pulp Pulp obtained by mild treatment of wood chips by any of the chemical pulping processes, which remove only part of the lignin from the wood chips, followed by mechanical treatment to complete the separation of individual cellulose fibers.

set A permanent or semipermanent deformation in wood caused by internal stresses.

setworks The mechanism on an edger, on a log carriage, or on twin and quad bandsaws for regulating the thickness of the wood being cut.

shake A rupture or separation along the grain. The term is most commonly applied to 'ring shakes,' which develop tangentially either within a given annual layer or at the boundary between two layers.

shakes In construction, a type of shingle usually hand cleft from a bolt and used for roofing or weatherboarding.

shear The displacement of woody tissues following fracture as a result of shearing stresses which cause the fibers to slide relative to one another.

shear, longitudinal Shearing stress that tends to cause the fibers to slide over each other lengthwise.

shear strength The capacity of a body to resist shearing stresses.

sheathing The structural covering, usually of boards, building fiberboards, waferboard, or plywood, placed over exterior studding or rafters of a structure.

shim stock Thin strips of wood used to level off irregularities under a hardwood floor.

shingles Thin, rectangular pieces of wood, sawn along the grain and tapering in thickness, used like tiles for roofing and weatherboarding.

shrinkage Contraction caused by drying wood below the fiber saturation point; it is greater in the wide face of flat-grain than in edge-grain lumber, and minimal in the longitudinal direction.

siding (cladding) The finish covering of the outside wall of a frame building, whether made of horizontal weatherboards, vertical boards with battens, shingles, or other material.

slab The exterior portion of a log removed in sawing lumber.

sliced veneer Veneer that is sliced off a log, bolt, or flitch with a knife.
soft rot A special type of decay developing under very wet conditions in the outer wood layers, caused by certain fungi that destroy the cellulose in the secondary cell walls; as a result the wood becomes soft. The decayed wood is similar in appearance to brown rot.
softwoods Generally, one of the botanical groups of trees that in most cases have needlelike or scalelike leaves (the conifers); also the wood produced by such trees. The wood does not contain pores. The term has no reference to the actual hardness of the wood.
spalt The pie-shaped portion of a shingle bolt that remains after processing.
specific gravity See *relative density.*
specific heat The heat in joules required to raise the temperature of one gram of wood 1°C.
spike grid A type of wood connector with teeth projecting from both surfaces that cut into the wood members as they are drawn together.
spike knot See *knots.*
spiral grain Wood in which the fibers take a spiral course about the stem of a tree. The spiral may extend toward the right or left around the tree stem. Spiral grain is a form of cross grain.
splints Trim from the edges of shingles and shakes.
splits Separations along the grain extending through a piece. Commonly caused by stresses set up in the wood during drying.
sporophore The fruiting body of a fungus; a conk.
springwood See *earlywood.*
stain A discoloration in wood that may be caused by such diverse agents as microorganisms, sunlight, metals, chemicals, and chemical interaction. The term also applies to materials used to impart color to wood.
- **blue stain** A bluish or grayish discoloration of the sapwood caused by the growth of certain dark-colored fungi on the surface and in the interior of the wood. Blue stain is made possible by the same conditions that favor the growth of other fungi.
- **brown stain** A dark brown discoloration of the sapwood of some pine logs that occurs during storage. Sometimes called 'coffee-brown stain,' it is caused by a fungus.
- **chemical brown stain** A brownish discoloration that may occur during the seasoning of certain softwoods, apparently caused by the concentration and oxidation of extractive chemicals.
- **hemlock brown stain** See *chemical brown stain.*
- **mineral stain** An olive to greenish-black or brown discoloration of undetermined cause in hardwoods.
- **sap stain** See *blue stain.*
- **sticker stain** A brown or blue stain that develops on lumber during seasoning where the stickers contact the boards.

star shake A number of heart shakes more or less in the form of a star.
stem The principal axis of a tree, capable of producing sawlogs, veneer logs, large poles, or pulpwood.
stickers Narrow wood strips used to separate the layers of lumber in a pile and thus improve air circulation.
straight grain Wood in which the fibers are aligned parallel to the axis of the piece.
strandwood A board made of long, narrow slices of softwood (mostly poplar)

bonded together in one direction in the horizontal plane, forming the middle layer, or core, in composite plywood.

strength The limit of ability of a member to sustain stress. Also, in a specific mode of test, the maximum stress sustained by a member loaded to failure.

strength ratio The hypothetical ratio of the strength of a structural member to that which it would have if it contained no strength-reducing characteristics (knots, cross grain, shake, etc.).

stress grades Lumber grades having assigned working stress and modulus of elasticity values in accordance with accepted principles of strength grading.

stress, working See *allowable unit stress.*

structural timbers Pieces of wood of relatively large size, the strength of which is the controlling element in their selection and use. Examples are trestle timbers (caps, posts, sills, bracing, bridge ties, guardrails); car timbers (car framing, including upper framing; car sills); framing for building (posts, sills, girders); ship timber (ship timbers, ship decking); and cross-arms for poles. See also *timbers.*

stud One of a series of slender wood structural members used as supporting elements in walls and partitions.

stump figure Figure produced by irregular grain in wood from the stump or base of a tree.

subfloor Boards or panel products used over floor joists as a working platform. See also *underlayment.*

sulfate process See *kraft (sulfate) process.*

sulfite process A chemical pulping process in which wood is cooked in aqueous acid sulfite solution containing free sulfur dioxide.

sulfite pulp A chemical wood pulp obtained by cooking wood chips in a bisulfite–sulfurous acid solution.

summerwood See *latewood.*

tall oil An oily material liberated from soap skimmings from sulfate pulping liquor. Used chiefly in the paint and lacquer industry.

tangential Strictly, coincident with a tangent at the circumference of a tree or log, or parallel to such a tangent. In practice, however, it often means roughly coincident with an annual layer. A tangential section is a longitudinal section through a tree or limb perpendicular to a radius. Flat-grain lumber is sawn tangentially.

tannins Complex, water-soluble phenolic extractives that precipitate gelatin and tan animal skins.

technical foliage All needles, leaves, and twigs below 0.6 cm diameter.

tension wood Reaction wood formed on the upper side of branches and inclined stems of hardwood trees. Tension wood is characterized anatomically by lack of cell-wall lignification and often by the presence of gelatinous fibers. It has excessive longitudinal shrinkage, and sawn surfaces usually have projecting fibers. Planed surfaces often are torn or have raised grain.

texture Refers to the size of the cellular components of wood; may also describe their relative uniformity in size. See also *grain.*

thermal conductivity A measure of the rate of heat flow through a material subjected to a temperature gradient, or the number of watts passing between the faces of a piece of wood 1 m^2 in area and 1 mm thick per 1°c temperature difference between the faces.

thermal diffusivity The ratio of the thermal conductivity to the product of density and specific heat. A measure of how quickly a material can absorb heat from its surroundings.

thinboard A particleboard made in thickness up to 6 mm (¼ in.) on a continuous rotary drum press. Differs from particleboard only by the pressing technique used in its manufacture.

timbers, rectangular Wood products (beams) 114 m (5 in.) thick or more with the width more than 38 mm (2 in.) greater than the thickness.

timbers, square Wood products (posts and timbers) 114 mm × 114 mm (5 in. × 5 in.) and larger with the width not more than 38 mm (2 in.) greater than the thickness.

torus Central thickened portion of a pit membrane of bordered pits.

tracheid An elongated cell with bordered pits and imperforate ends. Tracheids constitute the principal part of the cellular structure of softwoods. Tracheids are frequently referred to as fibers and are present in many hardwoods.

transverse Direction in wood perpendicular to that of the fibers. A transverse section is one that is cut across the grain at right angles to the fiber direction.

trimmer A battery of adjustable saws for trimming lumber to specific lengths or for removing defects.

truss An assembly of members, such as beams, bars, and rods, combined to form a rigid framework. All members are interconnected to form triangles of tension and compression members.

truss plates Steel plates in which nail-like teeth have been punched. Applied by a press, they combine the function of a gusset and of nails.

twin and quad bandsaws A twin bandsaw is an adjustable, double bandsaw headrig making two cuts simultaneously. A quad bandsaw makes four cuts simultaneously.

twist Distortion caused by the turning or winding of the edges of a board so that the four corners of any face are no longer in the same plane.

tyloses Ingrowths of parenchyma cells into the lumen of a vessel (or sometimes a fiber) occurring generally in the heartwood of certain hardwoods.

tylosoids Structures in resin ducts resembling tyloses in hardwoods.

undercure Incomplete cure of a chemical-setting adhesive, producing bonds of low strength.

underlayment A panel product used to provide an appropriate surface for a finished floor (tile, carpet, hardwood). It may also include the complete subfloor.

vapor pressure gradient A gradation in water vapor pressure established between the interior of wood and its surface during drying.

veneer A thin layer or sheet of wood. See also *rotary-cut*, *sawn*, and *sliced veneer*.

vertical grain See *edge grain*.

vessels Tubelike structures in porous woods (hardwoods only) made up of longitudinal series of relatively short, large-diameter cells having more or less open ends. Open vessels exposed on the surfaces of a piece of wood are known as pores.

waferboard A type of particleboard composed of wafers cut from roundwood bolts (mostly poplar) of uniform length and thickness resembling small pieces of veneer. The wafers are bonded together with resin binder, under heat and pressure, by a process similar to that by which particleboard is made.

wane Bark or lack of wood from any cause on any edge or corner of a piece of lumber.

warp Any variation from a true or plane surface. Warp includes bow, crook, cup, and twist, or any combination thereof.

wavy grain See *curly grain*.

weathering The mechanical or chemical disintegration and discoloration of the surface of wood caused by exposure to light, the action of dust and sand carried by winds, and the alternate shrinking and swelling of the surface fibers with the continual variation in moisture content brought about by changes in the weather. Weathering does not include decay.

western frame See *platform frame system*.

wet-bulb temperature The temperature when the thermometer bulb is kept moistened and hence cooled by evaporation. Because evaporation is greater in dry air, the thermometer will register a lower temperature at lower relative humidity.

white rot A condition caused by fungi attacking the cellulose and lignin in wood simultaneously, resulting in a whitish residue that may be spongy or stringy. May occur as pocket rot.

wood resin See *resin*.

xylem The wood portion of the tree stem, branches, and roots. It lies between the pith and the cambium.

Index